AF412813

MATHEMATICAL MODELING FOR SYSTEM ANALYSIS IN AGRICULTURAL RESEARCH

MATHEMATICAL MODELING FOR SYSTEM ANALYSIS IN AGRICULTURAL RESEARCH

Karel D. Vohnout

*Former Research Specialist
Inter-American Institute for
Cooperation in Agriculture*

**2003
ELSEVIER**

Amsterdam – Boston – London – New York – Oxford – Paris
San Diego - San Francisco – Singapore – Sydney – Tokyo

ELSEVIER SCIENCE B.V.
Sara Burgerhartstraat 25
P.O. Box 211, 1000 AE Amsterdam, The Netherlands

First edition 2003

Library of Congress Cataloging in Publication Data
A catalog record from the Library of Congress has been applied for.

British Library Cataloguing in Publication Data
A catalogue record from the British Library has been applied for.

ISBN: 0-444-51268-3

♾ The paper used in this publication meets the requirements of ANSI/NISO Z39.48-1992 (Permanence of Paper).
Printed in The Netherlands.

To my wife Cecilia

CONTENTS

5 CURVE FITTING AND EVALUATION

6 FRAMEWORK FOR MODELING AGRICULTURAL SYSTEMS

7 STOCHASTIC MODELS OF SYSTEMS

8 DETERMINISTIC MODELS OF DISCRETE SYSTEMS

9 DETERMINISTIC MODELS OF CONTINUOUS SYSTEMS

10 EXPERIMENTAL TESTS FOR A SYSTEM ANALYSIS PROBLEM

The word *system* may have different meanings for different people. In the scientific community, a system is usually said to be an arrangement, set, or collection of things connected in such a manner as to form an entirety or as to act as an integral unit. In a more mathematical sense, a system is defined as a physical or abstract object that, over the time scale, receives inputs from outside its boundaries, responding with changes of state and with outputs. An important implication of this definition is the awareness of a dynamic condition, the awareness of changes and evolution of the system over time.

With the new breakthroughs in computers and computer software for data processing, it is feasible and necessary to improve the existing thinking and research procedures. Thus, the objective of this work is to present a methodological foundation for research and analysis of biological systems, based on a system theoretical approach. It is intended as an aid to the scientist in his quest for improving the accuracy of research.

The book is essentially a proposal of procedures for mathematical modeling, as it applies to the design and analysis of biological systems. It is a guideline for the scientist to match mathematical models to working hypotheses, for selecting the most appropriate mathematical model and for matching experimental treatments and designs to models. For such, the book includes a mathematical background, theoretical definitions in system engineering and specific applications of mathematics to modeling procedures. It also includes procedures for the evaluation of mathematical models as they apply for experiments in agricultural systems. Specific features, related to the design of experimental tests in the research of agricultural systems and the processing and analyses of related experimental data, are also discussed.

The first chapter of this book is a general outlook of basic definitions and modeling procedures for system analysis in agricultural research. The next four chapters include selected topics from algebra, calculus and nonlinear curve fitting, which are directly related to the understanding of models in the subsequent chapters. Without the necessary skills in these topics, the manipulation of subjects in those chapters could become difficult and even frustrating.

The following four chapters contain specific applications of mathematics to the modeling and analysis of systems. Chapter 6 is a general outlook of the modeling process, aimed mainly at the utilization of linear mathematical models. Also explained in this chapter are procedures for interconnecting systems by means of differential or difference equations and the different types of responses expected from a system. Guidelines for defining stochastic models are presented in Chapter 7. Regular processes leading toward a steady state of the system, absorbing processes leading toward the extinction of some system states and the relationship between stochastic and deterministic models, are also evaluated in this chapter. Guidelines for defining deterministic models are presented in the eighth and ninth chapters. The main subjects examined in these two chapters are specific modeling procedures, the evaluation of deterministic models for conventional experiments in agricultural systems and methods for determining state equations from experimental data.

The last chapter includes very specific features related to the design of experimental tests in agricultural systems research and to processing and analysis of related experimental data. Relationships between the research problem and mathematical models of the working hypotheses and matching of experimental treatments to mathematical models, are the main subjects examined in this chapter.

Mathematics is used here as a tool and not as an end. Theoretical considerations are avoided and theorems and proofs are omitted. The lack of formality of this approach may not look attractive for mathematicians, but may not scare away agricultural research scientists, which are the targets of this book.

To be able to follow the material presented and discussed here, the reader needs college algebra and calculus and an acceptable background in statistics and experimental design.

This work is directed toward the utilization of linear mathematical models, represented by linear difference or differential equations. Agricultural Science is essentially an empirical science and there is seldom the necessity for going beyond linear models. Most topics and procedures outlined here are supported by examples and by case studies using actual data. Some equations, however, were computed from plotted data reported in the literature. Formal mathematical definitions follow examples and case studies. In this manner, it is expected that the reader can follow the full modeling process with a concrete image of the real problems, as they are projected into the abstract world. In addition and when appropriate, the proposed procedures are compared with traditional methods and accuracy of the use of data is evaluated.

It is the author's hope that, after the completion of this study, the reader should have the necessary skills and motivation to explore further and deeper into the subjects of this book. Graduate students and research professionals may both profit from the author's experience in the management and development of research procedures in agriculture, with a bias, however, toward animal science.

KAREL D. VOHNOUT
Tucson, Arizona

ACKNOWLEDGMENTS

I gratefully acknowledge the sincere comments of Professor Jorge Soria on the agricultural outlook of the book and his valuable efforts in detecting those mathematical topics that could be difficult to understand or follow up by agricultural scientists. It is with a definite sense of deep appreciation that I acknowledge the constructive criticism, helpful suggestions and careful review of the mathematical approach made by Professor Emeritus Wayne Wymore. The contributions of Dr. Soria and Dr. Wymore greatly improved the entire manuscript.

1

THE SCOPE OF SYSTEM ANALYSIS

The notion of a system and of system analysis may have different connotations to different people. Thus, the purpose of this chapter is to define these concepts objectively, as they relate to agricultural research in this book.

System analysis may be defined as the process of developing an abstract model of an existing system, such that the model would simulate the real system by means of a computer program. Then, the real system may be analyzed abstractly. The following steps may be included in this process:

- Statement of the research problem
- Defining the hypothesis about input-output relationships
- Designing a model of the system in terms consistent with the hypothesis
- Defining the system test plan
- Implementing the project and running system experiments
- Manipulation of the abstract model to simulate the real system

The statement of the research problem should answer the question, *what is the system supposed to do*. Answering this question leads to the notion of the input-output relationships. In nature, a production system has often countless input and output variables that need to be sorted out. This sorting leads to additional questions such as, what research on the input-output relationships is the most relevant and if such research is feasible as a research project. Conventionally, the statement of the research problem should be covered by the "Introduction" of the research project.

The hypothesis should answer the question *how are the inputs affecting the outputs*. A research team has often in mind some image of the system to answer intuitively this question. Thus, projecting this image into an abstract model of the system is possible. The idea of modeling includes the abstract model and the field model, which have a formal relationship called a homomorphic image. The field model has the experimental design as its basic blue prints. The experimental design is the floor plan of the field model. The abstract model should be consistent with the hypothesis on the selected input-output relationships and is also the result of a screening process over a set of proposed mathematical models.

At this time, a plan for testing the abstract model by means of field experiments on the field model should also be defined. Planing experiments on the field model should answer the question, *how is the abstract model going to be tested*. Answering this question determines the system test plan. The system test plan includes the field procedures and the proposed statistical analyses and is often called "Methods" or "Materials and Methods."

The field model, as implemented by the rules of experimental design, simulates the real system by means of system experiments. Results of field experiments are used for testing the abstract model. If not acceptable, adjustments to the abstract model and often also to the field model may be required. This iterative process may require long term projects and is often incompatible with flash type research.

Finally, if the mathematical model is accepted by statistical standards, it may be used for system experiments by means of computer simulations.

This chapter is a general outlook of basic definitions and procedures for modeling and evaluation of agricultural systems, as they are further presented and developed in this book. Thus, procedures for solving some problems in the examples that follow will be disclosed in the appropriate following chapters.

1.1 THE MATHEMATICAL CONCEPT OF A SYSTEM

The concept of a system is related to the notion of a dynamic physical or abstract object. This object is receiving inputs from outside its boundaries and is reacting to such inputs by state changes and by producing outputs. In the same manner as the state of the system depends on inputs, an output variable depends on the system states. This verbal definition can be translated into a precise mathematical theory.

A simple portrait of a system is that of a tank with a water admission pipe and an outlet with a pressure valve for collecting water from the tank. The content of the tank is emptied in proportion to the height of water in the tank. The change of the water level is the difference between admission and discharge of water. This prototype of a system is pictured in Fig. 1.1.1:

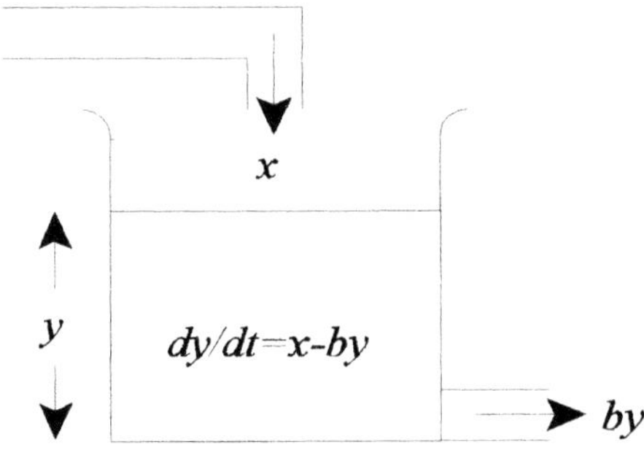

Figure 1.1.1

Note that the difference between the input and the output determines the change in the water level in the tank. Thus, the following first order linear differential equation represents the change of the water level:

$$\frac{dy}{dt} = x - by$$

where t is time, $x = f(t)$ is the water input, $y = g(t)$ is the height of water, b is the effect of the pressure valve and $by = z$ is the output, for $z = h(t)$. The above equation corresponds to a *continuous system*. The time variable of continuous systems is the set of nonnegative real numbers. An abstract representation of the system is the *black box* shown in Fig. 1.1.2:

Figure 1.1.2

A steady state is achieved when the input and the output are the same, that is $x=by$. Otherwise, the system is said to be in a transient state.

The simplest system is one where the input is a constant c. Then

$$\frac{dy}{dt} + by = c$$

This differential equation is easily solved by making the substitution $y = pq$, such that

$$\frac{dy}{dt} = p\frac{dq}{dt} + q\frac{dp}{dt}$$

Then

$$p\frac{dq}{dt} + \left(\frac{dp}{dt} + bp\right)q = c$$

The variable p is determined by solving $\frac{dp}{dt} + bp = 0$ in the above equation. Then

$$\ln p = -bt + \ln k_1$$
$$p = k_1 e^{-bt}$$

where k_1 is an integration constant. The variable q is determined by solving $p\dfrac{dq}{dt} = c$, such that

$$\int dq = \int \frac{c}{k_1} e^{bt} dt$$
$$q = \frac{c}{k_1 b} e^{bt} + k_2$$

where k_2 is an integration constant. Then

$$y = pq = k_1 e^{-bt}\left(\frac{c}{k_1 b} e^{bt} + k_2 \right)$$
$$= \frac{c}{b} + k_1 k_2 e^{-bt}$$

If the initial condition of the system is defined as y_0, then $k_1 k_2 = y_0 - c/b$ and the solution becomes

$$y = \frac{c}{b} + \left(y_0 - \frac{c}{b} \right) e^{-bt}$$

where c/b is an asymptotic value for the system at a steady state. This equation represents a *state trajectory* of the system. Note that changing the input c affects the steady state of the system. Note also that the initial state y_0 and the input $x = c$ determine the state y of the system at any time t. Clearly, if the system is started at a state y_0, is supplied by an input trajectory f and is run to some time t, then

$$y = u(f, y_0, t)$$

The above expression is called the *state transition function* of the system. The state transition function represents the dynamic behavior of the system.

From the above, it is clear that, given the initial conditions, a continuous system is completely determined by a differential equation or a set of interconnected differential equations and their solutions.

The output of the system was defined as $by = z$. Therefore, the expression

$$z = b\left[\frac{c}{b} + \left(y_0 - \frac{c}{b} \right) e^{-bt} \right]$$

represents an *output trajectory* of the system. Note that an output z is completely determined by the state y. Then

$$z = w(y)$$

The above expression is called the *output function* of the system.

　　　　When the state variables cannot be fractionalized, the system cannot be represented by differential equations. In such cases, the system is a *discrete system* and is represented by difference equations and their solutions. The time variable of discrete systems is the set of nonnegative integers. The same principles outlined for continuous systems apply also for discrete systems.

　　　　The following first order linear difference equation represents a discrete system, equivalent to the continuous system previously described[1]:

$$\Delta y_n = x_n - b y_n$$

where n is a discrete time value, x_n is the input and $b y_n$ is the output. The following is the corresponding difference equation in subscript notation:

$$y_{n+1} = x_n + (1 - b) y_n$$

where $\Delta y_n = y_{n+1} - y_n$. Note that the state y_{n+1} of the system at time $n+1$ is completely determined by the state y_n and by the input x_n at the discrete time n. Thus

$$y_{n+1} = v(y_n, x_n)$$

This expression is called the *next state function* of a discrete system.

　　　　If the input is constant, the difference equation of the system becomes

$$y_{n+1} = c + (1 - b) y_n$$

It follows that

---

[1]Readers that are unfamiliar with difference equations are referred to Chapter 3.

$$y_1 = c + (1-b)y_0$$
$$y_2 = c + (1-b)y_1 = c + (1-b)c + (1-b)^2 y_0$$
$$\vdots$$
$$y_n = c\left[1 + (1-b) + (1-b)^2 + \ldots + (1-b)^{n-1}\right] + (1-b)^n y_0$$

The terms within brackets represent a geometric series, such that

$$S_n = c\left[1 + (1-b) + (1-b)^2 + \ldots + (1-b)^{n-1}\right]$$
$$= \frac{1 - (1-b)^n}{1 - (1-b)}$$

Then, after replacing values and rearranging terms, the solution of the difference equation becomes

$$y_n = \frac{c}{b} + \left(y_0 - \frac{c}{b}\right)(1-b)^n$$

where y_0 is the initial condition of the system. This is a state trajectory of the discrete system for an input value of $x = c$. The above sequence shows that the initial state y_0 and the input $x_n = c$ determine the state of the system at any discrete time n. Clearly, if the system is started at a state y_0, is supplied by an input trajectory f and is run to some discrete time n, then

$$y_n = u(f, y_0, n)$$

represents the state transition function of the discrete system.

In conclusion, the mathematical definition of a system includes the following elements:

- The time variable, as a continuous or a discrete scale
- The states, as quantitative or qualitative variables
- The input variables
- The state transition function
- The output function, when appropriate

The full description of the above statement is presented in Chapter 6.

Summary

A system is a dynamic physical or abstract entity. It receives inputs from outside its boundaries and reacts to such inputs by state changes and by emission of outputs. The state y of a continuous system at any time t is completely determined by un input $x = f(t)$ and by the initial condition y_0. Thus, if the system is started at a state y_0, is supplied by an input trajectory f and is run to some time t, then $y = u(f, y_0, t)$. This expression is called the state transition function of the system. The output $z = f(t)$ of the system is determined only by the state y, that is $z = w(y)$. This expression is called the output function of the system. The state y_{n+1} of a discrete system at a period $n+1$ is completely determined by the input x_n and the state y_n at a period n, that is $y_{n+1} = v(y_n, x_n)$. This expression is called the next state function of the discrete system. Given the initial state, a linear system is completely determined by a differential or a difference equation or their solutions.

1.2 CLASSIFICATION OF AGRICULTURAL SYSTEMS

For practical purposes, the following factors were used for the classification criteria adopted for this book:

- The time scale of the system
- The uncertainties of events in the system
- Structure of the system

Within these classes, systems are named and classified according to the type of differential or difference equations of the mathematical model representing the system.

The Time Scale

There are two classes of systems as they relate to the time scale chosen for the mathematical model:

- Continuous systems
- Discrete systems

Continuous Systems. The time scale of continuous systems is the set of nonnegative real numbers. Continuous systems are called *differentiable systems* because they are represented by differential equations and their solutions.

Example 1.2.1 The following equation was fitted to the energy content of milk of a group of cows[2]:

---

[2]Computed from Lowman, B.G. et.al

$$y = 2.821 + 0.965e^{-0.0423t}$$

where y is the energy content of milk in MJoules/Kg and t is days after calving. Determine the corresponding differential equation.

Solution: The following is the first derivative of the state equation:

$$\frac{dy}{dt} = 0.965e^{-0.0423t}(-0.0423)$$

where $0.965e^{-0.0423t} = y - 2.821$. After replacing values, the following is the differential equation representing this system:

$$\frac{dy}{dt} = 0.1193 - 0.0423y$$

where 0.1193 is the energy input and $0.0423y$ is the energy output in MJoules/Kg/day.

Discrete Systems. The time scale of discrete systems is the set of nonnegative integers. Discrete state variables cannot be fractionalized, meaning that the system cannot be represented by differential equations. This is the case of state variables defined as number of individuals or as qualitative traits. Thus, the state changes are represented by difference equations.

Example 1.2.2 A rancher sells each month 3.6% of his feedlot steers and buys 90 new animals. The initial number of steers is 460. Define a mathematical model for the system.

Solution: This system is discrete because the state variable, number of steers, is discrete. The following difference equation represents the system:

$$y_{n+1} - y_n = 90 - 0.036y_n$$

where n is months, y_n is the number of steers corresponding to the present state of the system, y_{n+1} is the number of steers of the next state, $x = 90$ is the input as number of steers purchased and $z_n = 0.036y_n$ is the output as number of steers sold per month. The following is the corresponding state trajectory:

$$y_n = 2500 - 2040(0.964)^n$$

Example 1.2.3 It was found that, when the trees in a citrus farm are healthy, 20% get a disease within a year and when the trees are diseased, 30% of the trees recover. Define the mathematical model of the system.

Solution: This system can be modeled as a finite discrete system because the state variables are represented by two qualitative traits, the percent of healthy trees and the percent of diseased trees. The state transition matrix of the system is shown in the Table 1.2.11.The first row shows that the probability of healthy trees of remaining healthy in the next state is 80% and that the probability of becoming diseased is 20%. The second row shows that the probability of diseased trees of becoming healthy in the next state is 30% and that the probability of remaining diseased is 70%.

Table 1.2.1

Present State	Next State Probability	
	Healthy	**Diseased**
Healthy	0.80	0.20
Diseased	0.30	0.70

The state and output trajectories of the system are shown in Table 1.2.2. Note that the next state P_{n+1} of the system is defined by the product QP_n, where P_n is a state vector at time n and Q is the transition matrix. The initial condition of the system is assumed as $P_0 = (1,0)$, meaning that initially all the trees were healthy. By knowing the present state and the transition matrix, the next state of the system was predicted. Then, the next state of this system is completely determined by matrix Q and the present state.

As shown, the following matrix equation represents the system:

$$P_{n+1} = QP_n$$

where P_n is the set of states at time n and P_{n+1} is the set of states at time $n+1$.

Table 1.2.2.

Time	Present State P_n	Next State $P_{n+1} = QP_n$	Output Z_n
0	$(1, 0)$	$\begin{bmatrix} 0.8 & 0.2 \\ 0.3 & 0.7 \end{bmatrix}[1\ \ 0] = [0.8\ \ 0.2]$	$(-0.2, 0.2)$
1	$(0.8, 0.2)$	$\begin{bmatrix} 0.8 & 0.2 \\ 0.3 & 0.7 \end{bmatrix}[0.8\ \ 0.2] = [0.7\ \ 0.3]$	$(-0.1, 0.1)$
2	$(0.7, 0.3)$	$\begin{bmatrix} 0.8 & 0.2 \\ 0.3 & 0.7 \end{bmatrix}[0.7\ \ 0.3] = [0.65\ \ 0.35]$	$(-0.05, 0.05)$
3	$(0.65, 0.35)$	$\begin{bmatrix} 0.8 & 0.2 \\ 0.3 & 0.7 \end{bmatrix}[0.65\ \ 0.35] = [0.625\ \ 0.375]$	$(-0.025, 0.025)$
$\vdots$	$\vdots$	$\vdots$	$\vdots$

The state equation may also be written as

$$\Delta P_n = B - (I - Q)P_n$$

where B is the input of the system, I is an identity matrix and the output is the expression

$$Z_n = (Q - I)P_n$$

Note that the input is a null matrix, because no healthy or diseased trees are imported from outside the system. Then, the output of the system becomes the difference between the next state and the present state. The output shows the decrease of healthy trees and the increase of diseased trees, in percent units, at each new state of the system.

　　　If all the trees in the initial state were healthy, the following is the solution of the next state equation of the system[3]:

$$P_n = \frac{m}{5}\left[3 + 2(0.5)^n, \quad 2 - 2(0.5)^n\right]$$

for $P_n = (p_{1n}, p_{2n})$, where p_{1n} is healthy trees, p_{2n} is diseased trees, m is the total number of trees and n is years.

As shown in the next example, sometimes determining whether the system is continuous or discrete may be ambiguous.

Example 1.2.4 A forest area is chopped down and burned. After the first year, 20% of the burned area is the regrowth of trees and 30% is colonized by grasses. The remaining area stays as bare soil. Mortality of trees is 15% and mortality of grasses is 25%. Define the mathematical model of the system.

Solution: It may appear that grasses and trees are moving to colonize the bare soil. As such, this forest area may be defined as a continuous system, having the following mathematical model:

$$\frac{dY}{dt} = \begin{bmatrix} -(0.30+0.20) & 0.25 & 0.15 \\ 0.30 & -0.25 & 0 \\ 0.20 & 0 & -0.15 \end{bmatrix} Y$$

for $Y = (y_1, y_2, y_3)$, where Y is a column vector, y_1 is bare soil, y_2 is grasses, y_3 is trees and t is years. The solution of the above differential equations is the following set of state trajectories[4]:

$$y_1 = 0.2830 + 0.6937\,e^{-0.715t} + 0.0233\,e^{-0185t}$$
$$y_2 = 0.3396 - 0.4479\,e^{-0.715t} + 0.1083\,e^{-0.185t}$$
$$y_3 = 0.3773 - 0.2457\,e^{-0.715t} - 0.1316\,e^{-0.185t}$$

---

[3]See Chapters 4 and 9 for procedures for solving linear difference equations.

[4]See Chapter 4 and 10 for procedures for solving linear differential equations.

The area may also be defined as a finite discrete system with three states, percent of bare soil, percent grass and percent trees. Table 1.2.3 shows the corresponding transition matrix:

Table 1.2.3

Present State	Next State		
	Bare Soil	**Grasses**	**Trees**
Bare Soil	0.50	0.30	0.20
Grasses	0.25	0.75	0
Trees	0.15	0	0.85

The following set of next state equations represents the system:

$$Y_{n+1} = \begin{bmatrix} 0.50 & 0.30 & 0.20 \\ 0.25 & 0.75 & 0 \\ 0.15 & 0 & 0.85 \end{bmatrix} Y_n$$

As shown in the table and the above equations, 50% of bare soil may remain as bare soil, 30% may become grasses and 20% may become the regrowth of trees. It is also shown that 25% of the grasses may die out, reverting to bare soil and that 75% may remain as grasses. Fifteen percent of the trees may die and revert to bare soil and 85% may stay alive.

Note that the above transition matrix is the transpose of the matrix of the continuous model. Note also that Y_n is a row vector, whereas Y in the continuous system is a column vector.

After solving the above matrix equation, the following are the state trajectories when the initial state is bare soil[5]:

$$y_{1n} = 0.2830 + 0.6937(0.285)^n + 0.0233(0.815)^n$$
$$y_{2n} = 0.3396 - 0.4479(0.285)^n + 0.1083(0.815)^n$$
$$y_{3n} = 0.3773 - 0.2457(0.285)^n - 0.1316(0.815)^n$$

---

[5]See Chapters 4 and 9 for procedures.

Note that these expressions are identical to the state trajectories of the continuous model.

Very seldom is agricultural research data recorded as a continuous flow of information. Most often data of continuous systems is recorded at fixed Δt periods, determining the discretization of the time scale. By discretizing the time scale of a continuous system, the system is also discretized. An important question is here, how much information is lost within each Δt period, a question that must be taken into consideration when designing experiments. Clearly, Δt must be small enough as to prevent important information from being lost.

Uncertainties of Events

Most of the inputs reaching agricultural systems cannot be controlled and occur in a random pattern. Therefore, the operation of all agricultural systems is subject to some kind of uncertainties. Then, depending on whether these uncertainties are considered in the mathematical model or are ignored, two types of systems evolve:

- Stochastic systems
- Deterministic systems

Stochastic Systems. The basic feature of a stochastic model is that state variables are defined as probability distributions.

Example 1.2.5 Define the state probability distributions for the citrus trees in Example 1.2.3, assuming a binomial distribution of events.

Solution: The following was the state joint distribution expression defined for the system in Example 1.2.3, for an initial state $P_0 = (1, 0)$:

$$P_n = \frac{m}{5}\left[3 + 2(0.5)^n, \quad 2 - 2(0.5)^n\right]$$

where m is the total number of trees. This expression corresponds to a deterministic model of the system. It is assumed that the state variables have the binomial distribution, as shown below:

$$f_n(x_1, x_2) = \frac{m!}{x_1! x_2!} p_{1n}^{x_1} p_{2n}^{x_2}$$

where x_1 and x_2 are the number of healthy and diseased tress and p_1 and p_2 are the corresponding probabilities. Then, by replacing the P_n values in the binomial expression, it is possible to define the following state probability model of the system:

$$f_n(x_1,x_2) = \frac{m!}{x_1!\,x_2!}\left(\frac{3+2(0.5)^n}{5}\right)^{x_1}\left(\frac{2-2(0.5)^n}{5}\right)^{x_2}$$

The probability distribution curves of diseased trees are shown in Fig. 1.2.1. The total number of trees was assumed to be 10.

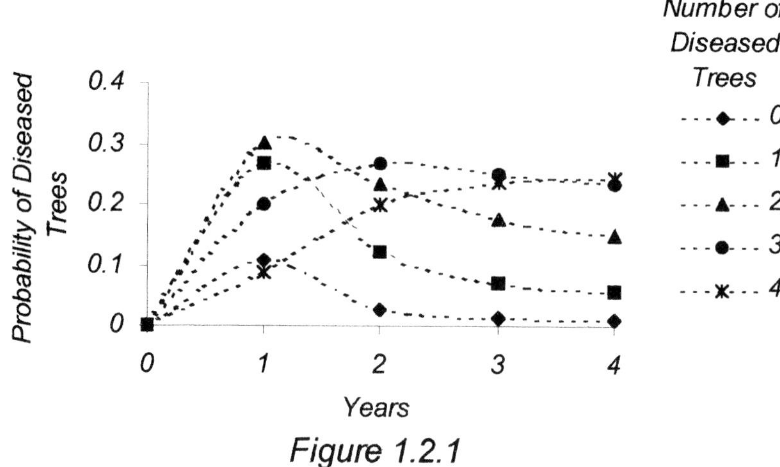

Figure 1.2.1

Deterministic Systems. In deterministic models, the states of the system are the expected values of the outcomes. Thus, deterministic models represent the expected or average behavior of the system. The first four examples in this section were all deterministic models.

A real system may be defined by a deterministic model or a stochastic model. Deterministic models are simpler and more widely used than stochastic models.

Example 1.2.6 The following is the deterministic model for the state equation of the previous example, expressed as expected values:

$$E_n(x_1,x_2) = \frac{m}{5}\left[3+2(0.5)^n, \quad 2-2(0.5)^n\right]$$

where x_1 is the expected number of healthy trees and x_2 is the expected number of diseased trees at time n. The graphic representation of expected values is shown in Fig. 1.2.2. The total number of trees is assumed to be 10.

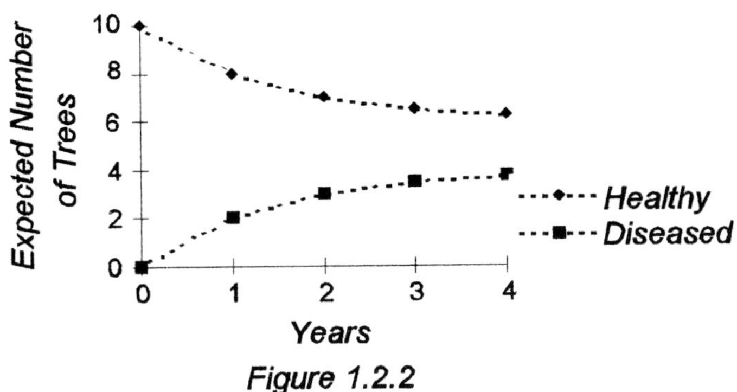

Figure 1.2.2

Structure of Systems

The notion of structure of a system is related to how *component systems* are coupled to form a more complicated system. The following classification was adopted here:

- Interactive coupled systems
- Conjunctive coupled systems

Interactive Coupled Systems. Interacting systems may be coupled by means of interconnected differential or difference equations, determining an *interactive coupling*. Interacting agricultural systems may be arranged in two groups:

- Compartmental systems
- Non compartmental systems

Compartmental Systems. Components of compartmental systems are called compartments. Such compartments work as chambers among which *some material is considered to move.*

Example 1.2.6 The movement of DDT from plant to soil is 25% per month, from soil to plant 2% and carried out with ground water 5%. Define the mathematical model of the system.

Solution: This system is represented in Fig. 1.2.3. The following set of differential equations defines the flow of DDT in the system[6]:

---

[6]See Chapters 4 and 10 for procedures.

$$\frac{dY}{dt} = \begin{bmatrix} -0.25 & 0.02 \\ 0.25 & -(0.02+0.05) \end{bmatrix} Y$$

for $Y = (y_p, y_s)$, where y_p is a state of the plant compartment, y_s is a state of the soil compartment and t is months. Coefficients with positive signs are the compartment inputs and coefficients with negative signs are the compartment outputs.

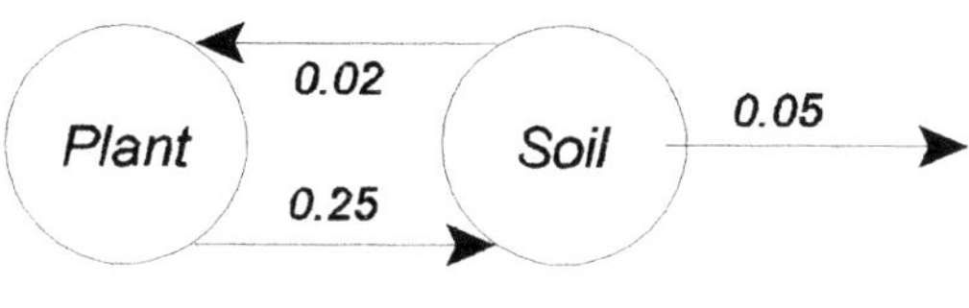

Figure 1.2.3

Non Compartmental Systems. Components of non compartmental systems sometimes are called black boxes, among which *some information is considered to move.*

Example 1.2.7 The following matrix equation defines the relationships between pasture yield and carrying capacity of a Kikuyu pasture field, as affected by rainfall[7][2]:

$$\frac{dY}{dt} = \begin{bmatrix} 0 & -2.9463 \\ 0.3338 & -2.5282 \end{bmatrix} Y + \begin{bmatrix} 0.1107 \\ 0.0119 \end{bmatrix} \frac{dx}{dt} + \begin{bmatrix} 0.0871 \\ 0 \end{bmatrix} x$$

for $Y = (y_1, y_2)$ where y_1 is leaf growth in kg of dry matter per ha/day, y_2 is the number of cows/ha and x is rainfall as mm/month. Determine the input and output of the system.

Solution: The mathematical model of the system has the form

$$\frac{dY}{dt} = AY + B\frac{dx}{dt} + Cx$$

The system input is rainfall, defined by the expression $Bdx/dt+Cx$ and the system output is AY. Fig. 1.2.4 shows the system black boxes exchanging information by input-output relationships.

---

[7]Computed from Murtagh, G.J. et.al.

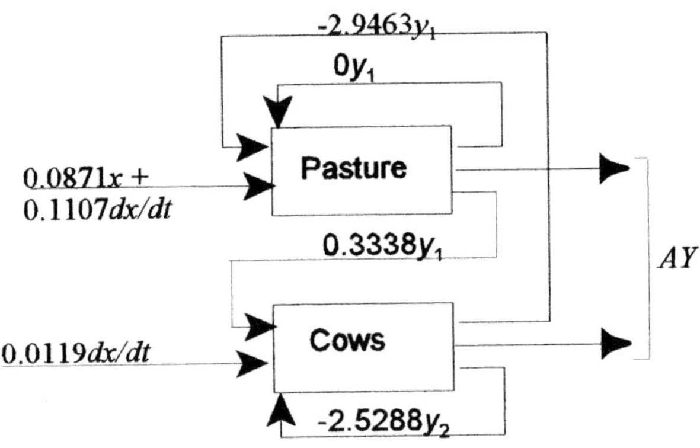

Figure 1.2.4

Conjunctive Coupled Systems. The idea of conjunctive coupling is that of a complex system where each of its components operates independently. This is the case, for example, of different plots or different experimental material, such that each plot is a component system and operates as a replication of the experiment. Grouping of experimental material determines the sources of variation in a typical analysis of variance.

As shown in the next example, the notion of conjunctive coupling is particularly helpful in factorial arrangements of treatments and in split-plot experimental designs.

Example 1.2.8 An experiment was designed to study how starch in the diet of cattle affects the digestibility of roughage. The experimental roughage were corn stalks, stems of the banana plant, sugarcane leaves and Stargrass hay. Different amounts of green bananas provided starch. *In vivo* digestibility procedures were carried out with six fistulated steers. Define the experiment as a conjunctive coupled system.

Solution: As shown in Fig. 1.2.5, each steer is a replication and a component system of the experiment as a system. Each roughage is a component within a fistulated steer as a system. Green bananas are the inputs.

Note that roughage and green bananas are both treatments, but only green bananas are inputs, because only green bananas are scheduled over the time variable. This is a very important factor in the design of experiments, because it determines the scaling of classes in the field design and in the analysis of variance. Inputs are placed always at the lowest end of the list of treatments and should never be confused with components.

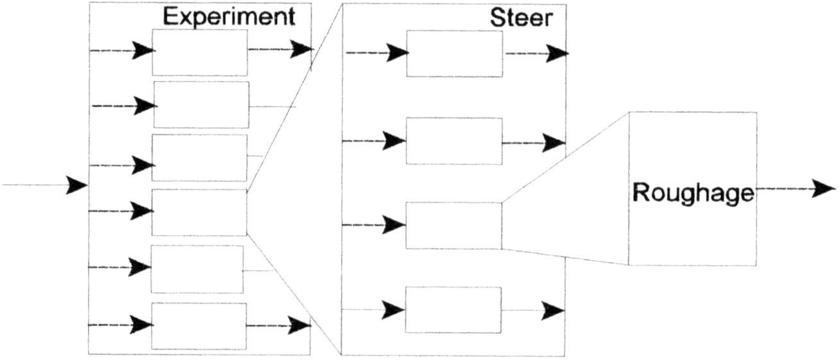

Figure 1.2.5

Summary

Depending on the time scale adopted for the mathematical model, systems are classified as continuous or discrete. Whether uncertainties are considered or are ignored, systems are stochastic or deterministic. Systems having interface relationships are called interactive coupled systems. If some material is moving among component systems, the system is called compartmental. Conversely, if components exchange only information, the system is non compartmental. Component systems having no interface relationships are called conjunctive coupled systems.

1.3 USING LINEAR MODELS IN AGRICULTURAL RESEARCH

Agricultural systems are very complex and are characterized by having multiple input variables of unknown or chaotic behavior. Thus, mathematical modeling in agricultural research is essentially an empirical process, with only few feasible theoretical considerations. Thus, a free choice of mathematical models of agricultural systems is possible. The simplest empirical option for modeling is using linear models. This was the approach taken in writing this book.

The following examples are presented here as an introduction to the use of linear models in agricultural systems. The first example is related to modeling and analyzing a deterministic system.

Example 1.3.1 The following are some selected experimental features from a research report on pasture production, as determined by pasture yield and pasture carrying capacity[8]:

Animals: Guernsey and Jersey cows with an average live weight of 364 kg.
Treatments: Nitrogen fertilized Kikuyu pastures (*Pennisetum clandestinum*), where more than 90% of the grass cover was Kikuyu and unfertilized Carpet grass (*Axonopus affinis*), where the average carpet grass cover was 59 %.
Measurements: Pasture yield, as dry green leaves in kg/ha/day and carrying capacity, as cows/ha. The pasture growing season was divided into ten 4-week periods. Data were collected for each period.

>   1) Define the mathematical models for pasture yield in both treatments.
>   2) Define a mathematical model for the relationship between yield and carrying capacity in the Kikuyu treatment.

Solution: The **first step** for defining the mathematical model of pasture yield is to arrange the data as a difference table. A difference table is a table that gives successive differences of $y_t = f(t)$, for $t = 1, 2, ...,$. For example,

$$\Delta y_t = y_{t+\Delta t} - y_t$$

is called a *first order finite difference,* where Δt is a time increment. Thus, a second order difference would be

$$\Delta^2 y_t = \Delta_{t+\Delta t} - \Delta y_t$$

and so on. The first entry in each column is called *the leading difference.*

Before defining a difference table for a continuous system, the data should be discretized. The derivative of a function is represented by the symbol dy/dt. The symbol dy is called *the differential of the state variable* and the symbol dt is called *the differential of the time variable.* The differential dt is always equal to the time increment Δt. Conversely, the differential dy is not equal to the finite difference Δy. However, if Δt is small enough, dy could be an acceptable approximation of Δy[9]. This was the accepted criterion for discretizing and fitting the Kikuyu data to the proposed mathematical model for the system.

The following is the difference table for the Kikuyu treatment, up to the second order difference. Note that this table was modified for a regression analysis, by having the

---

[8]Computed from Murtagh, G.J. et.al.

[9]This subject is discussed in Chapter 3.

leading differences aligned with the time scale values and not located in the traditional arrangement, between two successive entries of the preceding column. Note also that the time scale is discrete and that $\Delta t = 1 = \Delta t^2$.

Table 1.3.1

t	y	$\Delta y/\Delta t$	$\Delta^2 y/\Delta t^2$	x	$\Delta x/\Delta t$	$\Delta^2 x/\Delta t^2$
0.00	12.00	3.00	1.00	46.00	22.00	23.00
1.00	15.00	4.00	27.00	68.00	45.00	259.00
2.00	19.00	31.00	-33.00	113.00	304.00	-486.00
3.00	50.00	-2.00	25.00	417.00	-182.00	327.00
4.00	48.00	23.00	-28.00	235.00	145.00	-278.00
5.00	71.00	-5.00	-16.00	380.00	-133.00	-23.00
6.00	66.00	-21.00	0.00	247.00	-156.001	19.00
7.00	45.00	-21.00	18.00	91.00	-37.00	79.00
8.00	24.00	-3.00		54.00	42.00	
9.00	21.00			96.00		

The table includes the differences for the state variable y and the input variable x, where t is months, y is the leaf growth rate as kg of dry matter/ha/day and x is the rainfall input in mm/month.

The **second step** is defining the differential equation representing this system. The following linear second order difference equation was computed, by linear regression, from the difference table:

$$\frac{\Delta^2 y}{\Delta t^2} + 1.4567\frac{\Delta y}{\Delta t} + 0.6210y = 0.0972\frac{\Delta^2 x}{\Delta t^2} + 0.1975\frac{\Delta x}{\Delta t} + 0.1321x$$

If Δy and Δx are considered an acceptable approximation of dy and dx, then the following differential equation may be defined:

$$\frac{d^2 y}{dt^2} + 1.4567\frac{dy}{dt} + 0.6210y = 0.0972\frac{d^2 x}{dt^2} + 0.1975\frac{dx}{dt} + 0.1321x$$

As indicated by the statistics of the linear regression analysis shown in Table 1.3.2, the relationship between pasture yield and the rainfall input is significant. The coefficient of determination R^2 may be less than the above value, because this was a linear regression through the origin.

Table 1.3.2

R Square	0.99108
Adjusted R Square	0.97622
Standard Error	3.36442

Analysis of Variance

	DF	Sum of Squares	Mean Square
Regression	5	3774.04199	754.80840
Residual	3	33.95801	11.31934

Variable	B	SE B	t
$\Delta y/\Delta t$	-1.456682	0.361637	-4.028
y	-0.620985	0.194118	-3.199
$\Delta^2 x/\Delta t^2$	0.097204	0.008583	11.325
$\Delta x/\Delta t$	0.197499	0.037031	5.333
x	0.132113	0.041822	3.159

Each coefficient of the mathematical model was evaluated by a "t" test. If k_i is a coefficient of the state equation and k_0 is the corresponding hypothetical value, then where S_{k_i} is the standard error of the k_i coefficient. The null hypothesis is here $k_i - k_0 = 0$.

$$t = \frac{k_i - k_0}{S_{k_i}}$$

The **third step** is determining the solution of the system differential equation. Note that the above differential equation has the form:

$$\frac{d^2 y}{dt^2} + b_1 \frac{dy}{dt} + b_2 y = c_0 \frac{d^2 x}{dt^2} + c_1 \frac{dx}{dt} + c_2 x$$

The left hand may be represented by the polynomial $s^2 + b_1 s + b_2 = (s + \lambda_1)(s + \lambda_2)$, called the *characteristic equation* of the system[10]. Since pasture production often displays a cyclical pattern due to climatic conditions, then $\lambda = \alpha \pm i\beta$. Thus, the characteristic equation of the system is $s^2 + 1.4567s + 0.6210$, where $\lambda = -0.7284 \pm 0.3009i$ and $i = \sqrt{-1}$.

The following is the solution of the differential equation of the Kikuyu grass

---

[10]See Chapter 2 for definitions.

system[11]:

$$y = 13.36 - 13.36\cos 0.809t - 19.76\sin 0.809t$$
$$+ 31.11 + e^{-0.728t}\left[(13.01 - 31.11)\cos 0.301t\right]$$

where the terms $13.36 - 13.36\cos 0.809t - 19.76\sin 0.809t$ are the rainfall components of the state equation, 0.728 and 0.809 are the α and β terms of $\lambda = \alpha \pm i\beta$ in the characteristic equation of the system and the term 13.01 is the initial value y_0.

As indicated before, some error was introduced here by using values of the finite difference Δy for the differential operator dy and values of the difference Δx for the differential operator dx. Note, however, that this error is also implicit in the data, because the data was not recorded continuously. A non linear regression procedure was used to account for this discretization error. The summary of the non-linear regression statistics is shown below:

Table 1.3.3

Source	DF	Sum of Squares	Mean Square
Regression	4	17701.76629	4425.44157
Residual	6	271.23371	45.20562
Uncorrected Total	10	17973.00000	
(Corrected Total)	9	4208.90000	

R squared = 1 - Residual SS / Corrected SS = 0.93556

Parameter	Estimate	Asymptotic Std. Error	t
k_1	13.356942635	3.297376801	4.05
k_2	19.762102352	3.215554382	6.14
k_3	31.111270880	4.083738812	7.63
y_0	13.008102655	6.242228999	2.08

where

$$y = k_1 - k_1\cos 0.809t - k_2\sin 0.809t + k_3 + e^{-\alpha t}\left[(y_0 - k_3)\cos\beta t)\right]$$

Note that, by using the addition formula of sinuses and cosines, the state equation may be expressed in the form

---

[11]See Chapters 4 and 9 for procedures for solving linear differential equations.

$$y = k + ae^{-\alpha t}\cos[\beta(t-b)]$$

This form of the state equation is very useful and has a straightforward geometric interpretation:

- Parameter β modulates the frequency response of the system
- The term $ae^{-\beta t}$ modulates the amplitude response of the system
- Parameter b is the out-of-phase parameter
- Parameter k is the distance between the abscissa and the axes of the response curve
- A cycle is equal to $2\pi/\beta$

Note that when $\alpha<0$ the amplitude decreases over time, when $\alpha>0$ the amplitude increases and when $\alpha=0$ the amplitude is only determined by coefficient a. The above model is shown in Fig. 1.3.1 for $\alpha<0$.

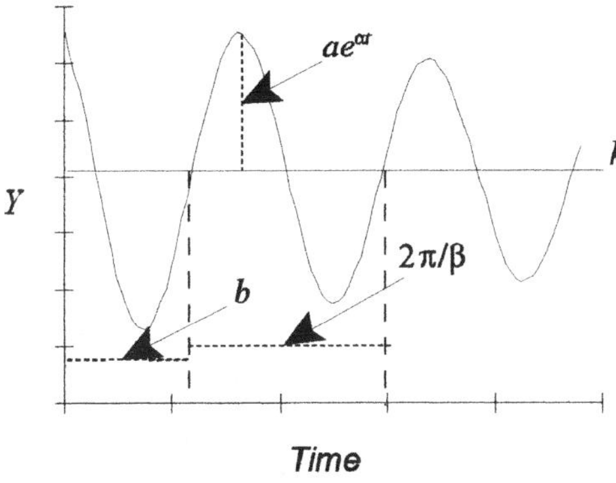

Figure 1.3.1

By the above transformation, the Kikuyu state equation becomes

$$y = 13.36 - 23.88\cos[0.809(t - 1.207)]$$
$$+ 31.11 + e^{-0.728t}[(13.01 - 31.11)\cos 0.301\,t]$$

This expression is the sum of two components of the state equation. The first component is called the *free response* of the system and is related only to the initial conditions y_0:

$$y_a = 13.01e^{-0.728t}\cos 0.301\,t$$

The second component is called the *forced response* of the system and is related only to the input $x = f(t)$:

$$y_b = 13.36 - 23.88\cos\left[0.809(t - 1.207)\right] + 31.11(1 - e^{-0.728t}\cos 0.301t)$$

The graph of the Kikuyu response functions is shown in Fig. 1.3.2:

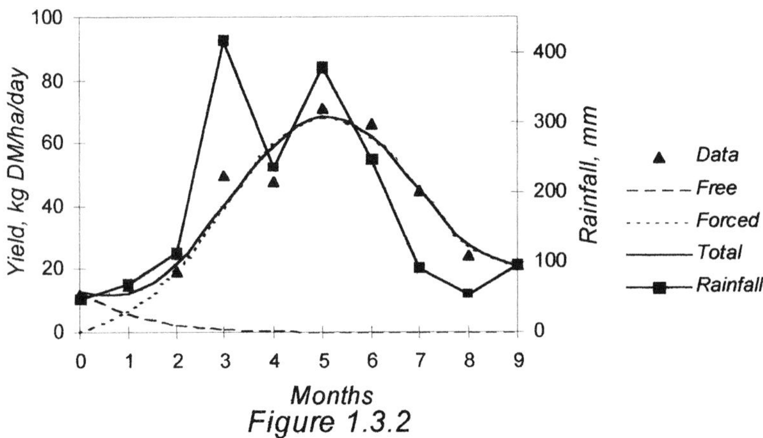

Figure 1.3.2

As expected, the free response fades away rapidly. In contrast, the forced response curve follows the shape of the rainfall data.

The following is the difference table for Carpet grass

Table 1.3.4

t	y	$\Delta y/\Delta t$	$\Delta^2 y/\Delta t^2$	x	$\Delta x/\Delta t$	$\Delta^2 x/\Delta t^2$
0.00	3.00	0.00	1.00	46.00	22.00	23.00
1.00	3.00	1.00	12.00	68.00	45.00	259.00
2.00	4.00	13.00	-12.00	113.00	304.00	-486.00
3.00	17.00	1.00	16.00	417.00	-182.00	327.00
4.00	18.00	17.00	-32.00	235.00	145.00	-278.00
5.00	35.00	-15.00	8.00	380.00	-133.00	-23.00
6.00	20.00	-7.00	2.00	247.00	-156.00	119.00
7.00	13.00	-5.00	-1.00	91.00	-37.00	79.00
8.00	8.00	-6.00		54.00	42.00	
9.00	2.00			96.00		

and the following is the corresponding difference equation:

$$\frac{\Delta^2 y}{\Delta t^2} + 1.9103\frac{\Delta y}{\Delta t} + 1.5624y = 0.02881\frac{\Delta^2 x}{\Delta t^2} + 0.06749\frac{\Delta x}{\Delta t} + 0.11321x$$

As indicated before, this difference equation was considered an acceptable approximation of the system differential equation, such that

$$\frac{d^2 y}{dt^2} + 1.9103\frac{dy}{dt} + 1.5624y = 0.02881\frac{d^2 x}{dt^2} + 0.06749\frac{dx}{dt} + 0.11321x$$

The characteristic equation of this system is $s^2 + 1.9103s + 1.5624$, where the roots are $\lambda = -0.9551 \pm 0.8063i$. The system statistics are given below:

Table 1.3.5

R Square	0.99722
Adjusted R Square	0.99259
Standard Error	1.23193

Analysis of Variance

	DF	Sum of Squares	Mean Square
Regression	5	1633.44703	326.68941
Residual	3	4.55297	1.51766

Variable	B	SE B	t
$\Delta y/\Delta t$	-1.910289	0.099422	-19.214
y	-1.562407	0.105729	-14.777
$\Delta^2 y/\Delta t^2$	0.028806	0.003481	8.274
$\Delta x/\Delta t$	0.067494	0.007415	9.103
x	0.113215	0.007590	14.917

The following is the solution of the differential equation for Carpet grass:

$$y = 7.52 - 7.52\cos 0.809t - 9.70\sin 0.809t + 7.57(1 - e^{-0.955t}\cos 0.806t)$$

where the terms $7.52 - 7.52\cos 0.809t - 9.70\sin 0.809t$ are components of the rainfall input and 0.955 and 0.806 are the numerical values α and β from the characteristic equation of the carpet grass system. Note that the coefficient for the initial value y_0 was not significant and was deleted from the state equation. Thus, there is no significant response to initial conditions for carpet grass and the above equation represents the forced response of the system.

The following is the statistical summary of the non-linear regression used to correct discretization errors for Carpet grass:

Table 1.3.6

Source	DF	Sum of Squares	Mean Square
Regression	3	2376.40104	792.13368
Residual	7	132.59896	18.94271
Uncorrected Total	10	2509.00000	
(Corrected Total)	9	996.10000	

R squared = 1 - Residual SS / Corrected SS = 0.86688

Parameter	Estimate	Asymptotic Std. Error	t
k_1	7.515671654	2.202448128	2.92
k_2	9.702182051	2.044905458	3.42
k_3	7.574700565	2.589750722	4.75

where

$$y = k_1 - k_1 \cos 0.809t - k_2 \sin 0.809t + k_3(1 - e^{\alpha t} \cos \beta t)$$

The graph of the state equations of the Kikuyu and the Carpet grasses is shown in Fig. 1.3.3.

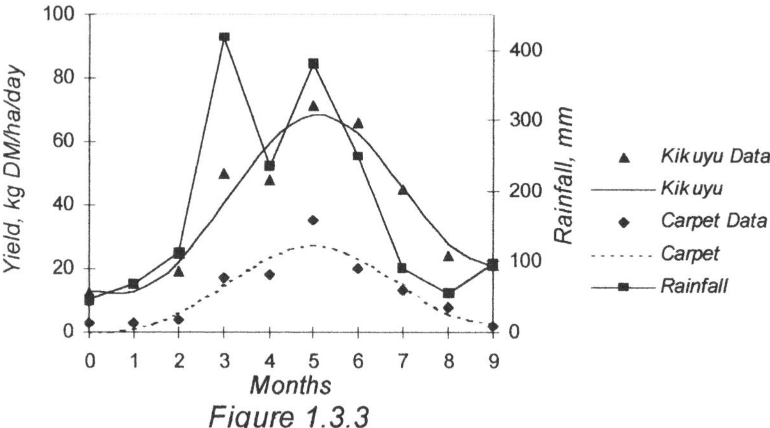

Figure 1.3.3

Differences between the Kikuyu and the Carpet treatments may be evaluated by the "t" test. If k_{i1} is a coefficient of the Kikuyu equation and k_{i2} is a coefficient of the Carpet equation, then

$$t = \frac{k_{i1} - k_{i2}}{\sqrt{S_{k_{i1}}^2 + S_{k_{i2}}^2}}$$

where $S_{k_{i1}}$ and $S_{k_{i2}}$ are standard errors. The null hypothesis is here $k_{i1} - k_{i2} = 0$.
The same approach may be used for evaluating relationships among different variables within a system, such as determining the relationships between pasture yield and carrying capacity in the Kikuyu pasture system.

The first step is arranging the data as a difference table, for $Y = (y_1, y_2)$, where y_1 is pasture yield, y_2 is carrying capacity and x is rainfall as mm/month:

Table 1.3.7

t	y_1	$\Delta y_1/\Delta t$	y_2	$\Delta y_2/\Delta t$	x	$\Delta x/\Delta t$
0.00	12.00	3.00	1.30	0.90	46.00	22.00
1.00	15.00	4.00	2.20	0.80	68.00	45.00
2.00	19.00	31.00	3.00	2.00	113.00	304.00
3.00	50.00	-2.00	5.00	1.60	417.00	-182.00
4.00	48.00	23.00	6.60	1.80	235.00	145.00
5.00	71.00	-5.00	8.40	0.90	380.00	-133.00
6.00	66.00	-21.00	9.30	-2.80	247.00	-156.00
7.00	45.00	-21.00	6.50	-2.80	91.00	-37.00
8.00	24.00	-3.00	3.70	-1.70	54.00	42.00
	9.00	21.00		2.00	96.00	

The **second step** is determining the mathematical model of the system. The following first order set of linear difference equations was fitted to the above data:

$$\frac{\Delta Y}{\Delta t} = \begin{bmatrix} 0 & -2.9463 \\ 0.3338 & -2.5288 \end{bmatrix} Y + \begin{bmatrix} 0.11066 \\ 0.01185 \end{bmatrix} \frac{\Delta x}{\Delta t} + \begin{bmatrix} 0.08708 \\ 0 \end{bmatrix} x$$

This system of equations shows that the number of cows affects negatively the change in pasture yield. These equations also indicates that the change in carrying capacity is affected positively by pasture yield and negatively by the number of cows. Again, these equations are considered as an acceptable approximation of the system differential equations. The statistics for the y_1 variable is shown in Table 1.3.8. Note that the coefficient for the y_1 variable was deleted because it was not statistically significant.

Table 1.3.8

R Square	0.95829		
Adjusted R Square	0.93744		
Standard Error	4.11405		

Variable	B	SE B	t
y_2	-2.946283	0.530966	-5.549
x	0.087075	0.013815	6.303
$\Delta x/\Delta t$	0.110656	0.009957	11.113

The following is the statistics for the y_2 variable:

Table 1.3.9

Multiple R	0.94348
R Square	0.89015
Adjusted R Square	0.83522
Standard Error	0.74886

Variable	B	SE B	t
y_1	0.333796	0.053107	6.285
y_2	-2.528848	0.401317	-6.301
$\Delta x/\Delta t$	0.011854	0.002060	5.753

The **third step** is finding the set of solutions of the system. The above set of equations has the form

$$\frac{\Delta Y}{\Delta t} = \begin{bmatrix} b_{11} & b_{12} \\ b_{21} & b_{22} \end{bmatrix} Y + \begin{bmatrix} c_1 \\ c_2 \end{bmatrix} \frac{\Delta x}{\Delta t} + \begin{bmatrix} c_1 \\ c_2 \end{bmatrix} x$$

where Y is the set of state variables. Again, it is assumed here that this model may represent the set of differential equations of the system. The characteristic equation of the system is given by the expansion of the following determinant, where $\lambda_1 = 0.4800$ and $\lambda_2 = 2.0488$:

$$|sI - B| = \begin{vmatrix} s - b_{11} & -b_{12} \\ -b_{21} & s - b_{22} \end{vmatrix} = \begin{vmatrix} s & 2.9463 \\ -0.3338 & s+2.5288 \end{vmatrix}$$

$$= s^2 + 2.5288s + 0.9835$$

$$= (s + 0.4800)(s + 2.0488)$$

Then, the following are the solutions of the system[12]:

$$Y_a = \begin{bmatrix} 80.89 & -68.93 \\ 0 & 1.23 \end{bmatrix} \begin{bmatrix} e^{-0.480t} \\ e^{-2.049t} \end{bmatrix}$$

This is the free response of the system. The following is the forced response:

$$Y_b = \begin{bmatrix} 0.427 & 0 \\ 0 & 0.339 \end{bmatrix} \begin{bmatrix} 72.24 \\ 12.46 \end{bmatrix} + \begin{bmatrix} -1.196(72.24) & 0.769(72.24) \\ 0.194(12.46) & -0.534(12.46) \end{bmatrix} \begin{bmatrix} e^{-0.480t} \\ e^{-2.049t} \end{bmatrix}$$
$$+ \begin{bmatrix} 1 & 0 \\ 0 & 1 \end{bmatrix} \begin{bmatrix} 13.57 \\ 1.073 \end{bmatrix} - \begin{bmatrix} 13.57 & 20.64 \\ 1.073 & 3.418 \end{bmatrix} \begin{bmatrix} \cos 0.809t \\ \sin 0.809t \end{bmatrix}$$

The total response Y is the sum $Y_a + Y_b$. Note that the terms

$$\begin{bmatrix} 1 & 0 \\ 0 & 1 \end{bmatrix} \begin{bmatrix} 13.57 \\ 1.073 \end{bmatrix} - \begin{bmatrix} 13.57 & 20.64 \\ 1.073 & 3.418 \end{bmatrix} \begin{bmatrix} \cos 0.809t \\ \sin 0.809t \end{bmatrix}$$

in the forced response are related to the rainfall input.

The coefficients of the free response are defined as follows:

$$Y_a = \begin{bmatrix} k_{12} & k_{13} \\ k_{22} & k_{23} \end{bmatrix} \begin{bmatrix} e^{-\lambda_1 t} \\ e^{-\lambda_2 t} \end{bmatrix}$$

---

[12]See Chapters 4 and 9 for procedures for solving systems of differential equations.

where k_{1i} stands for yield and k_{2i} stands for carrying capacity. The following are definitions of the forced response coefficients:

$$Y_b = \begin{bmatrix} \dfrac{b_{12}-b_{22}}{\lambda_1\lambda_2} & 0 \\[2ex] 0 & \dfrac{b_{21}-b_{11}}{\lambda_1\lambda_2} \end{bmatrix}\begin{bmatrix} k_{11} \\ k_{21} \end{bmatrix} + \begin{bmatrix} \dfrac{\lambda_1-b_{12}+b_{22}}{\lambda_1(\lambda_1-\lambda_2)}k_{11} & \dfrac{\lambda_2-b_{12}+b_{22}}{\lambda_2(\lambda_1-\lambda_2)}k_{11} \\[2ex] \dfrac{\lambda_1-b_{21}+b_{11}}{\lambda_1(\lambda_1-\lambda_2)}k_{21} & \dfrac{\lambda_2-b_{21}+b_{22}}{\lambda_2(\lambda_1-\lambda_2)}k_{21} \end{bmatrix}\begin{bmatrix} e^{-\lambda_1 t} \\ e^{-\lambda_2 t} \end{bmatrix}$$

$$+ \begin{bmatrix} 1 & 0 \\ 0 & 1 \end{bmatrix}\begin{bmatrix} k_{14} \\ k_{24} \end{bmatrix} - \begin{bmatrix} k_{14} & k_{15} \\ k_{24} & k_{25} \end{bmatrix}\begin{bmatrix} \cos 0.809\, t \\ \sin 0.809\, t \end{bmatrix}$$

The graph of pasture yield is shown in Fig. 1.3.4

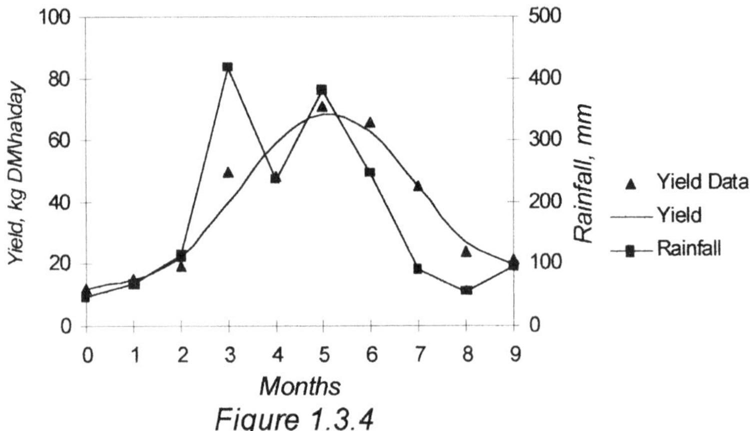

Figure 1.3.4

The following is corresponding summary of the non-linear regression, used to correct discretization errors for pasture yield:

Table 1.3.10

Source	DF	Sum of Squares	Mean Square
Regression	5	17708.15325	3541.63065
Residual	5	264.84675	52.96935
Uncorrected Total	10	17973.00000	
(Corrected Total)	9	4208.90000	

R squared = 1 - Residual SS / Corrected SS = 0.93707

Parameter	Estimate	Asymptotic Std. Error	t
k_{11}	72.236071842	10.742681413	6.73
k_{12}	80.894743168	17.793591477	4.55
k_{13}	68.925903214	19.888946475	3.47
k_{14}	13.570507544	3.690590480	3.68
k_{15}	20.639417786	4.228578006	4.88

The graph of carrying capacity is shown in Fig. 1.3.5

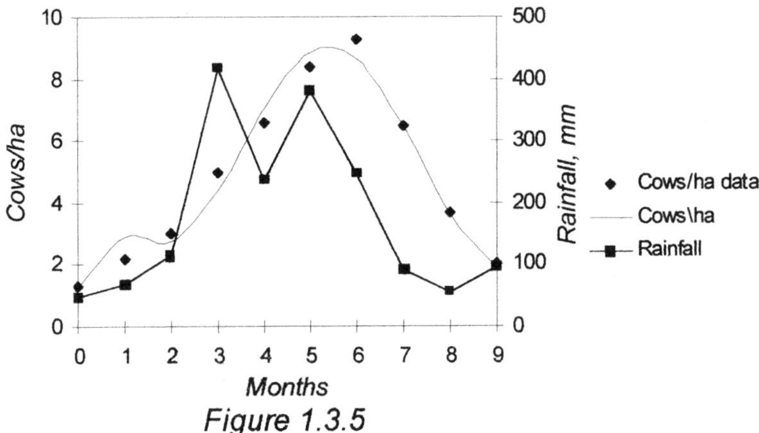

Figure 1.3.5

The following is the corresponding summary statistics for correcting discretization errors for carrying capacity:

Table 1.3.11

Source	DF	Sum of Squares	Mean Square
Regression	4	299.40021	74.85005
Residual	6	1.67979	0.27997
Uncorrected Total	10	301.08000	
(Corrected Total)	9	70.68000	

R squared = 1 - Residual SS / Corrected SS = 0.97623

Parameter	Estimate	Asymptotic Std. Error	t
k_{21}	12.461660491	0.847204071	14.71
k_{23}	1.229653345	0.526192538	2.33
k_{24}	1.073217885	0.265371083	4.05
k_{25}	3.417872472	0.253417049	13.50

The next example is related to a stochastic system.

Example 1.3.2 Determine the mathematical model and the state equations for the diseased trees in Example 1.2.3, assuming that some of the diseased trees may die.

Solution: The **first step** in defining the mathematical of the system is defining the data in as a transition table:

Table 1.3.12

Present State	Next State		
	Healthy	**Diseased**	**Dead**
Healthy	0.80	0.20	0
Diseased	0.30	0.10	0.6
Dead	0	0	1

The first row shows that the probability of healthy trees of remaining healthy in the next state is 80%, the probability of becoming diseased is 20% and the probability of dying is zero. The second row show that the probability of diseased trees of becoming healthy in the next state is 30%, the probability of remaining diseased is 10% and the probability of dying is 60%. The third row shows that dead trees would remain dead. The following probability matrix represents the above table:

$$Q = \begin{bmatrix} 0.8 & 0.2 & 0 \\ 0.3 & 0.1 & 0.6 \\ 0 & 0 & 1 \end{bmatrix}$$

The **second step** is defining the set of next state equations of the system. The state changes of the system are defined by the product QP_n, where P is a state probability vector. By knowing the present state P_n and the probability matrix Q, the next state QP_n of the system is predicted. Therefore, the following matrix equation represents the system:

$$P_{n+1} = P_n Q$$

where P_n is the set of states at time n and P_{n+1} is the set of states at time $n+1$.

The **third step** is defining the solution of the next state equation. The solution of the next state equations has the form $F(0)Q^n$, where $F(0)$ is the set of initial conditions[13]:

$$P_n = \begin{bmatrix} 1 & 0 & 0 \\ 0 & 1 & 0 \\ 0 & 0 & 1 \end{bmatrix}\begin{bmatrix} 0.8 & 0.2 & 0 \\ 0.3 & 0.1 & 0.6 \\ 0 & 0 & 1 \end{bmatrix}^n$$

Note that the first row in the initial state matrix shows that all the trees were healthy, the second row shows that all the trees in the initial state were diseased and the third row shows that all the trees in the initial state were dead. After solving the above power matrix and assuming that all the trees in the initial state were healthy, the solution is

$$P_n = [0.910(0.877)^n + 0.090(0.023)^n, \quad 0.234(0.877)^n - 0.234(0.023)^n,$$
$$- 1.144(0.877)^n + 0.144(0.023)^n + 1]$$

for $P_n = (p_{1n}, p_{2n}, p_{3n})$, where p_{1n} is the proportion of healthy trees, p_{2n} is the proportion of diseased trees, p_{3n} is the proportion of dead trees and n is years.

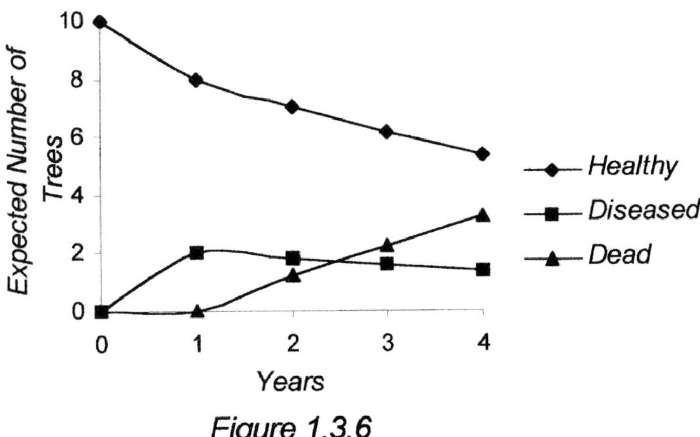

Figure 1.3.6

If the total number of trees is m, then the above equation becomes the expectation

---

[13]See Chapters 3 and 8 for solving stochastic models.

$$E(x_1,x_2,x_3) = m[0.910(0.877)^n + 0.090(0.023)^n, \quad 0.234(0.877)^n - 0.234(0.023)^n,$$
$$- 1.144(0.877)^n + 0.144(0.023)^n + 1]$$

where x_1 is the number of healthy trees, x_2 is the number of diseased trees and x_3 is the number of dead trees. This expression defines expected values and corresponds to a deterministic model of the system. The graphic representation of expected values is shown in Fig. 1.3.6. The total number of trees is assumed to be 10.

The **fourth step** is determining the stochastic model. The state variables may have a multinomial distribution, that is

$$f_n(x_1,x_2,x_3) = \frac{m!}{x_1!x_2!x_3} P_{1n}^{x_1} P_{2n}^{x_2} P_{3n}^{x_3}$$

where , x_2 and x_3 are the number of healthy, diseased and dead tress out of a total of m tx_1 rees. By replacing the P_n values in the multinomial equation, it is possible now to define the following state probability model of the system for an initial state $P_0=(1,0,0)$:

$$f_n(x_1,x_2,x_3) = \frac{m!}{x_1!x_2!x_3!}[0.910(0.877)^n + 0.090(0.023)^n]^{x_1}[0.234(0.877)^n - 0.234(0.023)^n]^{x_2}$$
$$[-1.144(0.877)^n + 0.144(0.023)^n + 1]^{x_3}$$

The probability distribution curve for eight healthy, two diseased and zero dead trees, is shown in Fig. 1.3.7.

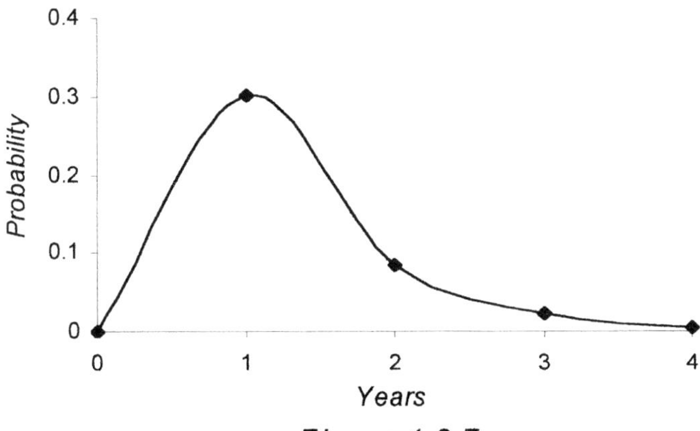

Figure 1.3.7

The total number of trees was assumed to be 10, resulting in 720 probability distribution curves, determined by the multinomial coefficient $m!/\, x_1!x_2!x_3!$.

The proposed modeling and analyses may improve or replace conventional procedures for evaluating a system and for comparing different systems.

Summary

The first step for defining the mathematical model and analyses of a deterministic system is arranging the data as a difference table. The difference table is used to determine the differential or difference equations representing the system. The solutions of the differential or difference equations are the state trajectories and parameters in these equations are evaluated by "t" tests. The first step for defining the mathematical model of a stochastic system is condensing the data in a state probability table, from which the next state equations of the system are determined. The next step is defining the corresponding deterministic solutions of the next state equations. The final step is applying a distribution function to the deterministic expressions.

2

CHARACTERISTIC VALUES

The selection of topics for this chapter has been aimed at gaining a basic understanding and proficiency in the manipulation of linear equations, as this subject relates to further chapters. Matrix techniques are applied here for solving linear equations and for determining characteristic equations, roots and vectors. Some readers may wish to skip some familiar material of the first two sections of this chapter.

2.1 SYSTEMS OF LINEAR EQUATIONS

As disclosed in the previous chapter, system analysis is related to the process of developing an abstract model of a system, such that the model would simulate the real system by means of a computer program. It was also indicated that the approach selected for modeling a system was the use of linear models. Therefore, some basic definitions and concepts are presented here for a clear understanding of the subject.

Linear Combinations

The concept of *linear combination* is related to the straight line. For the particular case of the X, Y plane, it is possible to define a linear combination of the variables x and y, such that

$$ax + by = c$$

The above equation is called linear, because it represents a straight line. Then, the following definition stands for linear combination:

Definition 2.1.1 If $y_1, y_2, ..., y_n$ are m-component vectors and $k_1, k_2, ..., k_n$ are scalars, then vector $y = k_1 y_1 + k_2 y_2 + ... + k_n y_n$ is called a linear combination of $y_1, y_2, ..., y_n$.

This concept is illustrated in the following example.

Example 2.1.1 Given the scalars k_1, k_2, k_3 and vectors $y_1 = (1,2,3)$, $y_2 = (4,-3,7)$ and $y_3 = (-2,7,-1)$ define a linear combination of vectors y_1, y_2, y_3.

Solution: All linear combinations of y_1, y_2 and y_3 are of the form

$$y = k_1(1,2,3) + k_2(4,-3,7) + k_3(-2,7,1)$$
$$= (k_1 + 4k_2 - 2k_3,\ 2k_1 - 3k_2 + 7k_,\ 3k + 7k_2 - k_3)$$

where k_1, k_2 and k_3 are arbitrary scalars. The following is the matrix form of the above expression:

$$y = \begin{bmatrix} 1 & 4 & -2 \\ 2 & -3 & 7 \\ 3 & 7 & -1 \end{bmatrix} \begin{bmatrix} k_1 \\ k_2 \\ k_3 \end{bmatrix}$$

Linear combinations are either *linearly dependent* or *linearly independent*. Linear dependence is defined as follows:

Definition 2.1.2 The m-component vectors $y_1, y_2, \ldots, y_n$ are said to be linearly dependent, if there are scalars $k_1, k_2, \ldots, k_n$, not all equal to zero, such that $k_1 y_1 + k_2 y_2 + \ldots + k_n y_n = 0$

Consider the following example:

Example 2.1.2 If y_1 represents the daily average consumption of dry matter by a group of cattle, y_2 represents the daily average consumption of crude protein and 0.115 is the crude protein content of food on a dry basis, define a linear combination.

Solution: The two variables are related by equation, $0.115y_1 = y_2$. This relationship can be rewritten as follows:

$$0.115y_1 - y_2 = 0$$

It is clear here that y_1 and y_2 are related, because $k_1 = 0.115$ and $k_2 = -1$ are different from zero. Therefore, y_1 and y_2 are linearly dependent.

The following definition stands for linear independence:

Definition 2.1.3 The m-component vectors $y_1, y_2, \ldots, y_n$ are said to be linearly independent, if there are scalars $k_1 = k_2 = \ldots = k_n = 0$, such that $k_1 y_1 + k_2 y_2 + \ldots + k_n y_n = 0$

Consider the next example:

Example 2.1.3 Given vectors $y_1 = (1, 0, -2)$, $y_2 = (-4, 3, 5)$ and $y_3 = (1, 2, 3)$ and scalars

k_1, k_2, and k_3, show that y_1, y_2 and y_3 are linearly independent.

Solution: The system may be represented by the following matrix expression:

$$y = \begin{bmatrix} 1 & 0 & -2 \\ -4 & 3 & 2 \\ -2 & 5 & 3 \end{bmatrix} \begin{bmatrix} k_1 \\ k_2 \\ k_3 \end{bmatrix} = 0$$

No solutions to the above system can be found other than $k_1 = k_2 = k_3 = 0$. Thus, y_1, y_2 and y_3 are linearly independent. The reader is encouraged to test this result

Linear Systems

As indicated, any equation of the form $ax + by = c$, where a, b and c are constants and (x, y) are variables, is a linear equation because it represents a straight line in the X, Y plane. The equation is *satisfied* whenever an ordered pair (α,β) of real numbers is substituted for (x, y) and $a\alpha + b\beta = c$. This is a linear combination of the variables x and y. The pair (α,β) is called a *solution* of the equation. The set of all solutions of an equation is its *solution set*.

Two or more linear equations in the same variables are said to form a system, when the equations are satisfied by a given solution. The system is then called *a system of linear equations*. Thus, two equations form a system if an ordered pair of real numbers (α,β) is found to satisfy both equations. Then (α,β) is a solution of the system.

A system of equations may have a single solution, multiple solutions or no solution at all. The system in the following example has a single solution:

Example 2.1.4 A system with a single solution:

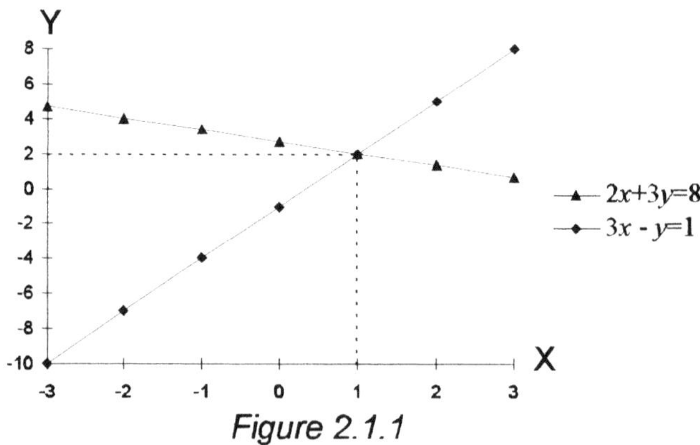

Figure 2.1.1

The solution of this system is the ordered pair (1, 2), since both equations are satisfied when $x = 1$ and $y = 2$. As shown in Fig. 2.1.1, the solution is the point of intersection between the two straight lines.

A system with multiple solutions is illustrated in the following example:

Example 2.1.5 A system with infinite solutions:

$$\begin{aligned} 2x + 3y + z &= 14 \\ x + y + z &= 6 \\ 3x + 5y + z &= 22 \end{aligned}$$

A set of some solutions of this system is shown in the following table:

Table 2.1.1

Variables	Solutions				
	S_1	S_2	S_3	S_4	S_5
x	4	6	8	2	0
y	2	1	0	3	4
z	0	-1	2	1	2

The reader is encouraged to check these solutions.

When there are fewer equations than unknowns, there are infinite number of solutions, as shown in the following example:

Example 2.1.6 A system with infinite number of solutions and fewer equations than unknowns:

$$\begin{aligned} y_1 + y_2 - y_3 &= 6 \\ 3y_1 + 6y_2 - 3y_3 &= 16 \end{aligned}$$

The solutions of this system are

$$y_1 = \frac{20}{3} + y_3 \quad ; \quad y_2 = -\frac{2}{3}$$

As indicated in the next example, not all systems of linear equations have solutions. When the lines are parallel, there is no intersection and the system has no solutions. A linear system that has no solutions is said to be *inconsistent*.

Example 2.1.7 A system with parallel lines:

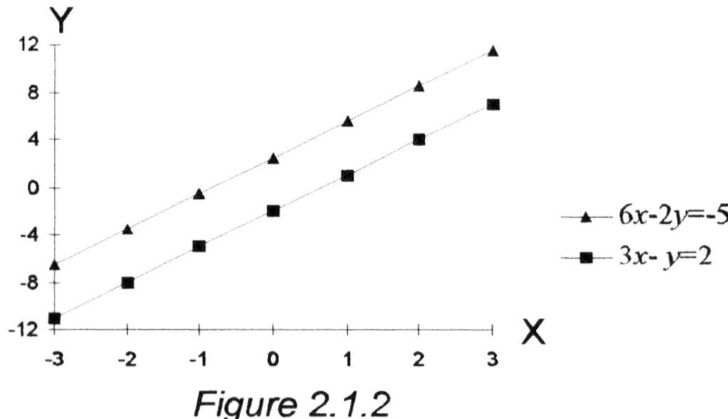

Figure 2.1.2

Note that solutions for a set of linear equations exist only if the system is consistent. Thus, the following definition is here set forth:

Definition 2.1.4 A system represented by the matrix equation $Ay = b$ is said to be *consistent* if it has at least one solution.

Whether the system has one solution, multiple solutions or no solution at all, the general linear system of m equations and n unknowns can be represented as follows:

$$a_{11}y_1 + a_{12}y_2 + \dots + a_{1n}y_n = b_1$$
$$a_{21}y_1 + a_{22}y_2 + \dots + a_{2n}y_n = b_2$$
$$\vdots \qquad \vdots \qquad \qquad \vdots \qquad \vdots$$
$$a_{m1}y_1 + a_{m2}y_2 + \dots + a_{mn}y_n = b_m$$

The following definition applies here:

Definition 2.1.5 A system represented by the matrix equation $Ay=b$ is *homogeneous* if $b_1 = b_2 = \dots = b_m = 0$. Otherwise, the system is said to be *nonhomogeneous*.

In all the above examples, the systems were nonhomogeneous. A homogeneous system of equations is illustrated in the next example:

Example 2.1.8 A homogeneous system:

$$
\begin{aligned}
2y_1 + y_2 + y_3 &= 0 \\
5y_1 + 2y_2 + 2y_3 &= 0 \\
y_1 - y_2 + y_3 &= 0
\end{aligned}
$$

Note that homogeneous systems are never inconsistent. All such systems have the solution $(0, 0, \dots ,0)$. This is called a *trivial* solution. Thus, the problem in a homogeneous system is to determine whether there is a non trivial solution. Note also that a system with fewer equations than unknowns, always has a *non trivial* solution. Otherwise, to avoid trivial cases, at least one a_{ij} coefficient must be non-zero. In addition, systems with a single non trivial solution are always nonhomogeneous. Conversely, systems with many solutions may be either homogeneous or nonhomogeneous.

Order and Rank of a Matrix

A matrix is defined by its *order* and by its *rank*. The order of a matrix is determined by the number of rows and columns and is defined as follows:

Definition 2.1.6 The order, called also *the matrix dimension*, represents the size of a matrix

Thus, a matrix A with r rows and c columns has order $r \times c$ and can be written as $A_{r \times c}$. When the number of rows and columns are equal, the matrix is referred as a *squared matrix* and is describer as being of order r.

Example 2.1.9 A matrix of order 2 x 3:

$$
A_{2x3} = \begin{bmatrix} 4 & 0 & -3 \\ -7 & 3 & 1 \end{bmatrix}
$$

Example 2.1.10 A matrix of order 3:

$$
A = \begin{bmatrix} 1 & -4 & 11 \\ 0 & 3 & -6 \\ -2 & 5 & -16 \end{bmatrix}
$$

The rank of a matrix is defined as follows:

Definition 2.1.7 The rank $r(A)$ of a matrix A represents the number of linearly independent rows or columns of the matrix.

Example 2.1.11 Determine the rank of the following matrix:

$$A = \begin{bmatrix} 1 & 0 & -1 & 2 \\ 3 & 1 & 4 & 2 \\ 5 & 2 & 9 & 2 \end{bmatrix}$$

Solution: This is a rectangular matrix of order 3 x 4 and, therefore $r(A) \le 3$. Note that the third and the fourth columns are linear combinations of the first two:

$$\begin{bmatrix} -1 \\ 4 \\ 9 \end{bmatrix} = -\begin{bmatrix} 1 \\ 3 \\ 5 \end{bmatrix} + 7\begin{bmatrix} 0 \\ 1 \\ 2 \end{bmatrix} \quad ; \quad \begin{bmatrix} 2 \\ 2 \\ 2 \end{bmatrix} = 2\begin{bmatrix} 1 \\ 3 \\ 5 \end{bmatrix} - 4\begin{bmatrix} 0 \\ 1 \\ 2 \end{bmatrix}$$

The third row is a linear combination of the first two:

$$\begin{bmatrix} 5 & 2 & 9 & 2 \end{bmatrix} = 2\begin{bmatrix} 3 & 1 & 4 & 2 \end{bmatrix} - 4\begin{bmatrix} 1 & 0 & -1 & 2 \end{bmatrix}$$

Thus, matrix A has only two independent columns and two independent rows, therefore, $r(A) = 2$.

From the above, it is clear that rank and order are related in the following manner:

- The rank of a square matrix is equal or less than its order
- The rank of a m x n rectangular matrix is equal or less than the smaller value of m and n

Defining the order of a matrix is straightforward. However, defining the rank requires determining the number of linearly independent rows or columns. Since the number of linearly independent rows or columns is also the order of the largest minor determinant whose value is different from zero, finding the order of such minors is also a straightforward procedure to find the rank of a matrix. Thus, the following definition applies here:

Definition 2.1.8 The rank $r(A)$ of a matrix A is the order of the largest minor of the matrix determinant, whose value is different from zero.

The reader is reminded that a determinant can be expanded as a linear function of minor order determinants derived from it. Thus, a third order determinant can be expanded as a linear function of three second order determinants. The minor order determinants are simply called *minors*.

Example 2.1.12 Find the rank of matrix A in Example 2.1.11 using determinants.

Solution: Matrix A was defined as

$$A = \begin{bmatrix} 1 & 0 & -1 & 2 \\ 3 & 1 & 4 & 2 \\ 5 & 2 & 9 & 2 \end{bmatrix}$$

By definition, a determinant is a polynomial of the elements of a square matrix. However, matrix A can be partitioned, such that a group of three column vectors may be selected into a square matrix B. Then, the rank of B is determined by evaluating $|B|$:

$$|B| = \begin{vmatrix} 1 & 0 & -1 \\ 3 & 1 & 4 \\ 5 & 2 & 9 \end{vmatrix} = \begin{vmatrix} 1 & 4 \\ 2 & 9 \end{vmatrix} - \begin{vmatrix} 3 & 1 \\ 5 & 2 \end{vmatrix} = 1 - 1 = 0$$

Note that the value of $|B|$ is zero, but the value os its minors is different from zero. Thus, $r(A) = 2$, because the order of the minors is two. Thus

- If any of the rows or columns of a matrix are linearly dependent, its determinant is zero.
- If all the rows or columns of a matrix are linearly independent, then the determinant is not zero.

The following properties of the rank of a matrix are summarized here:

- The rank of a square matrix of order n is equal or less than its order
- The rank of a rectangular matrix or order $m \times n$ is equal or less than the smaller value of m and n
- If $r(A) = r$, there are at least one minor of order r whose value is different from zero and all minors of order greater than r are zero

Summary

A vector $y = k_1 y_1 + k_2 y_2 + ... + k_n y_n$ is called a linear combination of vectors $y_1, y_2, ..., y_n$, where $k_1, k_2, ..., k_n$ are scalars. Vectors $y_1, y_2, ..., y_n$ are said to be linearly dependent, if there are scalars $k_1, k_2, ..., k_n$, not all zero, such that $k_1 y_1 + k_2 y_2 + ... + k_n y_n = 0$. Vectors $y_1, y_2, ..., y_n$ are said to be linearly independent, if there are scalars $k_1 = k_2 = ... = k_n = 0$, such that $k_1 y_1 + k_2 y_2 + ... + k_n y_n = 0$. A system represented by the matrix equation $Ay = b$ is said to be consistent if it has at least one solution. It is said to be homogeneous if $b_1 = b_2 = ... = b_m = 0$. Homogeneous systems are never inconsistent. Solutions of the form $(0,0,...,0)$ are called trivial. Systems with fewer equations than unknowns have always non trivial solutions. Systems with a single non trivial solution are always nonhomogeneous, while systems with multiple solutions may be homogeneous or nonhomogeneous. The rank of a square matrix is equal or less than its order and the rank of a $m \times n$ rectangular matrix is equal or less than the smaller value of m and n.

2.2 SOLVING LINEAR SYSTEMS

As defined before, a system represented by the matrix equation $Ay = b$ may have a single solution, multiple solutions or no solutions. Some selected procedures for solving linear equations are described in this section.

Single Solution Systems

Two procedures will be used for solving single solution systems, a determinant procedure and a matrix inversion procedure.

A system with two variables will be utilized as a model to explain the procedure for solving linear equations using determinants. Thus, the system

$$a_{11} y_1 + a_{12} y_2 = b_1$$
$$a_{21} y_1 + a_{22} y_2 = b_2$$

is written in the following matrix form:

$$\begin{bmatrix} a_{11} & a_{12} \\ a_{21} & a_{22} \end{bmatrix} \begin{bmatrix} y_1 \\ y_2 \end{bmatrix} = \begin{bmatrix} b_1 \\ b_2 \end{bmatrix}$$

Note that this system can be easily solved by any of the well known algebraic procedures based on successive elimination, such that

$$y_1 = \frac{a_{22}b_1 - a_{12}b_2}{a_{11}a_{22} - a_{21}a_{12}} \quad ; \quad y_2 = \frac{a_{11}b_2 - a_{21}b_1}{a_{11}a_{22} - a_{21}a_{12}}$$

These procedures may become cumbersome if the system has more than two variables. However, the above solutions can be expressed easily as determinants. Then, the denominator of the above solutions is

$$a_{11}a_{22} - a_{21}a_{12} = \begin{vmatrix} a_{11} & a_{12} \\ a_{21} & a_{22} \end{vmatrix}$$

The numerator of y_1 is obtained by replacing the first column of matrix A by vector b:

$$a_{22}b_1 - a_{12}b_2 = \begin{vmatrix} b_1 & a_{12} \\ b_2 & a_{22} \end{vmatrix}$$

The numerator of y_2 is obtained by replacing the second column of matrix A by vector b:

$$a_{12}b_2 - a_{21}b_1 = \begin{vmatrix} a_{11} & b_1 \\ a_{21} & b_2 \end{vmatrix}$$

Thus, the solution, in terms of determinants must be

$$y_1 = \frac{\begin{vmatrix} b_1 & a_{12} \\ b_2 & a_{22} \end{vmatrix}}{\begin{vmatrix} a_{11} & a_{12} \\ a_{21} & a_{22} \end{vmatrix}} \quad ; \quad y_2 = \frac{\begin{vmatrix} a_{11} & b_1 \\ a_{21} & b_2 \end{vmatrix}}{\begin{vmatrix} a_{11} & a_{12} \\ a_{21} & a_{22} \end{vmatrix}}$$

Example 2.2.1 Solve the system

$$3x + 2y = -12$$
$$2x - 3y = 5$$

Solution:

$$x = \frac{\begin{vmatrix} -12 & 2 \\ 5 & -3 \end{vmatrix}}{\begin{vmatrix} 3 & 2 \\ 2 & -3 \end{vmatrix}} = -2 \quad ; \quad y = \frac{\begin{vmatrix} 3 & -12 \\ 2 & 5 \end{vmatrix}}{\begin{vmatrix} 3 & 2 \\ 2 & -3 \end{vmatrix}} = -3$$

Example 2.2.2 Solve the system

$$\begin{aligned} x + y &= 1 \\ 2x + 2y &= 2 \end{aligned}$$

Solution: This system has an infinite number of solutions, because the second equation is a multiple of the first. Thus, the determinant of the denominator is zero and matrix A has no inverse:

$$\begin{vmatrix} 1 & 1 \\ 2 & 2 \end{vmatrix} = 0$$

Example 2.2.3 Solve the system

$$\begin{aligned} 3x - y &= 2 \\ 6x - 2y &= 3 \end{aligned}$$

Solution: This system has no solutions. The two lines are parallel, because the slope is the same, as shown below:

$$\begin{aligned} y &= -2 + 3x \\ y &= -3/2 + 3x \end{aligned}$$

As in the previous example, the determinant of the denominator is also zero:

$$\begin{vmatrix} 3 & -1 \\ 6 & -2 \end{vmatrix} = 0$$

Example 2.2.4 Solve the system

$$2x + 8y + 6z = 20$$
$$4x + 2y - 2z = -2$$
$$3x - y - z = 11$$

Solution: Denominator:

$$\begin{vmatrix} 2 & 8 & 6 \\ 4 & 2 & -2 \\ 3 & -1 & 1 \end{vmatrix} = 2\begin{vmatrix} 2 & -2 \\ -1 & 1 \end{vmatrix} - 8\begin{vmatrix} 4 & -2 \\ 3 & 1 \end{vmatrix} + 6\begin{vmatrix} 4 & 2 \\ 3 & -1 \end{vmatrix} = -140$$

The numerator of the x solution is

$$\begin{vmatrix} 20 & 8 & 6 \\ -2 & 2 & -2 \\ 11 & -1 & 1 \end{vmatrix} = 20\begin{vmatrix} 2 & -2 \\ -1 & 1 \end{vmatrix} - 2\begin{vmatrix} 8 & 6 \\ -1 & 1 \end{vmatrix} + 11\begin{vmatrix} 8 & 6 \\ 2 & -2 \end{vmatrix} = -280$$

The numerator of the y solution is

$$\begin{vmatrix} 2 & 20 & 6 \\ 4 & -2 & -2 \\ 3 & 11 & 1 \end{vmatrix} = -20\begin{vmatrix} 4 & -2 \\ 3 & 1 \end{vmatrix} - 2\begin{vmatrix} 2 & 6 \\ 3 & 1 \end{vmatrix} - 11\begin{vmatrix} 2 & 6 \\ 4 & -2 \end{vmatrix} = 140$$

The numerator of the z solution is

$$\begin{vmatrix} 2 & 8 & 20 \\ 4 & 2 & -2 \\ 3 & -1 & 11 \end{vmatrix} = 20\begin{vmatrix} 4 & 2 \\ 3 & -1 \end{vmatrix} + 2\begin{vmatrix} 2 & 8 \\ 3 & -1 \end{vmatrix} + 11\begin{vmatrix} 2 & 8 \\ 4 & 2 \end{vmatrix} = -560$$

Then

$$x = -280/-140 = 2$$
$$y = 140/-140 = -1$$
$$z = -560/-140 = 4$$

Solving linear equations using determinants may not be practical for systems with many variables. An alternate procedure is based in the following relationship:

$$A^{-1}Ay = A^{-1}b$$

Since $A^{-1}A = I$, where I is the identity matrix and $Iy = y$, then

$$y = A^{-1}b$$

By this way, the problem of finding solutions for a linear system is the problem of finding the inverse A^{-1}.

Several procedures are available for inverting matrices. The matrix inversion by elementary operations is used in this book, because this procedure can be applied to find the system solutions when a matrix is either invertible or non invertible. It can also be used when the system is homogeneous or nonhomogeneous. The procedure changes, by elementary row operations, the original system of equations $Ay = Ib$ to $Iy = A^{-1}b$. The row operations that transform matrix A to the identity matrix I, also transforms I to A^{-1}. Matrix A and the adjoined identity matrix I are called an *augmented matrix*. The following example illustrate the procedure.

Example 2.2.5 Solve the system

$$\begin{bmatrix} 1 & 2 \\ 2 & 3 \end{bmatrix}\begin{bmatrix} x_1 \\ x_2 \end{bmatrix} = \begin{bmatrix} 0 \\ 1 \end{bmatrix}$$

Solution: Perform the following row operations:

Augmented Matrix	Row Operations

$$\begin{bmatrix} 1 & 2 & \vdots & 1 & 0 \\ 2 & 3 & \vdots & 0 & 1 \end{bmatrix}$$

$$\begin{bmatrix} 1 & 2 & \vdots & 1 & 0 \\ 0 & -1 & \vdots & -2 & 1 \end{bmatrix} \qquad (-2) \textit{ times the first plus the second row}$$

$$\begin{bmatrix} 1 & 2 & \vdots & 1 & 0 \\ 0 & 1 & \vdots & 2 & -1 \end{bmatrix} \qquad (-1) \textit{ times thesecond row}$$

$$\begin{bmatrix} 1 & 0 & \vdots & -3 & 2 \\ 0 & 1 & \vdots & 2 & -1 \end{bmatrix} \qquad (-2) \textit{ times the second plus the first row}$$

Thus

$$x = \begin{bmatrix} -3 & 2 \\ 3 & -1 \end{bmatrix} \begin{bmatrix} 0 \\ 1 \end{bmatrix} = \begin{bmatrix} 2 \\ -1 \end{bmatrix}$$

Note that $A A^{-1} = I$:

$$A A^{-1} = \begin{bmatrix} 1 & 2 \\ 2 & 3 \end{bmatrix} \begin{bmatrix} -3 & 2 \\ 3 & -1 \end{bmatrix} = \begin{bmatrix} 1 & 0 \\ 0 & 1 \end{bmatrix}$$

Example 2.2.6 A diet for heifers uses corn, soybean meal and sorghum silage as ingredients. The nutrient composition and daily requirements of the heifers are summarized in Table 2.2.1, where DM = Dry Matter DE = Digestible Energy and CP = Crude Protein. Find how much of each ingredient is needed to balance the diet for the above requirements.

Table 2.2.1

Nutrients	Ingredients			Requirements
	Corn	Soybean Meal	Sorghum Silage	
DM, Kg/Kg	0.89	0.89	0.29	2.80
DE, Mcal/Kg	3.45	3.17	0.7	7.45
CP, Kg/Kg	0.089	0.441	0.024	0.32

Solution: This problem may be defined as a system of linear equations, where y represents the quantity of each ingredient in the food:

$$0.89y_1 + 0.89y_2 + 0.29y_3 = 2.80$$
$$3.45y_1 + 3.17y_2 + 0.70y_3 = 7.45$$
$$0.089y_1 + 0.441y_2 + 0.024y_3 = 0.320$$

Note that matrix A represents the nutrient composition of the ingredients. The quantity of each ingredient required to meet the heifers' requirements is vector y. Vector b is the heifers' daily requirement. Elementary row operations for inverting matrix A are as indicated below.

$$\begin{bmatrix} 0.89 & 0.89 & 0.29 \\ 3.45 & 3.17 & 0.70 \\ 0.089 & 0.441 & 0.024 \end{bmatrix} \begin{bmatrix} y_1 \\ y_2 \\ y_3 \end{bmatrix} = \begin{bmatrix} 2.80 \\ 7.45 \\ 0.320 \end{bmatrix}$$

First: 1/ 0.89 times row 1
Second: 0.089 times row 2 - 3.45 times row 3
Third: row 3 - 0.089 times row 1
Fourth: row 1 - 1/0.352 times row 3
Fifth: 0.005 times row 2 - 0.0205 times row 3
Sixth: 0.0134 times row 3 + 0.352 times row 2
Seventh: -1/0.0134 times row 2
Eight: 1/0.00006706 times row 3
Ninth: row 1 - 0.34 times row 3

Then, the equivalent augmented matrices from the above operations are

$$\begin{bmatrix} 0.89 & 0.89 & 0.29 & | & 1 & 0 & 0 \\ 3.45 & 3.17 & 0.70 & | & 0 & 1 & 0 \\ 0.089 & 0.441 & 0.024 & | & 0 & 0 & 1 \end{bmatrix} \dots \approx \dots \begin{bmatrix} 1 & 0 & 0 & | & -1.7439 & 0.7948 & -2.2086 \\ 0 & 1 & 0 & | & -1.5284 & -0.0332 & 2.8145 \\ 0 & 0 & 1 & | & 9.2687 & -2.3373 & -1.8597 \end{bmatrix}$$

The reader may want to check the above operations. The following is the resulting inverse:

$$A^{1} = \begin{bmatrix} -1.7439 & 0.7948 & -2.2086 \\ -0.1528 & -0.0332 & 2.8145 \\ 9.2687 & -2.3373 & -1.8597 \end{bmatrix}$$

Thus, the solution of the problem is

$$\begin{bmatrix} -1.7439 & 0.7948 & -2.2086 \\ -0.1528 & -0.0332 & 2.8145 \\ 9.2687 & -2.2373 & -1.8597 \end{bmatrix} \begin{bmatrix} 2.80 \\ 7.45 \\ 0.32 \end{bmatrix} = \begin{bmatrix} y_1 \\ y_2 \\ y_3 \end{bmatrix}$$

Then, the following is the amount of each ingredient needed to balance the diet for the heifers:

$$y_1 = 0.332 \text{ Kg of corn/day}$$
$$y_2 = 0.225 \text{ Kg of soybean meal/day}$$
$$y_3 = 7.944 \text{ Kg of sorghum silage/day}$$

Multiple Solutions Systems

The procedures described previously are valid only for systems with solutions of the form $y = A^{-1}b$, when matrix A is invertible. An invertible matrix is *nonsingular* and the system is nonhomogeneous. A matrix is said to be nonsingular if its row vectors and its column vectors are linearly independent. Otherwise, the matrix is said to be *singular*. In multiple solutions systems, matrix A is not invertible. Then, an alternative procedure is needed.

As indicated previously, elementary matrix operations can be used to find solutions when matrix A is either invertible or non invertible and also when the system is either homogeneous or nonhomogeneous.

Example 2.2.6 Find solutions for the following system:

$$\begin{bmatrix} 2 & 3 & 1 \\ 1 & 1 & 1 \\ 3 & 5 & 1 \end{bmatrix} \begin{bmatrix} x \\ y \\ z \end{bmatrix} = \begin{bmatrix} 14 \\ 6 \\ 22 \end{bmatrix}$$

Solution: This system has n = 3 unknowns and rank r = 2. A set of solutions was defined in Example 2.1.5. At that stage, however, it was not explained how those solutions were obtained. The following row operations were performed here on the augmented matrix:

First: 2 times row 3 - 3 times row 1
Second: row 2 - 0.5 times row 1
Third: row 3 + 2 times row 2
Fourth: 0.5 times row 1 + row 2
Fifth: 2 times row 2

$$\begin{bmatrix} 2 & 3 & 1 & | & 14 \\ 1 & 1 & 1 & | & 6 \\ 3 & 5 & 1 & | & 22 \end{bmatrix} \approx \begin{bmatrix} 2 & 3 & 1 & | & 14 \\ 1 & 1 & 1 & | & 6 \\ 0 & 1 & -1 & | & 2 \end{bmatrix} \approx \begin{bmatrix} 2 & 3 & 1 & | & 14 \\ 0 & -1/2 & 1/2 & | & -1 \\ 0 & 1 & -1 & | & 2 \end{bmatrix} \approx$$

$$\begin{bmatrix} 2 & 3 & 1 & | & 14 \\ 0 & -1/2 & 1/2 & | & -1 \\ 0 & 0 & 0 & | & 0 \end{bmatrix} \approx \begin{bmatrix} 1 & 1 & 1 & | & 6 \\ 0 & -1/2 & 1/2 & | & -1 \\ 0 & 0 & 0 & | & 0 \end{bmatrix} \approx \begin{bmatrix} 1 & 1 & 1 & | & 6 \\ 0 & -1 & 1 & | & -2 \\ 0 & 0 & 0 & | & 0 \end{bmatrix}$$

The equivalent system is now

$$\begin{bmatrix} 1 & 1 & 1 \\ 0 & -1 & 1 \\ 0 & 0 & 0 \end{bmatrix} \begin{bmatrix} x \\ y \\ z \end{bmatrix} = \begin{bmatrix} 6 \\ -2 \\ 0 \end{bmatrix}$$

Then

$$x + y + z = 6$$
$$-y + z = -2$$

By solving the above equations, it was found that

$$y = z + 2$$
$$x = -2z + 4$$

If z is made a constant k, then the set V of solutions becomes

$$V = \begin{vmatrix} -2k + 4 \\ k + 2 \\ k \end{vmatrix}$$

A set of some explicit solutions of this system is given in the table below. Note that these solutions are the same shown in Example 2.1.5.

Table 2.2.2

Variables	Solutions				
	$s_1,\ k=0$	$s_2,\ k=-1$	$s_3,\ k=-2$	$s_4,\ k=1$	$s_5,\ k=2$
x	4	6	8	2	0
y	2	1	0	3	4
z	0	-1	-2	1	2

Example 2.2.7 Find the set of solutions for the following homogeneous system:

$$2x + 3y + z = 0$$
$$x + y + z = 0$$
$$3x + 5y + z = 0$$

This is the same system of equations of the previous example but made homogeneous. Therefore, writing the last equivalent form of the elementary transformations in the previous example is easy. Then

$$\begin{bmatrix} 1 & 1 & 1 \\ 0 & -1 & 1 \\ 0 & 0 & 0 \end{bmatrix} \begin{bmatrix} x \\ y \\ z \end{bmatrix} = \begin{bmatrix} 0 \\ 0 \\ 0 \end{bmatrix}$$

By solving these equations, it is found that

$$\begin{aligned} x &= -2z \\ y &= z \end{aligned}$$

If z is made k, the set V of solutions will be

$$V = \begin{bmatrix} -2k \\ k \\ k \end{bmatrix}$$

Example 2.2.8 Solve the following system:

$$\begin{bmatrix} 1 & 2 & -1 & 9 \\ 2 & 4 & 3 & 3 \\ -1 & -2 & 6 & -24 \\ 1 & 2 & 4 & -6 \end{bmatrix} \begin{bmatrix} y_1 \\ y_2 \\ y_3 \\ y_4 \end{bmatrix} = \begin{bmatrix} 4 \\ 13 \\ 1 \\ 9 \end{bmatrix}$$

Solution: The following elementary operations were performed here:

First: row 3 + row 1
Second: row 2 - 2 times row 1
Third: row 4 - row 1
Fourth: row 3 - row 2
Fifth: row 4 - row 2
Sixth: 1/5 times row 2
Seventh: row 1 + row 2

$$\begin{bmatrix} 1 & 2 & -1 & 9 & | & 4 \\ 2 & 4 & 3 & 3 & | & 13 \\ -1 & -2 & 6 & -24 & | & 1 \\ 1 & 2 & 4 & -6 & | & 9 \end{bmatrix} \approx \ \dots \ \approx \begin{bmatrix} 1 & 2 & 0 & 6 & | & 5 \\ 0 & 0 & 1 & -3 & | & 1 \\ 0 & 0 & 0 & 0 & | & 0 \\ 0 & 0 & 0 & 0 & | & 0 \end{bmatrix}$$

The reader is encouraged to check the above operations. The equivalent system is now

$$\begin{bmatrix} 1 & 2 & 0 & 6 \\ 0 & 0 & 1 & -3 \\ 0 & 0 & 0 & 0 \\ 0 & 0 & 0 & 0 \end{bmatrix}\begin{bmatrix} y_1 \\ y_2 \\ y_3 \\ v_4 \end{bmatrix} = \begin{bmatrix} 5 \\ 1 \\ 0 \\ 0 \end{bmatrix} \quad ; \quad \begin{aligned} y_1 &= -2y_2 - 6y_4 + 5 \\ y_3 &= 3y_4 + 1 \end{aligned}$$

By making $y_2 = k_2$ and $y_4 = k_4$, the set V of solutions is

$$V = \begin{bmatrix} y_1 \\ y_2 \\ y_3 \\ v_4 \end{bmatrix} = \begin{bmatrix} 5 - 2k_2 - 6k_4 \\ k_2 \\ 1 + 3k_4 \\ k_4 \end{bmatrix}$$

Summary

For single solution systems of the form $Ay = b$, each solution for y can be obtained by determinant procedures. However, if the system has many variables, solving the system by matrix inversion procedures may be more practical. Elementary operations can be used to find solutions when matrix A is either invertible or non invertible and when the system is either homogeneous or non homogeneous.

2.3 CHARACTERISTIC EQUATION, ROOTS AND VECTORS

If a system of linear equations is defined as $Ay=b$, where A is a square matrix of order n, a question arises on whether there exists any vector such that Ay is a constant multiple of y. This question leads to the concepts of *characteristic equation, characteristic roots,* known also as *latent roots* or *eigenvalues* and *characteristic vectors,* called also *invariant vectors, latent vectors* or *eigenvectors*.

The Characteristic Equation

If A is a square matrix of order n and y is a column vector, such that $Ay=b$, then a question is posed whether a vector y exists, such that Ay is a constant multiple of y. Then

$$Ay = \lambda y$$

where λ is a scalar value. If λ and y satisfy this equation, then λ is said to be a characteristic root of matrix A, corresponding to the characteristic vector y of A.

The above equation can be rewritten as $Ay - \lambda y = 0$, which is equivalent to the matrix equation

$$(A - \lambda I)y = 0$$

where I is an identity matrix and 0 represents a null vector. A homogeneous equation of this form has a non trivial solution for vector y, only if its determinant is zero. Then

$$|A - \lambda I| = 0$$

When matrix A is of order n, the expansion of this determinant yields a polynomial equation of degree n in λ, which is known as the *characteristic equation* of the system. The polynomial is called *characteristic polynomial*.

Example 2.3.1 Given matrix

$$A = \begin{bmatrix} 3 & -6 \\ 2 & -5 \end{bmatrix}$$

find the characteristic equation of the system.

Solution: The expansion of the determinant is

$$|A - \lambda I| = \begin{vmatrix} 3-\lambda & -6 \\ 2 & -5-\lambda \end{vmatrix} = (3 - \lambda)(-5 - \lambda) + 12$$

Then, the characteristic equation is $\lambda^2 + 2\lambda - 3 = 0$

Example 2.3.2 Find the characteristic equation of the system represented by the following matrix:

$$A = \begin{bmatrix} 1 & 3 & 0 \\ 7 & 1 & 5 \\ 0 & -4 & 1 \end{bmatrix}$$

Solution: The expansion of the determinant is here

$$|A - \lambda I| = \begin{vmatrix} 1-\lambda & 3 & 0 \\ 7 & 1-\lambda & 5 \\ 0 & -4 & 1-\lambda \end{vmatrix}$$

$$-\lambda^3 + \lambda^2(1+1+1) - \lambda\left[\begin{vmatrix} 1 & 5 \\ -4 & 1 \end{vmatrix} + \begin{vmatrix} 1 & 0 \\ 0 & 1 \end{vmatrix} + \begin{vmatrix} 1 & 3 \\ 7 & 1 \end{vmatrix}\right] + |A$$

Then, the characteristic equation is $\lambda(\lambda^2 - 3\lambda + 2) = 0$.

The same rules for determining the characteristic equation of system of linear equations, having the form $Ay = b$, apply for determining the characteristic equation of a system of linear differential equations of the form $dY/dt = AY$, where $Y = (y_1, y_2,...,y_n)$.

Example 2.3.3 Determine a characteristic equation of the following system of differential equations:

$$\frac{dy_1}{dt} = 2y_1 - 3y_2$$

$$\frac{dy_2}{dt} = -4y_1 + 3y_2$$

Solution: The above system can be written in the following matrix form:

where $Y = (y_1, y_2)$. Then, the matrix equation is here

$$\begin{bmatrix} 2-\lambda & -3 \\ -4 & 3-\lambda \end{bmatrix} Y = \begin{bmatrix} 0 \\ 0 \end{bmatrix}$$

The characteristic equation is the expansion of the determinant of the system:

$$|A - \lambda I| = \begin{vmatrix} 2-\lambda & -3 \\ -4 & 3-\lambda \end{vmatrix} = \lambda^2 - 5\lambda - 6 = 0$$

Note that the characteristic equation is a second degree polynomial because there are two variables in the system of differential equations.

Example 2.3.4 The movement of DDT from a group of orange trees to soil is 25% per month and from soil to trees is 30%. The movement of the insecticide from soil to weeds is 15% per month and from weeds to soil is 20%. The following set of differential equations represents the system:

$$\frac{dY}{dt} = \begin{bmatrix} -0.25 & 0 & 0.30 \\ 0 & -015 & 0.20 \\ 0.25 & 0.15 & -0.50 \end{bmatrix} Y$$

for $Y = (y_1, y_2, y_3)$, where y_1 is the amount of DDT in the trees, y_2 is the amount of DDT in the weeds and y_3 is the amount of DDT in soil. Determine the characteristic equation of the system.

Solution: The matrix equation of the system is here

$$\begin{bmatrix} -(0.25+\lambda) & 0 & 0.30 \\ 0 & -(0.15+\lambda) & 0.20 \\ 0.25 & 0.15 & -(0.50+\lambda) \end{bmatrix} Y = \begin{bmatrix} 0 \\ 0 \\ 0 \end{bmatrix}$$

The characteristic equation is found by expanding the determinant of the system:

$$\begin{vmatrix} -(0.25+\lambda) & 0 & 0.30 \\ 0 & -(0.15+\lambda) & 0.20 \\ 0.25 & 0.15 & -(0.50+\lambda) \end{vmatrix} = \lambda^3 + 0.9\lambda^2 + 0.1325\lambda = 0$$

Since the system is represented by three variables in y, the characteristic equation is a third degree polynomial.

In general terms, if a_{ij} is the ij element of matrix A, for $i,j = 1,2,...,n$, then the characteristic equation of A is represented by the diagonal expansion of the following determinant:

$$|A-\lambda I| = \begin{vmatrix} a_{11}-\lambda & a_{12} & ... & a_{1n} \\ a_{21} & a_{22}-\lambda & ... & a_{2n} \\ \vdots & \vdots & \vdots & \vdots \\ a_{n1} & a_{n2} & ... & a_{nn}-\lambda \end{vmatrix} = (-\lambda)^n + s_1(-\lambda)^{n-1} + ... + s_{n-1}(-\lambda) + s_n = 0$$

where the s coefficients are the sums of certain minor determinants of A. Specifically, by a diagonal expansion, $s_1 = a_{11} + a_{22} + ... + a_{nn}$ and $s_n = |A|$.

Characteristic Roots and Vectors

If matrix A is of order n, the characteristic equation of A is a polynomial of degree n in the variable λ. A polynomial of degree n has n solutions, that is

$$\lambda_1, \lambda_2, ..., \lambda_n$$

Each of these solutions is called a characteristic root of A. For each solution, equation $\square\square = \lambda y$ must hold true. It is expected that, corresponding to the n solutions of the characteristic equation, there are n linearly independent vectors

$$v_1, v_2, ..., v_n$$

which are the characteristic vectors of A. Then

$$Av_i = \lambda_i v_i \quad ; \quad i = 1,2,...,n$$

Derivation of the characteristic roots and generation of the corresponding vectors is shown in the following examples. These examples are related only to solving quadratic and cubic equations in λ. This task is not difficult, but solving higher-order polynomials becomes increasingly more complicated as the degree of the polynomial increases. Several numerical methods are available for determining these roots. These methods can be handled fast and easily by personal computers and even by some pocket size calculators.

Example 2.3.5 The characteristic equation of matrix A in Example 2.3.1 was

$$|A - \lambda I| = \begin{vmatrix} 3-\lambda & -6 \\ 2 & -5-\lambda \end{vmatrix} = \lambda^2 + 2\lambda - 3 = 0$$

Find the corresponding characteristic roots and vectors.

Solution: If the characteristic equation is $\lambda^2 + 2\lambda - 3 = (\lambda + 3)(\lambda - 1) = 0$, then the characteristic roots of the system are the following solutions of the characteristic equation:

$$\lambda_1 = -3 \quad \text{and} \quad \lambda_2 = 1$$

For a matrix equation $Ay = \lambda y$, where $y = (y_1, y_2)$, the characteristic vector v_1 for root $\lambda_1 = -3$ satisfies equation $Ay = -3y$. That is

$$\begin{bmatrix} 3 & -6 \\ 2 & -5 \end{bmatrix} \begin{bmatrix} y_1 \\ y_2 \end{bmatrix} = \begin{bmatrix} -3y_1 \\ -3y_2 \end{bmatrix} \quad ; \quad \begin{aligned} 3y_1 - 6y_2 &= -3y_1 \\ 2y_1 - 5y_2 &= -3y_2 \end{aligned}$$

By solving the above equations, it is found that $y_1 = y_2$. Then, any vector

$$v_1 = k_1 \begin{bmatrix} 1 \\ 1 \end{bmatrix}$$

where k_1 is any constant value, is a characteristic vector of A and a solution of the system. The characteristic vector v_2 corresponding to root $\lambda_2 = 1$ satisfies equation $Ay=y$, that is

$$\begin{bmatrix} 3 & -6 \\ 2 & -5 \end{bmatrix} \begin{bmatrix} y_1 \\ y_2 \end{bmatrix} = \begin{bmatrix} y_1 \\ y_2 \end{bmatrix} \quad ; \quad \begin{aligned} 3y_1 - 6y_2 &= y_1 \\ 2y_1 - 5y_2 &= y_2 \end{aligned}$$

By solving the above equations, it is found that $y_1 = 3y_2$. Then, any vector

$$v_2 = k_2 \begin{bmatrix} 3 \\ 1 \end{bmatrix}$$

where k_2 is any constant value, is another characteristic vector of A and also a solution of the system.

Defining a characteristic matrix V of solutions is now possible:

$$V = \begin{bmatrix} 1 & 3 \\ 1 & 1 \end{bmatrix} \begin{bmatrix} k_1 \\ k_2 \end{bmatrix}$$

Note that the characteristic vectors v_1 and v_2 are linearly independent, because $|V| \neq 0$.

Example 2.3.6 The characteristic equation in Example 2.3.2 was $\lambda(\lambda^2 - 3\lambda + 2) = 0$. Find the corresponding characteristic roots and vectors.

Solution: The characteristic roots of the above equation are $\lambda_1 = 0$, $\lambda_2 = 1$ and $\lambda_3 = 2$. The corresponding characteristic vector for root $\lambda_1 = 0$ satisfies equation $Ay = 0$, that is

$$\begin{bmatrix} 1-0 & 3 & 0 \\ 7 & 1-0 & 5 \\ 0 & -4 & 1-0 \end{bmatrix} \begin{bmatrix} y_1 \\ y_2 \\ y_3 \end{bmatrix} = \begin{bmatrix} 0 \\ 0 \\ 0 \end{bmatrix} \quad ; \quad \begin{aligned} y_1 + 3y_2 \qquad\quad &= 0 \\ 7y_1 + y_2 + 5y_3 &= 0 \\ -4y_2 + y_3 &= 0 \end{aligned} \quad ; \quad \begin{aligned} y_1 &= -3y_2 \\ y_2 &= y_2 \\ y_3 &= 4y_2 \end{aligned}$$

Then, any vector

$$v_1 = k_1 \begin{bmatrix} -3 \\ 1 \\ 4 \end{bmatrix}$$

is a characteristic vector and a solution of the system. The corresponding characteristic vector for root $\lambda_2 = 1$ satisfies equation $(A - I)y = 0$, that is

$$\begin{bmatrix} 1-1 & 3 & 0 \\ 7 & 1-1 & 5 \\ 0 & -4 & 1-1 \end{bmatrix}\begin{bmatrix} y_1 \\ x_2 \\ y_3 \end{bmatrix} = \begin{bmatrix} 0 \\ 0 \\ 0 \end{bmatrix} \quad ; \quad \begin{array}{rcl} y_2 & = & 0 \\ 7y_1 + 5y_3 & = & 0 \\ -4y_2 & = & 0 \end{array} \quad ; \quad \begin{array}{rcl} 7y_1 & = & -5y_3 \\ y_2 & = & 0 \end{array}$$

Thus, any vector

$$v_2 = k_2 \begin{bmatrix} 5 \\ 0 \\ -7 \end{bmatrix}$$

is also a characteristic vector and a solution of the system. The corresponding characteristic vector for root $\lambda_3 = 2$ satisfies equation $(A - 2I)y = 0$, that is

$$\begin{bmatrix} 1-2 & 3 & 0 \\ 7 & 1-2 & 5 \\ 0 & -4 & 1-2 \end{bmatrix}\begin{bmatrix} y_1 \\ y_2 \\ y_3 \end{bmatrix} = \begin{bmatrix} 0 \\ 0 \\ 0 \end{bmatrix} \quad ; \quad \begin{array}{rcl} -y_1 + 3y_2 & = & 0 \\ 7y_1 - y_2 + 5y_3 & = & 0 \\ -4y_2 - y_3 & = & 0 \end{array} \quad ; \quad \begin{array}{rcl} y_1 & = & 3y_2 \\ y_2 & = & y_2 \\ y_3 & = & -4y_2 \end{array}$$

Also any vector

$$v_3 = k_3 \begin{bmatrix} 3 \\ 1 \\ -4 \end{bmatrix}$$

is a characteristic vector and a solution of the system. Then, the characteristic matrix V of the system is

$$V = \begin{bmatrix} -3 & 5 & 3 \\ 1 & 0 & 1 \\ 4 & -7 & -4 \end{bmatrix} \begin{bmatrix} k_1 \\ k_2 \\ k_3 \end{bmatrix}$$

The vectors of the above characteristic matrix are linearly independent.

So far, the examples presented here are related to systems with characteristic roots all different. If the characteristic equation has multiple roots, the problem is to determine a set of linearly independent characteristic vectors. In the following example the system has multiple characteristic roots.

Example 2.3.7 Determine the characteristic roots and vectors of the following matrix:

$$A = \begin{bmatrix} -1 & -2 & -2 \\ 1 & 2 & 1 \\ -1 & -1 & 0 \end{bmatrix}$$

Solution: The characteristic equation is here

$$|A - \lambda I| = \begin{vmatrix} -1-\lambda & -1 & -2 \\ 1 & 2-\lambda & 1 \\ -1 & -1 & -\lambda \end{vmatrix} = (\lambda - 1)(\lambda^2 - 1) = 0$$

and the roots are $\lambda_1 = -1$, $\lambda_2 = 1$, $\lambda_3 = 1$. Note that $\lambda_2 = \lambda_3$. Hence, $\lambda = 1$ is a multiple root with multiplicity 2. For root $\lambda_1 = -1$, the corresponding vector satisfies equation $(A+I)y=0$. Thus

$$\begin{bmatrix} -1+1 & -2 & -2 \\ 1 & 2+1 & 1 \\ -1 & -1 & 1 \end{bmatrix} \begin{bmatrix} y_1 \\ y_2 \\ y_3 \end{bmatrix} = \begin{bmatrix} 0 \\ 0 \\ 0 \end{bmatrix} \quad ; \quad \begin{aligned} -2y_2 - 2y_3 &= 0 \\ y_1 + 3y_2 + y_3 &= 0 \\ -y_1 - y_2 + y_3 &= 0 \end{aligned} \quad ; \quad \begin{aligned} y_1 &= 2y_3 \\ y_2 &= -y_3 \\ y_3 &= y_3 \end{aligned}$$

Then, any vector

$$v_1 = k_1 \begin{bmatrix} 2 \\ -1 \\ 1 \end{bmatrix}$$

is a characteristic vector and a solution of the system. For multiple roots $\lambda = 1$, the corresponding vectors satisfy equation $(A - I)y = 0$:

$$\begin{bmatrix} 1-1 & -2 & -2 \\ 1 & 2-1 & 1 \\ -1 & -1 & -1 \end{bmatrix} \begin{bmatrix} y_1 \\ y_2 \\ y_3 \end{bmatrix} = \begin{bmatrix} 0 \\ 0 \\ 0 \end{bmatrix} \quad ; \quad \begin{array}{rcl} -2y_1 - 2y_2 - 2y_3 &=& 0 \\ y_1 + y_2 + y_3 &=& 0 \\ -y_1 - y_2 - y_3 &=& 0 \end{array} \quad ; \quad \begin{array}{rcl} y_1 &=& y_1 \\ y_2 &=& 0 \\ y_3 &=& -y_1 \end{array}$$

However, the solution must provide here two linearly independent vectors from the same root. Note that the above equations are all represented by expression $y_1 + y_2 + y_3 = 0$. Then, any vector v, such that $y_1 + y_2 + y_3 = 0$, is an appropriate solution. The following are two possibilities:

$$v_2 = k_2 \begin{bmatrix} 1 \\ 0 \\ -1 \end{bmatrix} \quad ; \quad v_3 = k_3 \begin{bmatrix} 1 \\ -1 \\ 0 \end{bmatrix}$$

Thus, the characteristic matrix is as follows:

$$V = \begin{bmatrix} 2 & 1 & 1 \\ -1 & 0 & -1 \\ 1 & -1 & 0 \end{bmatrix} \begin{bmatrix} k_1 \\ k_2 \\ k_3 \end{bmatrix}$$

Summary

The characteristic equation of a system of linear equations is a polynomial of degree n in the λ_i characteristic roots and is obtained by expanding determinant $|A - \lambda I| = 0$. Then, equation $(A - \lambda I)y = 0$ must hold true for each of the characteristic roots. Corresponding to each characteristic root there is a characteristic vector v_i, such that $(A - \lambda_i I) = 0$.

3

THE CALCULUS FOUNDATION OF MODELING

The concept of a system is related to the notion of change. Depending on wether the system is discrete or is continuous, change is usually expressed as difference or as differential equations.

The objective of this chapter is to present a conceptual overview of difference and differential equations. Because the calculus of finite differences is frequently overlooked in formal mathematical training, selected topics of series and finite differences are also included.

3.1 SERIES

The notion of a series is derived from the summation of the terms of a *sequence*. A sequence is a succession of terms formed according to a fixed rule or law. For example,

$$1,\ 4,\ 9,\ 16,\ 25,\ \dots\ ,n^2$$

is a sequence, and

$$1\ +\ 4\ +\ 9\ +\ 16\ +\ 25\ +\ \dots\ +\ n^2$$

is a series. Then, a series is defined as follows:

Definition 3.1.1 A series S_n is the sum of the terms of a sequence. When the number of terms is limited, the series is said to be a *finite series*. When the number of terms is unlimited, the series is called an *infinite series*.

The variable S_n is a function of n, the number of terms. When n increases without a limit, the series is said to be *convergent* if S_n approaches a finite limit, that is

$$\lim_{n \to \infty} S_n\ =\ u$$

where u is a positive real number. When S_n approaches infinity as a limit, the series is said to be *divergent*:

$$\lim_{n \to \infty} S_n = \infty$$

Example 3.1.1 Find the value of S_n in the following geometric series.

$$S_n = 90 \,(1 + 0.964 + 0.964^2 + ... + 0.964^{n-1})$$

Solution: For simplicity, define $a = 90$ and $r = 0.964$. Then

$$\begin{aligned}
S_n &= a(1 + r + r^2 + ... + r^{n-1}) \\
rS_n &= a(r + r^2 + r^3 + ... + r^n) \\
S_n - rS_n &= a(1 - r^n) \\
S_n &= \frac{a(1 - r^n)}{1 - r}
\end{aligned}$$

If $|r| < 1$, then r^n decreases in value as n increases. Therefore

$$\lim_{n \to \infty} S_n = \frac{a}{1 - r}$$

The following is obtained when the numerical values are replaced in the above result:

$$\lim_{n \to \infty} S_n = \frac{90}{1 - 0.964} = 2500$$

Example 3.1.2 Show that the geometric series is divergent when $|r| > 1$.

Solution: S_n can be written as follows:

$$S_n = \frac{a(r^n - 1)}{r - 1}$$

If $|r| > 1$, then r^n increases to infinity as n increases indefinitely. Thus, S_n will become infinite.

Example 3.1.3 It was found that consumption of molasses reduces pasture consumption of dairy heifers, as defined by the following equation [1]:

$$y = 30.1 - 0.501x$$

where y is pasture consumption in Mcal/day of digestible energy (DE) and x is a molasses supplement in Mcal/day, also as digestible energy. This effect is illustrated in Fig. 3.1.1.

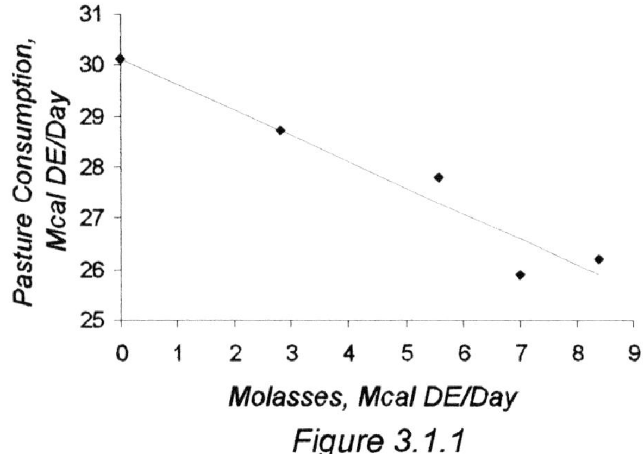

Figure 3.1.1

Find how much digestible energy as molasses the heifers need for a target total energy consumption of 34.6 Mcal/day.

Solution: The difference between the target energy consumption of 34.6 and 30.1 and the pasture energy consumption when the molasses value is zero is 4.5 Mcal/day. Note that, for each Mcal of molasses, the pasture consumption of the heifers decreases by 0.501 Mcal/day. Supplementing 4.5 Mcal of molasses will decrease pasture consumption in 2.25

---

[1]Computed from Beaudouin, J.

Mcal of pasture. Therefore, an additional supplementation of 2.25 Mcal of molasses is needed, that will decrease pasture consumption by an additional 1.13 Mcal/day and so on. Thus, the following geometric series can be defined:

$$S_n = 4.5(1 + 0.501 + 0.501^2 + 0.501^3 + ... + 0.501^n)$$

Then, the solution is

$$\lim_{n \to \infty} S_n = \frac{4.5}{1 - 0.501} = 9.02 \text{ Mcal/day}$$

This solution is shown in Fig 3.1.2.

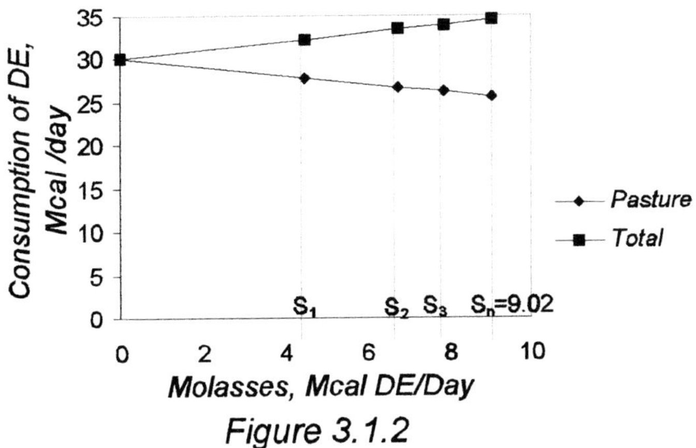

Figure 3.1.2

Note that this system may also be defined by the following set of linear equations:

$$y = 30.1 - 0.501x$$

$$y + x = 34.6$$

Then, $x = 9.02$ and $y = 25.6$.

It is often important to determine the limit of a convergent series when n increases without a limit. Conversely, for divergent series, it is only possible to determine the value of S_n for a finite value of n. Thus, it is essential to have the means for testing if the series is convergent or divergent.

One available procedure to determine the convergence of a series is called the *test-ratio* method and may be applied as follows to any series. Let $u_1 + u_2 + ... + u_{n+1} + ...$ be an infinite series. By considering consecutive terms, the following ratio can be defined:

$$R = \frac{u_{n+1}}{u_n}$$

Let ρ be the limit of R when n becomes infinite. Then

$$\rho = \lim_{n \to \infty} \frac{u_{n+1}}{u_n}$$

The following rules are here set forth without proof:

- If $\rho < 1$, the series is convergent
- If $\rho > 1$, the series is divergent
- If $\rho = 1$, the test fails

Example 3.1.4 Test the following series:

A. $\quad 1 + \dfrac{1}{1!} + \dfrac{1}{2!} + \dfrac{1}{3!} + ...$
B. $\quad \dfrac{1!}{10} + \dfrac{2!}{10^2} + \dfrac{3!}{10^3} + ...$
C. $\quad \dfrac{1}{1*2} + \dfrac{1}{3*4} + \dfrac{1}{5*6} + ...$

Solution:

A. $\quad \dfrac{u_{n+1}}{u_n} = \dfrac{(n-1)!}{n!} = \dfrac{1}{n} \quad ; \quad \rho = \lim_{n \to \infty} \dfrac{1}{n} = 0 \qquad$ The series is convergent

B. $\quad \dfrac{u_{n+1}}{u_n} = \dfrac{(n+1)!}{10^{n+1}} * \dfrac{10^n}{n!} = \dfrac{n+1}{10} \quad ; \quad \rho = \lim_{n \to \infty} \dfrac{n+1}{10} = \infty \qquad$ The series is divergent

C. $\quad u_n = \dfrac{1}{(2n-1)2n} \quad ; \quad u_{n+1} = \dfrac{1}{(2n+1)(2n+2)} \quad ; \quad \rho = \lim_{n \to \infty} \dfrac{4n^2 - 2n}{4n^2 + 6n + 2} = 1$

Note that series C may be either convergent or divergent, meaning that the test failed. Thus, another testing procedure is needed. Often determining if a series is convergent or divergent is possible by comparing it term by term with another series whose convergence or divergence was previously determined.

Example 3.1.5 Determine if series C in the previous example is convergent or divergent.

Solution: Compare series C term by term with the following convergent geometric series:

$$\text{Geometric Series:} \quad 1 + \frac{1}{2} + \frac{1}{2^2} + \frac{1}{2^3} + \ldots \quad ; \quad \text{Series C:} \quad \frac{1}{1(2)} + \frac{1}{3(4)} + \frac{1}{5(6)} + \ldots$$

The terms of the geometric series are never less than the corresponding terms of the series being tested. Thus, series C is convergent.

Summary

A sequence is a succession of terms arranged according to a fixed rule and a series is the sum of such terms. When the number of terms is finite, the series is said to be a finite series. Conversely, if the number of terms is infinite, the series is called an infinite series. Infinite series are said to be convergent if the sum of its terms approaches a finite limit when the number of terms increases without a limit. Otherwise, the series are said to be divergent. Series may be tested for convergence by the test-ratio procedure. If this method fails, the series may be compared term by term with another series whose convergence or divergence was previously determined.

3.2 FINITE DIFFERENCES

The topics covered in this section have been aimed at getting a basic knowledge on finite differences, as related to the manipulation of difference equations in Chapter 6.

Definition of a Finite Difference

Consider the function $y = f(t)$ depicted in Fig.3.2.1. Then, a finite difference is defined as follows:

Definition 3.2.1 If Δy is the difference between two values of $f(t)$ and Δt is the increment in the independent variable, then $y + \Delta y = f(t + \Delta t)$.

Thus

$$\Delta y = f(t + \Delta t) - f(t)$$

Δy is called a *finite difference*. The symbol Δ is called the *difference operator*. The geometrical interpretation of a finite difference is shown in Fig. 3.2.1.

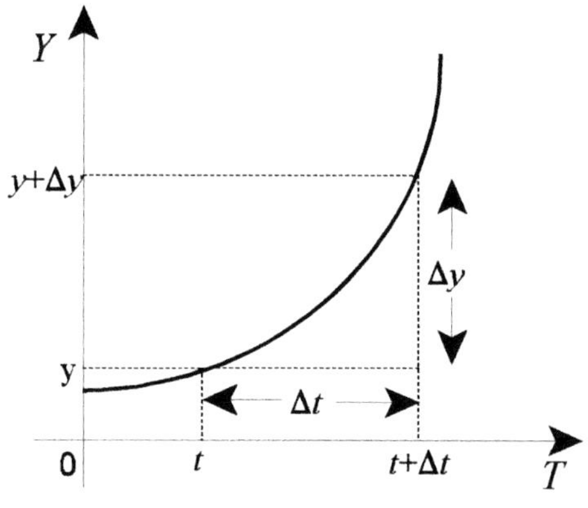

Figure 3.2.1

Example 3.2.1 Find a finite difference expression for equation $y = 2x^2 + 3x$.

Solution:

$$
\begin{aligned}
y &= 2x^2 + 3x \\
y + \Delta y &= 2(x + \Delta x)^2 + 3(x + \Delta x) \\
\Delta y &= 2(x + \Delta x)^2 + 3(x + \Delta x) - 2x^2 + 3x \\
&= \Delta x(4x + 2\Delta x + 3)
\end{aligned}
$$

If $\Delta x = 2$, then $\Delta y = 2(4x + 7)$.

From the definition of a finite difference, it is clear that a relationship exists between finite differences and derivatives. A derivative is determined if the finite difference Δy is divided by Δt and the limit of this ratio is taken when Δt approaches zero:

$$
\lim_{\Delta t \to 0} \frac{\Delta y}{\Delta t} = \frac{dy}{dt} = \lim_{\Delta t \to 0} \frac{f(t + \Delta t) - f(t)}{\Delta t}
$$

It is possible to perceive more clearly the analogy between the difference calculus and the differential calculus by comparing the general rules for finite differences and the rules for differentiation. The general rules of the difference calculus are presented below:

$$\Delta k = 0 \quad ; \quad k = \text{constant}$$
$$\Delta[kf(t)] = k\Delta f(t)$$
$$\Delta[f(t) + g(t)] = \Delta f(t) + \Delta g(t)$$
$$\Delta[f(t)g(t)] = f(t)\Delta g(t) + g(t+\Delta t)\Delta f(t) = g(t)\Delta f(t) + f(t+\Delta t)\Delta g(t)$$
$$= f(t)\Delta g(t) + g(t)\Delta f(t) + \Delta f(t)\Delta g(t)$$
$$\Delta\left[\frac{f(t)}{g(t)}\right] = \frac{g(t)\Delta f(t) - f(t)\Delta g(t)}{g(t)g(t+\Delta t)}$$

The following are the general rules of differentiation:

$$\frac{d}{dt}k = 0 \quad ; \quad k = \text{constant}$$
$$\frac{d}{dt}[kf(t)] = k\frac{d}{dt}f(t)$$
$$\frac{d}{dt}[f(t)+g(t)] = \frac{d}{dt}f(t) + \frac{d}{dt}g(t)$$
$$\frac{d}{dt}[f(t)g(t)] = f(t)\frac{d}{dt}g(t) + g(t)\frac{d}{dt}f(t)$$
$$\frac{d}{dt}\left[\frac{f(t)}{g(t)}\right] = \frac{g(t)\frac{d}{dt}f(t) - f(t)\frac{d}{dx}g(x)}{[g(t)]^2}$$

The differential calculus would give the results of the difference calculus in the special case when Δt approaches zero as a limit.

Example 3.2.2 Show that $\Delta[f(t)g(t)] = f(t)\Delta g(t) + g(t)\Delta f(t) + \Delta f(t)\Delta g(t)$.

Solution: According to the definition of a finite difference,

$$\Delta[f(t)g(t)] = f(t+\Delta t)g(t+\Delta t) - f(t)g(t)$$
$$f(t+\Delta t) = f(t) + \Delta f(t)$$
$$g(t+\Delta t) = g(t) + \Delta g(t)$$

After the proper replacements, the following is obtained:

$$\Delta[f(t)g(t)] = [f(t)+\Delta f(t)][g(t)+\Delta g(t)] - f(t)g(t)$$
$$= \Delta f(t)\Delta g(t) + g(t)\Delta f(t) + f(t)\Delta g(t)$$

Note that it is possible to obtain the derivative of the product of two functions by dividing the difference of the product by Δt and making this to approach zero as a limit:

$$\lim_{\Delta t \to 0} \frac{\Delta[f(t)g(t)]}{\Delta t} = \frac{d}{dt}[f(t)g(t)]$$
$$= g(t)\frac{d}{dt}f(t) + f(t)\frac{d}{dt}g(t)$$

The following are the differences of special functions, which are also analogous to their derivative counterparts:

$$\Delta[k^x] = k^x(k^{\Delta x} - 1) \quad ; \quad k = \text{constant}$$
$$\Delta[e^{kx}] = e^{kx}(e^{kx} - 1)$$
$$\Delta[\sin\ kx] = 2\sin(k\Delta x/2)\cos k(x + \Delta x/2)$$
$$\Delta[\cos\ kx] = -2\sin(k\Delta x/2)\sin k(x + \Delta x/2)$$
$$\Delta[\ln\ x] = \ln(1 + \Delta x/x)$$
$$\Delta[\log_b\ x] = \log_b(1 + \Delta x/x)$$

Example 3.2.3 Show that $\Delta[\sin kx] = 2\sin(k\Delta x/2)\cos k(x + \Delta x/2)$.

Solution: According to the definition of a finite difference

$$\Delta[\sin\ kx] = \sin k(x + \Delta x)\ -\ \sin\ kx$$
$$= \sin A\ -\ \sin B$$

This difference formula may be written as[2]

$$\Delta[\sin\ kx] = 2\sin\frac{(A - B)}{2}\cos\frac{(A + B)}{2}$$
$$= 2\sin(k\Delta x/2)\ \cos k(x + \Delta x/2)$$

Subscript Notation

Consider the function $f(t)$ and make the transformation $t = a + n\Delta t$ for variable t. Defining a new function y_n is now possible, such that $y_n = f(a + n\Delta t)$. Then, the finite

---

[2]See any manual of mathematical tables

difference can be written as

$$\Delta y_n = \Delta f(a+n\Delta t)$$
$$= f(a+n\Delta t+\Delta t) - f(a+n\Delta t)$$
$$= f(a+(n+1)\Delta t) - f(a+n\Delta t)$$

If $a=0$ and $\Delta t=1$, then $t=n$ and the new independent variable becomes $n \in N[0,\infty)$. Now the finite difference can be redefined as follows:

$$\Delta y_n = f(n+1) - f(n)$$
$$= y_{n+1} - y_n$$

Thus, Δy_n is the difference between two values of the dependent variable.

Note that, because $\Delta t=1$, the difference Δy_n is equivalent to the first derivative dy/dt in the differential calculus. Note also that y_n represents a sequence of values of the dependent variable, defined over the discrete time scale N. As shown below, the rules of the difference calculus are not changed by the subscript transformation:

$$\Delta[k] = 0 \quad ; \quad k = \text{constant}$$
$$\Delta[ky_n] = k\Delta y_n$$
$$\Delta[y_n+z_n] = \Delta y_n + \Delta z_n$$
$$\Delta[y_n z_n] = y_n\Delta z_n + z_{n+1}\Delta y_n$$
$$\Delta\left[\frac{y_n}{z_n}\right] = \frac{y_n\Delta z_n - z_n\Delta y_n}{z_n z_{n+1}}$$
$$\Delta[k^n] = k^n (k-1)$$
$$\Delta[e^{kn}] = e^{kn} (e^k-1)$$
$$\Delta[\ln n] = \ln (1+1/n)$$
$$\Delta[\log_b n] = \log_b (1+1/n)$$
$$\Delta[\sin kn] = 2 \sin (k/2) \cos k(n+1/2)$$
$$\Delta[\cos kn] = -2 \sin (k/2) \sin k(n+1/2)$$

Example 3.2.4 Show that $\Delta y_n z_n = y_n\Delta z_n + z_{n+1}\Delta y_n$.

Solution: According to the definition of a finite difference, $\Delta y = y_{n+1} - y_n$. Then

$$\Delta y_n z_n = y_{n+1}z_{n+1} - y_n z_n$$
$$z_n = z_{n+1} - \Delta z_n$$

Thus

$$\Delta y_n z_n = y_{n+1} z_{n+1} - y_n(z_{n+1} - \Delta z_n)$$
$$= z_{n+1}(y_{n+1} - y_n) + y_n \Delta z_n$$
$$= z_{n+1} \Delta y_n + y_n \Delta z_n$$

Example 3.2.5 The rate of change of an insect population is described by the following difference equation:

$$y_n - y_{n-1} = 500(2^{-n})$$

where y is number of insects and n is periods in weeks. Find the population growth equation.

Solution: It follows that:

$$y_n = y_{n-1} + 500(2^{-n})$$
$$y_1 = y_0 + 500(2^{-1})$$
$$y_2 = y_1 + 500(2^{-2}) = y_0 + 500(2^{-1}) + 500(2^{-2})$$
$$\vdots$$
$$y_n = y_0 + 500(2^{-1}) + 500(2^{-2}) + \ldots + 500(2^{-n})$$

This solution includes the series

$$S_n = 2^{-1} + 2^{-2} + \ldots + 2^{-n}$$
$$2S_n = 1 + 2^{-1} + \ldots + 2^{-(n+1)}$$
$$S_n - 2S_n = -1 + 2^{-n}$$
$$S_n = 1 - 2^{-n}$$

Then

$$y_n = y_o + 500(1 - 2^{-n})$$

This is a *general solution* for the given difference equation, where y_o is the summation constant. If a particular value is assigned to y_o , then the solution is called a *particular solution.* A particular solution for $y_o = 1000$ is shown in Figure 3.2.

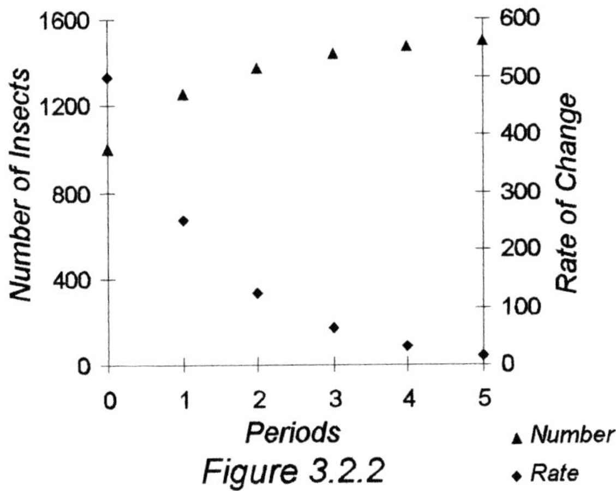

Figure 3.2.2

Example 3.2.6 A rancher sells each month 3.6 % of his feedlot steers and buys 90 new animals. Define a difference equation and the state equation for this system.

Solution: The difference equation is given by the difference of what the rancher sells and what he buys:

$$\Delta y_n = y_{n+1} - y_n = -0.036 y_n + 90$$

where y is number of individuals and n is periods in months. Then, it follows that:

$$y_{n+1} = (1 - 0.036)\, y_n + 90 = 0.964\, y_n + 90$$
$$y_1 = 0.964\, y_0 + 90$$
$$y_2 = 0.964\, y_1 + 90 = (0.964)^2\, y_0 + 0.964(90) + 90$$
$$\vdots$$
$$y_n = (0.964)^n\, y_0 + 90[(0.964)^{n-1} + (0.964)^{n-2} + \ldots + (0.964)^{n-n}]$$

The following geometric series is included in this solution:

$$S_n = 90[1 + 0.964 + 0.964^2 + \ldots + 0.964^{n-1})$$
$$= \frac{90(1 - 0.964^n)}{1 - 0.964}$$
$$= 2500(1 - 0.964^n)$$

Then

$$y_n = (0.964)^n \, y_0 + 2500(1 - 0.964^n)$$
$$= 2500 + (y_0 - 2500) \, 0.964^n$$

The constant y_0 is the initial number of animals in the ranch. A particular solution for $y_o = 1000$ and the corresponding difference equation are shown in Figure 3.2.3.

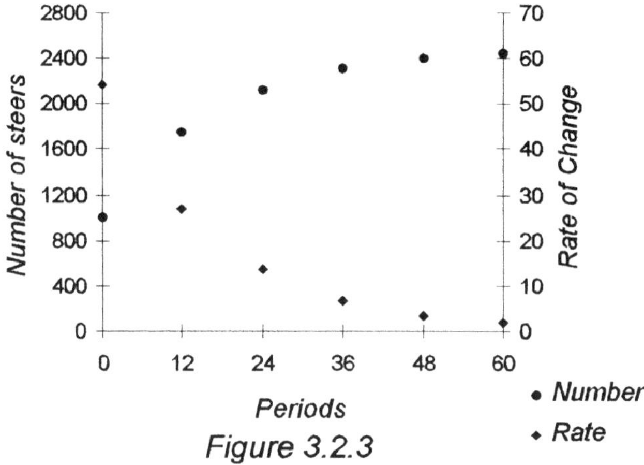

Figure 3.2.3

In the work that follows only the subscript notation will be used.

Summary

A finite difference is defined as $\Delta y = f(t + \Delta t) - f(t)$. The limit of the finite difference divided by Δt is a derivative when Δt approaches zero. A finite difference can be redefined in a subscript notation as $\Delta y_n = y_{n+1} - y_n$.

3.3 DIFFERENTIALS

The derivative of the function $y = f(x)$ is represented by the symbol dy/dx. This symbol should not be considered as an ordinary fraction with dy as numerator and dx as denominator but as a representation of the limit of the quotient $\Delta y/\Delta x$ as Δx approaches zero as a limit. As will be shown, giving a geometrical meaning to dy and dx separately is important.

As exposed in Fig. 3.3.1, $dy/dx = \tan \beta = RT/PR$. Note that segment $RT=dy$ and segment $PR=dx$. Note also that segment $RQ=\Delta y \neq dy$, except for the particular case of the straight line $y=a+bx$. Conversely, always $dx = \Delta x$. Then

$$dy = f'(x)dx = f'(x)\Delta x$$

The symbol dy is called the *differential of f*(x), the symbol dx is called the *differential of the independent variable x* and the symbol d is called the differential operator.

As illustrated above, the differential dy is not equal to the increment of the

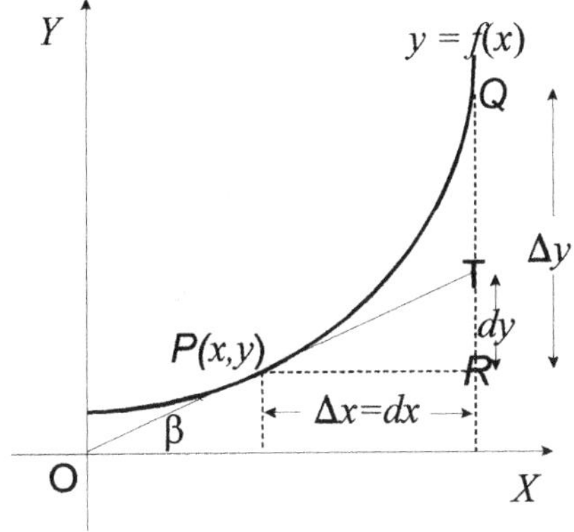

Figure 3.3.1

function Δy. However, if Δx is small enough, dy could be an acceptable approximation for Δy. Usually, calculating differentials and using this value rather than the corresponding increment is simpler. This is shown in the following example.

Example 3.3.1   The daily average consumption of forage by a group of steers, as a function of crude protein in forage, is given by the following equation [3]:

---

[3]Vohnout, K., Unpublished.

$$y = \frac{x}{0.0701 + 0.0102x}$$

The variable y is dry matter consumption in grams per unit of metabolic body weight ($W^{0.75}$) and the variable x is percent crude protein. Find the difference between the differential of the function and the corresponding consumption increment, when crude protein is increased from 5 to 10%, from 10 to 11% and from 10 to 15%.

Solution: The first derivative of the mathematical model is

$$\frac{dy}{dx} = \frac{0.0701}{(0.0701 + 0.0102x)^2}$$

If $dy = f'x\,\Delta x$ and $\Delta \hat{y} = f(x_1) - f(x_2)$ then, when protein is increased from 5 to 10% for $\Delta x = 5$, the following results are obtained:

$$dy = \left[\frac{0.0701}{[0.0701 + 0.0102(5)]^2}\right]5 = 23.90$$

$$\Delta y = \frac{10}{0.0701 + 0.0102(10)} - \frac{5}{0.0701 + 0.0102(5)} = 16.82$$

Thus, $dy - \Delta y = 7.08$. When protein is increased from 10 to 11% for a $\Delta x = 1$, the following relations are obtained:

$$dy = \frac{0.0701}{(0.0701 + 0.1020)^2} = 2.37$$

$$\Delta y = \frac{11}{0.0701 + 0.1122} - \frac{10}{0.0701 + 0.1020} = 2.23$$

Thus, $dy - \Delta y = 0.14$. When protein is increased from 10 to 15% for a $\Delta x = 5$, the following expressions are obtained:

$$dy = \frac{0.3505}{(0.0701 + 0.1020)^2} = 11.83$$

$$\Delta y = \frac{15}{0.0701 + 0.1530} - \frac{10}{0.0701 + 0.1020} = 9.12$$

Thus, $dy - \Delta y = 2.71$. Note that the difference between dy and Δy increases with the increase of Δx. Note also that, because of the diminishing returns nature of the function, as the independent variable increases, the error introduced by the dy estimate decreases. These relationships are shown in Fig. 3.3.2.

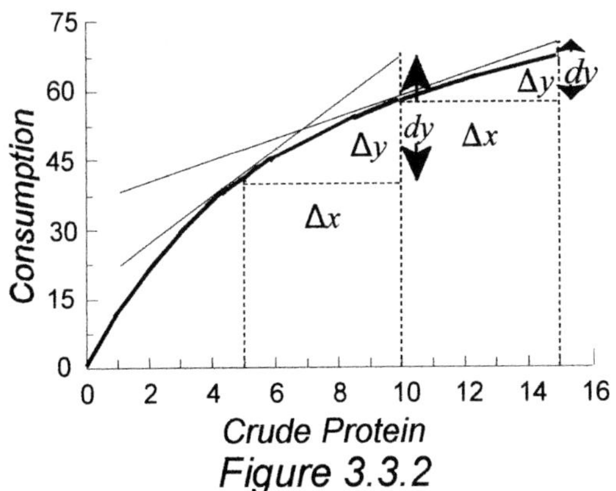

Figure 3.3.2

Example 3.3.2 The following equation was fitted to the lactation curve of a group of dairy cows[4]:

$$y = e^{-484t}(298 + 411t)$$

where y is milk production, Kg/month and t is months. Determine the difference between the differential of the function and the corresponding increment in milk production between 0.5 and 1 month and between 0.5 and 1.5 months in the lactation curve.

Solution: The first derivative of the lactation curve equation is

$$\frac{dy}{dt} = e^{-0.484t}(267 - 199t)$$

Then, the difference in milk production between 0.5 and 1 month is given by the following expressions:

---

[4] Vohnout, K., Unpublished

$$dy = 0.5e^{-0.484(0.5)}[267 - 199(0.5)] = 65.7$$

$$\Delta y = e^{-0.484}[298 + 411] - e^{-484(0.5)}[298 + 411(0.5)] = 41.69$$

Thus, $dy - \Delta y = 24.0$. The difference in milk production between 0.5 and 1.5 months is

$$dy = e^{-0.484(0.5)}[267 - 199(0.5)] = 131.4$$

$$\Delta y = e^{-484(0.15)}[298 + 411(1.5)] - e^{-484(0.5)}[298 + 411(0.5)] = 47.19$$

Thus, $dy - \Delta y = 84.2$.

Data are seldom recorded continuously. Most often, experimental data are recorded at given intervals of time. Thus, information between data points is lost. As shown, understanding differentials and increments is important in the design of system experiments, in relation to the manipulation of errors introduced by the intermittent collection of data and in relation to the analysis of results.

Summary

The symbol dy/dx is not an ordinary fraction, but the representation of the derivative of a function. However, dy and dx have a geometrical meaning if considered separately that is, always $dx = \Delta x$ but $dy \neq \Delta y$, except for the straight line case. Since obtaining differentials is simpler than obtaining increments, when Δx is small, dy could be an acceptable approximation for Δy. Data is seldom recorded continuously. Thus, the difference between dy and Δy reflects the amount of information lost between increments in the independent variable.

3.3.4 DIFFERENCE EQUATIONS

The reader has already been introduced in the previous section to the notion of difference equations. Thus, it should come as no surprise that a relationship exists between differential equations and difference equations. If a differential equation is an equation involving derivatives or differentials, then a difference equation is one involving finite differences.

A differential equation is the limit of a difference equation when the time increment Δt approaches zero as its limit. This relationship is shown here using a first order constant coefficients linear equation:

$$\lim_{\Delta t \to 0} \left(\frac{\Delta y}{\Delta t} + ay = b \right) = \frac{dy}{dt} + ay = b$$

where $\Delta y = y_{n+1} - y_n$. Then, the first order difference equation may be written as

$$\frac{y_{n+1} - y_n}{\Delta t} - ay_n = b$$
$$y_{n+1} - (1 - a\Delta t)y_n = b\Delta t$$

As will be shown, the solution of a differential equation is also the limit of a difference equation when the time increment Δt approaches zero as its limit.

The solution of the first order constant coefficients differential equation is

$$y = \left(f(0) - \frac{b}{a} \right) e^{-at} + \frac{b}{a}$$

and the solution of the corresponding difference equation is

$$y_n = \left(f(0) - \frac{b}{a} \right) (1 - a\Delta t)^{t/\Delta t} + \frac{b}{a}$$

where $t/\Delta t = n$. The following definitions are from the infinitesimal calculus:

$$\lim_{n \to \infty} \left(1 + \frac{1}{n} \right)^n = e = 2.71828...$$

If $n = 1/\Delta t$, then

$$\lim_{\Delta t \to 0} (1 + \Delta t)^{1/\Delta t} = e \quad ; \quad \lim_{\Delta t \to 0} (1 + \Delta t)^{t/\Delta t} = e^t \quad ; \quad \lim_{\Delta t \to 0} (1 - a\Delta t)^{t/\Delta t} = e^{-at}$$

Thus

$$\lim_{\Delta t \to 0}\left[\left(f(0) - \frac{b}{a}\right)(1 - a\Delta t)^{t/\Delta t} + \frac{b}{a}\right] = \left(f(0) - \frac{b}{a}\right)e^{-at} + \frac{b}{a}$$

For the purpose of this book, difference equations are classified according to the following criteria:

- The order and degree of the equation
- Linearity or non linearity
- Inclusion or non inclusion of the dependent variable in each term of the equation
- Inclusion or non inclusion of the time variable in one or more terms of the equation

Combinations of all the above factors are possible. Of the above list, order and linearity are the most relevant for further chapters and are also the most often used to name a discrete system.

Note that two or more difference equations may form a system. In such case, the system is called multidimensional.

Order and Degree

The *order* of difference equations is determined by the sequence of successive differences of a function and is defined as follows:

Definition 3.4.1 The order of a difference equation is the difference between the largest and the smallest argument of the function involved. If $n+m$ is the largest argument and n is the smallest, then the order is $(n+m)-n = m$.

Thus

$$\Delta y_n = y_{n+1} - y_n$$
$$\Delta^2 y_n = \Delta y_{n+1} - \Delta y_n$$
$$= y_{n+2} - y_{n+1} - (y_{n+1} - y_n)$$
$$= y_{n+2} - 2y_{n+1} + y_n$$
$$\Delta^3 y_n = \Delta^2 y_{n+1} - \Delta^2 y_n$$
$$= y_{n+3} - 2y_{n+2} + y_{n+1} - (y_{n+2} - 2y_{n+1} + y_n)$$
$$= y_{n+3} - 3y_{n+2} + 3y_{n+1} - y_n$$

Successive differences as the above can be represented as a *difference table*. A difference table is a table that gives successive differences of $y = f(n)$, for $n = 1, 2, \ldots ,$. The entries in each column after the second are located between two successive entries of the preceding column and are equal to the difference between these entries. A difference table is completely determined when only one entry in each column beyond the first is known.

As shown in the difference table, entry $\Delta^2 y_1$ in the fourth column is located between entries Δy_1 and Δy_2 of the third column, which means that $\Delta^2 y_1 = \Delta y_2 - \Delta y_1$, $\Delta^3 y_2 = \Delta^2 y_3 - \Delta^2 y_2$ and so on. The first entry in each column is called a *leading difference* for the column. Thus, the leading difference for the successive columns are $\Delta y_0, \Delta^2 y_0, \dots, \Delta^5 y_0$.

Note that the first value of n in the difference table is 0. However, the first value can be any integer number, either positive or negative. Note also that the subscript in each entry relates that entry to the corresponding value of the independent variable n.

Table 3.4.1

n	y_n	Δy_n	$\Delta^2 y_n$	$\Delta^3 y_n$	$\Delta^4 y_n$	$\Delta^5 y_n$
0	y_0					
		Δy_0				
1	y_1		$\Delta^2 y_0$			
		Δy_1		$\Delta^3 y_0$		
2	y_2		$\Delta^2 y_1$		$\Delta^4 y_0$	
		Δy_2		$\Delta^3 y_1$		$\Delta^5 y_0$
3	y_3		$\Delta^2 y_2$		$\Delta^4 y_1$	
		Δy_3		$\Delta^3 y_2$		
4	y_4		$\Delta^2 y_3$			
		Δy_4				
5	y_5					

Example 3.4.1 Define the n difference in equation $y_n = n^3 - n^2$ and the corresponding difference table.

Solution:

$$\Delta y_n = (n+1)^3 - (n+1)^2 - (n^3 - n^2) = 3n^2 + n$$
$$\Delta^2 y_n = 3(n+1)^2 + (n+1) - (3n^2 + n) = 6n + 4$$
$$\Delta^3 y_n = 6(n+1) + 4 - (6n+4) = 6$$

Table 3.4.2

n	y_n	Δy_n	$\Delta^2 y_n$	$\Delta^3 y_n$
1	0			
		4		
2	4		10	
		14		6
3	18		16	
		30		
4	48			

The concept of order is further illustrated in the following examples.

Example 3.4.2 Determine the order of the following equations:

$$y - y_{n-1} = 2^{-n} \qquad \textit{Order } n-(n-1) = 1$$
$$y_{n+1} = 2^{-n} y_n + (y_{n-1})^2 \qquad \textit{Order } n+1 - (n-1) = 2$$
$$2y_{n+2} + 3y_{n+1} = \sin(y_{n+1}) \qquad \textit{Order } n+2 - (n+1) = 1$$
$$2y_{n+2} + 3y_{n+1} + y_n = 0 \qquad \textit{Order } n+2 - n = 2$$

Example 3.4.3 The evolution of a population of birds is given by the following state equation:

$$y_n = 780 - 1015(0.573)^n + 265(0.240)^n$$

The variable y is the number of individuals and the variable n is periods in years. Define the corresponding second order difference equation.

Solution: The following is the first order difference equation:

$$y_{n+1} - y_n = 780 - 1015(0.575)^{n+1} + 265(0.240)^{n+1} - [780 - 1015(0.575)^n + 265(0.240)^n]$$
$$= 1015[(0.575)^n(1 - 0.575)] - 265[(0.240)^n(1 - 0.240)]$$

The following is the second order difference equation:

$$y_{n+2} - y_n = 780 - 1015(0.575)^{n+2} + 265(0.240)^{n+2} - [780 - 1015(0.575^n + 265(0.240)^n]$$
$$= 1015[(0.575)^n(1 - 0.575^2)] - 265[(0.240)^n(1 - 0.240^2)]$$

This second difference equation must now be restructured. For such, the state equation and the two difference equations can be expressed as follows:

$$y_n = c - A + B$$
$$\Delta y_n = a_1 A - b_1 B$$
$$\Delta^2 y_n = a_2 A - b_2 B$$

By selecting any two equations in the system, solving for the A and B unknowns, replacing these values in the third equation and replacing the numerical values of c, A, B, a_1, a_2, b_1, b_2, the second order difference equation looks now as follows:

$$y_{n+2} - 0.815 y_{n+1} + 0.138 y_n = 252$$

The reader is encouraged to check the above calculations. This is a tedious procedure. However, a shortcut is available. Defining a characteristic equation from the state equation of the system is possible:

$$(\lambda - 0.575)(\lambda - 0.240) = \lambda^2 - 0.815\lambda + 0.138$$

where λ is a characteristic root. Then $y_{n+2} - 0.815 y_{n+1} + 0.138 y_n = c$. To find the numerical value for c, some numerical values for y_{n+2}, y_{n+1} and y_n are needed. For $n = 0$, the following equations can be defined:

$$
\begin{aligned}
y_n &= 780 - 1015 + 265 = 30 \\
y_{n+1} &= 1015(1-0.575) - 265(1-0.240) + y_n = 260 \\
y_{n+2} &= 1015(1-0.575^2) - 265(1-0.240^2) + y_n = 460
\end{aligned}
$$

Then $c = 460 - 0.815(260) + 0.138(30) = 252$

As referred earlier, systems are named according to the order of the difference equations of their mathematical model. Thus, the system of Example 3.3.3 is a second order system.

The following is the definition for *degree*:

Definition 3.4.2 The degree of a difference equation is the value of the largest exponent affecting the term of largest order in the equation.

Example 3.4.4 Determine the degree of the following equations:

$$
\begin{aligned}
(y_{n+2})^2 &= (1 + y_{n+1})^2 &&\textit{Second degree} \\
y_{n+1} + y_n &= 0 &&\textit{First degree} \\
(y_{n+3})^3 - y_n &= (y_{n+1})^4 &&\textit{Third degree}
\end{aligned}
$$

Linearity

A *linear difference equation* may be defined as follows:

Definition 3.4.3 A linear difference equation is one in which the dependent variable and any of its differences are of no degree greater than one.

The above definition implies that the dependent variable should not be expressed as products, logarithms, trigonometric functions or any other non linear terms. If a difference equation contains a non linear term, it is called a *non linear difference equation*. As indicated before, discrete systems are named accordingly. Thus, a linear difference equation of order m is an equation having the form

$$
g_0(n)y_{n+m} + g_1(n)y_{n+m-1} + \ldots + g_m(n)y_n = f(n)
$$

where $g_i(n)$ and $f(n)$ are functions of the independent variable or constants.

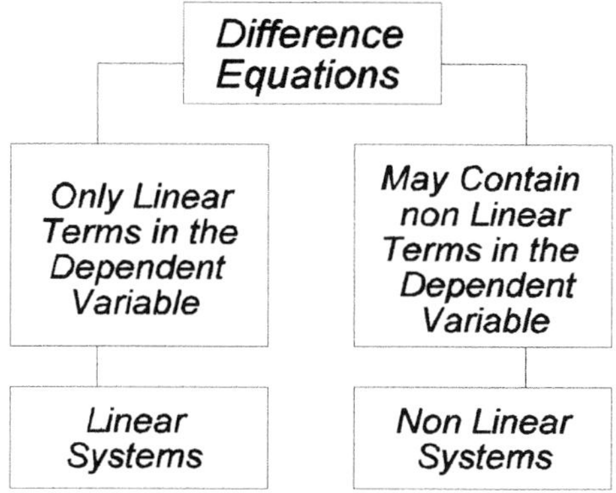

Figure 3.4.1

Example 3.4.5 Determine the linearity of the following equations:

$$y_{n+3} - 5y_{n+2} + 6y_{n+1} + 3y_n = 0 \qquad Linear$$
$$y_n y_{n+1} = y_{n-1}^2 \qquad Non\ Linear$$
$$y_{n+1} - y_n + ny_{n+1}y_n = 0 \qquad Non\ Linear$$
$$y_{n+1} - a(n)y_n = r(n) \qquad Linear$$
$$y_{n+2}y_n^2 = y_{n+1}^3 \qquad Non\ Linear$$

In some cases non linear equations can be linearized by proper transformations, as shown in the next example.

Example 3.4.6 Linearize the non linear equations in Example 3.4.5.

Solution: For equation $y_n y_{n+1} = y_{n-1}^2$:

$$\ln y_n + \ln y_{n+1} - 2\ln y_{n-1} = 0$$
$$v_{n+1} + v_n - 2v_{n-1} = 0 \quad ; \quad v_n = \ln y_n$$

For equation $y_n y_{n+1} - y_n + ny_{n+1}y_n = 0$:

$$\frac{1}{y_n} - \frac{1}{y_{n+1}} + n = 0$$

$$v_{n+1} - v_n = n \quad ; \quad v_n = 1/y_n$$

For equation $y_{n+2}y_n^2 = y_{n+1}^3$:

$$\ln y_{n+2} + 2\ln y_n = 3\ln y_{n+1}$$

$$v_{n+2} - 3v_{n+1} + 2v_n = 0 \quad ; \quad v_n = \ln y_n$$

Example 3.4.7 The population of a protected bird is represented by the following state equation:

$$y_n = (0.606)^n(450n + 250)$$

where y is number of birds and n is periods in years. Find out if the corresponding difference equation is linear or non linear.

Solution: The graphs of the state equation is shown in Fig. 3.4.2 and the corresponding difference equation follows.

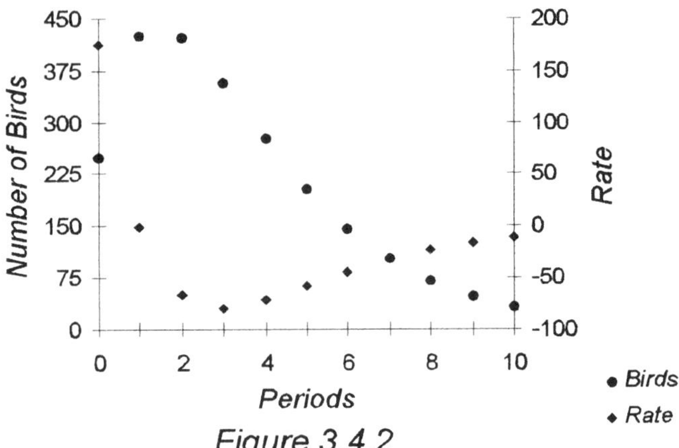

Figure 3.4.2

$$y_n = (0.606)^n(450n + 250)$$
$$y_{n+1} = (0.606)^{n+1}[450(n+1) + 25]$$
$$= 0.606(0.606)^n(450n + 250) + 450(0.606)^{n+1}$$
$$= 0.606y_n + 450(0.606)^{n+1}$$

Then $y_{n+1} - 0.606y_n = 272.7(0.606)^n$. This is a first order linear equation.

Example 3.4.8 The number of colonies of some bacteria species in a Petri dish, was found to grow according to the following state equation:

$$y_n = \frac{1}{0.0169 + 0.0279(0.9449)^n}$$

where y is the number of colonies and n is periods in hours. Determine if the corresponding difference equation is linear or non linear.

Solution: The following is a simple procedure:

$$y_{n+1} = \frac{1}{0.0169 + 0.0279(0.9449)^{n+1}} = \frac{1}{0.0169 + 0.0279(0.9.449)(0.9449)^n}$$

From the state equation, it is found that $0.0279(0.9449)^n = (1 - 0.0169y_n)/y_n$. By replacing this value in the above equation, the following is obtained:

$$y_{n+1} = \frac{y_n}{0.000931y_n + 0.9449}$$

Thus

$$y_{n+1} - \frac{y_n}{0.000931y_n + 0.9449} = 0$$

This difference equation is non linear. The state equation and its corresponding difference equation are shown in Fig. 3.4.3.

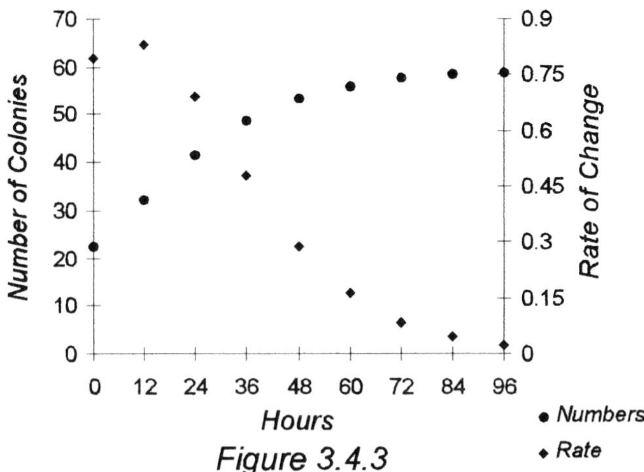

Figure 3.4.3

Homogeneity

The property of homogeneity refers to the distribution of the dependent variable in the difference equation. Consider a linear difference equation of the form

$$g_0(n)y_{n+m} + g_1(n)y_{n+m-1} + \ldots + g_m(n)y_n = f(n)$$

where $g_i(n)$ represents functions of the independent variable or constants, $f(n)$ is a function of time or a constant and $n+m$ is the order of the difference equation.

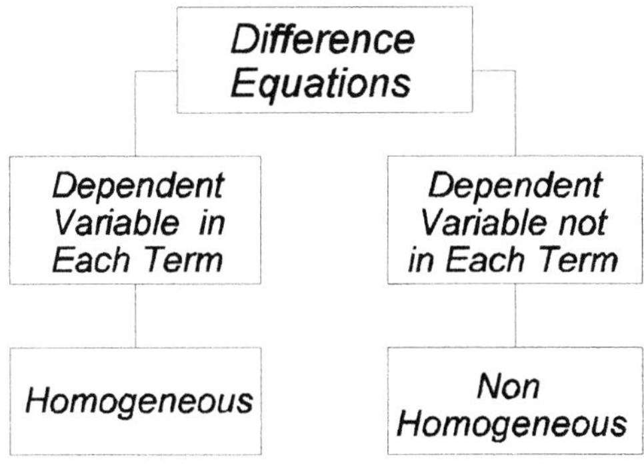

Figure 3.4.4

Then the concept of homogeneity is defined as follows:

Definition 3.4.4 A difference equation is called *homogeneous* if the dependent variable appears exactly once in each term of the equation and the term $f(n) = 0$.

 Definition 3.4.5 If some terms of the equation do not contain the dependent variable, the equation is *non homogeneous*.

Example 3.4.9 A sample of homogeneous and non homogeneous equations.

$$
\begin{aligned}
y_{n+2} + n y_{n+1} + 4 n_n &= 0 & &\textit{Non homogeneous} \\
\sin n\, y_{n+1} &= -\cos n\, y_n & &\textit{Homogeneous} \\
y_{n+1} + y_n &= 4 & &\textit{Non homogeneous} \\
y_{n+2} + y_{n+1} + n &= 0 & &\textit{Non homogeneous} \\
y_{n+1} + \frac{b}{n^2} y_n + \frac{c}{n^2} &= 0 & &\textit{Non homogeneous}
\end{aligned}
$$

A complementary definition applies here:

Definition 3.4.6 If all the $g_i(n)$ terms of the equation are constants, the equation is called a *difference equation with constant coefficients*.

Example 3.4.10 An insect control program was tested for one year in a pasture field. The following state equation was fitted to the data:

$$ y_n = 2193(0.6686)^n - 1943(0.5359)^n $$

where y is the number of bugs per square meter and n is months. Determine if the system is homogeneous or non homogeneous.

Solution: The following is the system difference equation:

$$ y_{n+2} - 1.2045 y_{n+1} + 0.3583 y_n = 0 $$

The dependent variable appears exactly once in each term of the equation. Therefore, the system is homogeneous with constant coefficients. The reader is encouraged to check that the above difference equation corresponds to the state equation.

Example 3.4.11 Determine if the difference equation in Example 3.3.7 is homogeneous or non homogeneous.

Solution: Equation $y_{n+1} - 0.606y_n = 272.7(0.606)^n$ is non homogeneous because the dependent variable does not appear exactly once in each term. Note that the term $272.7(0.606)^n \neq 0$ and does not contain the dependent variable.

The Time Variable

Time is always the independent variable of a system and, unless otherwise specified, is the only independent variable of the system. The following definitions are related to the time variable:

Definition 3.4.7 When one or more terms of the equation depend explicitly on the time variable, the equation is called a *time variant difference equation.*

Definition 3.4.8 If none of the terms of the equation depends explicitly on the variable time, the equation is called a *time invariant difference equation.*

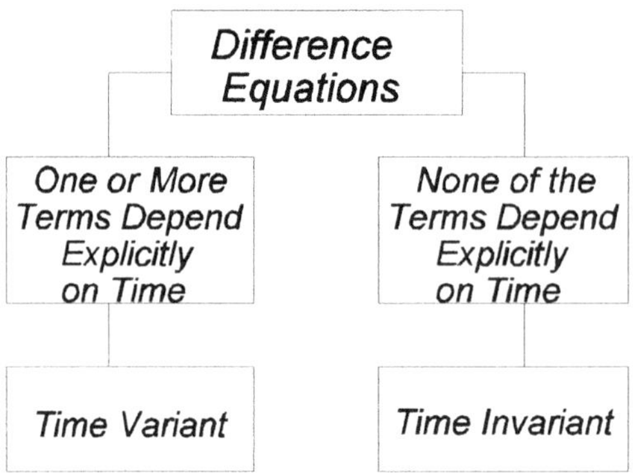

Figure 3.4.5

Example 3.4.12 Some time variant and time invariant difference equations:

$$y_{n+2} + ny_{n+1} + 4n_n = 0 \qquad Time\ variant$$
$$\sin ny_{n+1} = -\cos ny_n \qquad Time\ variant$$
$$y_{n+1} + y_n = 4 \qquad Time\ invariant$$
$$y_{n+2} + y_{n+1} + y_n = 0 \qquad Time\ invariant$$
$$y_{n+1} + \frac{b}{n^2}y_n + \frac{c}{n^2} = 0 \qquad Time\ variant$$

Example 3.4.13 An insect population was monitored for one year in a pasture field. The following is the state equation fitted to the data:

$$y_n = 152 + 149\,n^{2.28}(0.440)^n$$

where y is the number of bugs per square meter and n is months. Determine if the corresponding difference equation is time variant or time invariant.

Solution: The above expression has the following first order difference equation

$$
\begin{aligned}
y_{n+1} &= 152 + 149(n+1)^{2.28}(0.440)^{n+1} \\
&= 152 + (y_n - 152)\left(\frac{n+1}{n}\right)^{2.28}(0.440) \\
&= 152 + 0.440\left(\frac{n+1}{n}\right)^{2.28} y_n - 152(0.440)\left(\frac{n+1}{n}\right)^{2.28}
\end{aligned}
$$

Thus

$$y_{n+1} - 0.440\left(\frac{n+1}{n}\right)^{2.28} y_n = 152 - 66.9\left(\frac{n+1}{n}\right)^{2.28}$$

This is a first order, non homogeneous time variant difference equation.

Example 3.4.14 The population of an animal species is represented by the following state equation:

$$y_n = 309 + 957\left(\frac{1}{2}\right)^n \cos\left[\frac{2\pi}{3}(n - 0.3648)\right]$$

where y is the number of animals and n is the number of generations. Determine if the corresponding difference equation is time variant or time invariant.

Solution: The following is the difference equation representing this system:

$$y_{n+2} + 0.5y_{n+1} + 0.25y_n = 540$$

This is a second order, non homogeneous time invariant equation.

Summary

Difference equations are expressions involving finite differences. The order of a difference equation is the difference between the largest and the smallest argument of the function involved. If $n+m$ is the largest argument and n is the smallest, then the order is $(n+m)-n = m$. The degree is the value of the largest exponent corresponding to the largest order term. A linear difference equation is one in which the dependent variable and any of its differences are not of degree greater than one. Otherwise, the equation is non linear. A difference equation is called homogeneous if the dependent variable appears exactly once in each term of the equation, otherwise is non homogeneous. If one or more terms depend explicitly on the variable time, the equation is time variant. If none of the terms depends explicitly on the time variable, the equation is called time invariant. Discrete systems are often named according to the type of difference equation of the mathematical model assigned to define the system.

3.5 DIFFERENTIAL EQUATIONS

A differential equation is an equation involving derivatives or differentials. In this book, difference equations are classified according to the same criteria defined previously for difference equations:

- The order and degree of the equation
- Linearity or non linearity
- Inclusion or non inclusion of the dependent variable in each term of the equation
- Inclusion or non inclusion of the time variable in one or more terms of the equation

Continuous systems are usually named according to the type of differential equations of the mathematical model of the system.

Order and Degree

The *order* of differential equations is defined as follows:

Definition 3.5.1 The order of a differential equation is that of the derivative of highest order in the expression.

The order of a differential equation is determined by the sequence of successive derivatives of a function. Thus, the first derivative of a function determines a first order differential equation. The second derivative determines a second order differential equation and so on.

Example 3.5.1 Find the successive derivatives and the order of equation $y = 3x^4$.

Solution:

$$\frac{dy}{dx} = 12x^3 \qquad \textit{First order}$$

$$\frac{d}{dx}\left(\frac{dy}{dx}\right) = \frac{d^2y}{dx^2} = 36x^2 \qquad \textit{Second order}$$

$$\frac{d}{dx}\left(\frac{d}{dx}\left(\frac{dy}{dx}\right)\right) = \frac{d^3y}{dx^3} = 72x \qquad \textit{Third order}$$

$$\frac{d}{dx}\left(\frac{d}{dx}\left(\frac{d}{dx}\left(\frac{dy}{dx}\right)\right)\right) = \frac{d^4y}{dx^4} = 72 \qquad \textit{Fourth order}$$

Example 3.5.2 The reader may want to confirm the order of the following equations:

$$\frac{dy}{dx} = 2x \qquad \textit{First order}$$

$$\frac{d^2y}{dx^2} + 4\left(\frac{dy}{dx}\right)^3 + 4y = 0 \qquad \textit{Second order}$$

$$\frac{d^3y}{dx^3} + y^2(1 + x^4) = 0 \qquad \textit{Third order}$$

$$\left(\frac{d^2y}{dx^2}\right)^2 = \left(1 + \left(\frac{dy}{dx}\right)^2\right)^3 \qquad \textit{Second order}$$

In general terms, the order of a differential equation is related to the dimension and complexity of a system. This statement is illustrated in the following example.

Example 3.5.3 The growth of the population for two species of insects is given by the following state equations:

$$y_1 = 305e^{0.10t} \qquad \textit{First species}$$

$$y_2 = 563e^{0.08t} \qquad \textit{Second species}$$

The variable y is the number of individuals and the variable t is time. Find the differential equations for each of the two species of insects and for the two species combined as a single system.

Solution: The following are the first order differential equations for the above species:

$$\frac{dy_1}{dt} = 305e^{0.10t}(0.10)$$

$$\frac{dy_2}{dt} = 563e^{0.08t}(0.08)$$

These equations can also be expressed as follows:

$$\frac{dy_1}{dt} - 0.10y_1 = 0$$

$$\frac{dy_2}{dt} - 0.08y_2 = 0$$

The state equation, for the two species combined, is given by joining the state equations of each single species. This state equation and its first and second derivatives are shown below:

$$y = 305e^{0.10t} + 563e^{0.08t}$$

$$\frac{dy}{dt} = 305e^{0.10t}(0.10) + 563e^{0.08t}(0.08)$$

$$\frac{d^2y}{dt^2} = 305e^{0.10t}(0,10)^2 + 563e^{0.08t}(0.08)^2$$

The second derivative must now be restructured. For such, the above system of equations can be expressed in the following terms:

$$y = A + B$$

$$\frac{dy}{dt} = aA + bB$$

$$\frac{d^2y}{dt^2} = (a)^2A + (b)^2B$$

By selecting any two equations in the system, solving the A and B unknowns and replacing these values in the third equation, the second order differential equation becomes

$$\frac{d^2y}{dt^2} - (a+b)\frac{dy}{dt} + aby = 0$$

Thus, the second order differential equation is

$$\frac{d^2y}{dt^2} - 0.18\frac{dy}{dt} + 0.008y = 0$$

The second order differential equation represents the two insect populations combined, which is a larger and more complex system than any of the single species represented by the first order differential equations. The reader is encouraged to check all the above operations.

Note that it is possible to define a characteristic equation from the state equation of the system. By defining the characteristic equation of the system, it is possible to determine its differential equation directly from the state equation.

The following definition stands for *degree*:

Definition 3.5.2 The degree of a differential equation is the value of the largest exponent affecting the largest order differential term.

Example 3.5.4 Determine the degree of the following equations:

$$\left(\frac{d^2y}{dx^2}\right)^2 = \left(1+\frac{dy}{dx}\right)^3 \qquad \textit{Second degree}$$

$$\frac{dy}{dx} = -\frac{x}{y} \qquad \textit{First degree}$$

$$\frac{d^2y}{dx^2} + y = 0 \qquad \textit{First degree}$$

$$\left(\frac{d^2y}{dx^2}\right)^3 = \left(y+\frac{dy}{dx}\right)^4 \qquad \textit{Third degree}$$

Linearity

A *linear differential equation* is defined as follows:

Definition 3.5.3 A linear differential equation is one in which the dependent variable and any of its derivatives are of no degree greater than one.

The above definition can be represented by the following expression:

$$g_o(t)\frac{d^n y}{dt^n} + g_1(t)\frac{d^{n-1}y}{dt^{n-1}} + \dots + g_n(t)y = f(t)$$

where $g_i(t)$ represents functions of the independent variable or constants and n is the order of the differential equation. This means that the dependent variable should not be expressed as products, logarithms, trigonometric functions or any other non linear terms, such as

$$\left(\frac{dy}{dt}\right)^2, \quad y\left(\frac{dy}{dt}\right), \quad y^2, \quad \sin y, \quad y\ln y$$

If the differential equation contains a non linear term, it is called a *non linear differential equation.* As indicated in Fig. 3.5.1, systems are named accordingly.

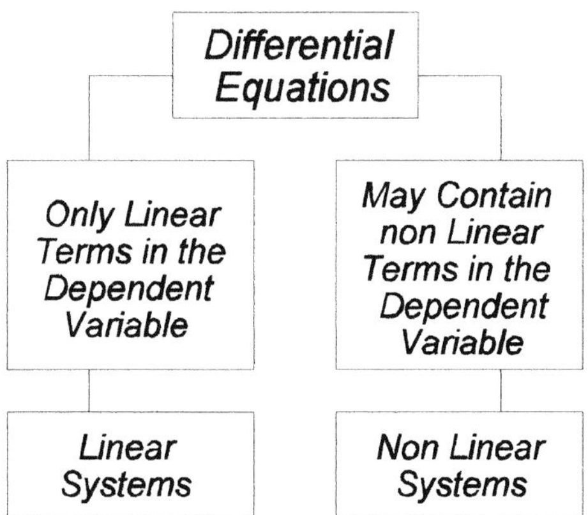

Figure 3.5.1

Example 3.5.5 A sample of linear and non linear differential equations:

$$\frac{d^2x}{dt^2} = 8x \qquad \textit{Linear}$$

$$\frac{dy}{dx} + 2xy = e^{-x^2} \qquad \textit{Linear}$$

$$\frac{dy}{dt} = y^2 \qquad \textit{Non linear}$$

$$\frac{dx}{dt} - t^2x = 0 \qquad \textit{Linear}$$

$$\left(\frac{dy}{dx}\right)^2 + ye^y = 1 + x \qquad \textit{Non linear}$$

$$x\frac{d^2x}{dt^2} = t^3 \qquad \textit{Non linear}$$

Example 3.5.6 The following state equations were obtained from "in situ" digestibility data of forage samples[5]:

$$y = 74.9 - 30.6e^{-0.0113t} - 38.7e^{-0.0928t}$$

$$y = \frac{1}{0.0157 + 0.0335e^{-0.0743t}}$$

where y is percent digestion and t is time in hours. Determine if the differential equations of the proposed state equations are linear or non linear.

Solution: The first and second derivatives for the first state equation are shown below:

$$\frac{dy}{dt} = 0.0113(30.6)e^{-0.0113t} + 0.0928(38.7)e^{-0.0928t}$$

$$\frac{d^2y}{dt^2} = -(0.0113)^2(30.6)e^{-0.0113t} - (0.0928)^2(38.7)e^{-0.928t}$$

---

[5]Computed from San Martin, F.A.

The state equation and its derivatives can be expressed as follows:

$$y - c = A + B$$
$$\frac{dy}{dt} = aA + bB$$
$$\frac{d^2y}{dt^2} = -a^2A - b^2B$$

where A and B are the exponential expressions. By selecting any two of the above equations, solving the A and B terms and replacing these values in the third equation, the differential equation representing the system is obtained:

$$\frac{d^2y}{dt^2} + (a + b)\frac{dy}{dt} + aby = abc$$
$$\frac{d^2y}{dt^2} - 0.1051\frac{dy}{dt} + 0.00105y = 0.0785$$

The first mathematical model is linear. Note that the characteristic equation of the system is here

$$(\lambda + a)(\lambda + b) = \lambda^2 + (a + b)\lambda + ab$$
$$(\lambda - 0.0113)(\lambda - 0.0928) = \lambda^2 - 0.1051\lambda + 0.00105$$

Note also that the characteristic equation can be determined directly from the state equation. As disclosed before, determining the characteristic equation becomes a shortcut for determining the system differential equation.

The following is the first derivative of the second state equation:

$$\frac{dy}{dt} = \frac{0.0743(0.0335)e^{-0.0743t}}{(0.0157 + 0.0335e^{0.0743t})^2}$$

The reader is encouraged to establish that this equation can also be expressed as follows:

$$\frac{dy}{dt} + 0.00117y^2 - 0.0743y = 0$$

The second mathematical model is non linear and is known as the *logistic equation*. The logistic equation is widely used to describe growth processes, mainly in bacterial populations. "In vitro" and "in situ" processes are related to bacterial digestion. As illustrated in Fig.3.5.2, both models fit the data accurately, however the linear model is a first choice.

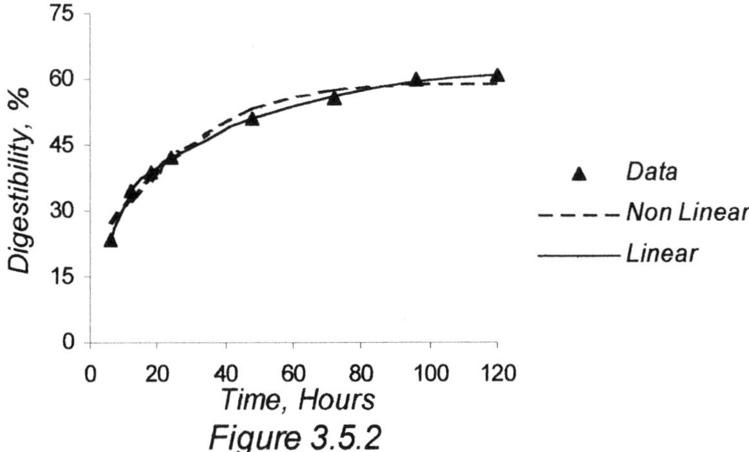

Figure 3.5.2

The statistical results for the two models are shown in the following table:

Table 3.5.1

Model	R^2	$S_{y.x}$
Linear	0.999	0.782
Non Linear	0.974	2.87

As disclosed in the example, linear mathematical models can often describe agricultural data at least as accurately as non linear models. Since agricultural research is an empirical science and linear models are easier to manipulate than non linear, the linear approach may be the first modeling choice.

Homogeneity

Given a differential equation of the form, where $g_i(t)$ represents functions of the independent variable or constants, $f(t)$ is a function of time or a constant and n is the order of the differential equation

$$g_o(t)\frac{d^n y}{dt^n} + g_1(t)\frac{d^{n-1}y}{dt^{n-1}} + \dots + g_n(t)y = f(t)$$

the following definitions apply:

Definition 3.5.4 A linear differential equation is called *homogeneous* when the dependent variable appears exactly once in each term of the equation and the term $f(t) = 0$.

Definition 3.5.5 When some terms in the equation do not contain the dependent variable, the equation is called *non homogeneous*.

Definition 3.4.6 If the $g_i(x)$ expressions represents only constant terms, the equation is called a *differential equation with constant coefficients*.

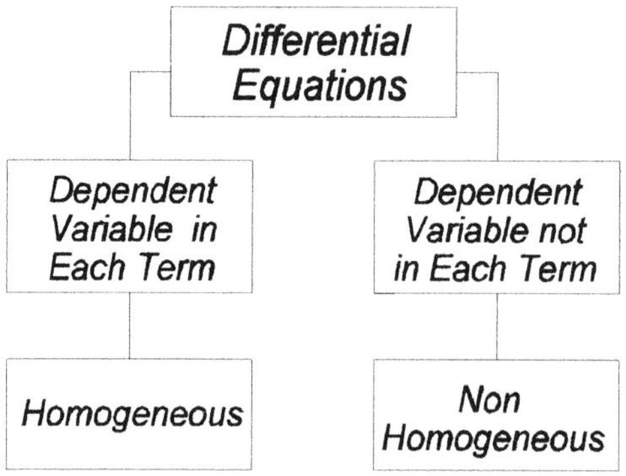

Figure 3.5.3

Example 3.5.7 A sample of homogeneous and non homogeneous differential equations:

$$\frac{d^2y}{dt^2} + t\frac{dy}{dt} + 4y = 0 \qquad\qquad \textit{Homogeneous}$$

$$\sin x\frac{dy}{dx} = -\cos x\, y \qquad\qquad \textit{Homogeneous}$$

$$\frac{d^2y}{dx^2} + \frac{dy}{dx} + x = 0 \qquad\qquad \textit{Non homogeneous}$$

$$\frac{dy}{dx} - \frac{b}{x^2}y + \frac{c}{x^2} = 0 \qquad\qquad \textit{Non homogeneous}$$

Example 3.5.8 The difference between the number of cattle, adjusted by the method of put and take and the carrying capacity of a pasture land is given by the state equation

$$y = e^{-0.69t}\left[10.0\cos(2.09t) + 14.4\sin(2.09t)\right]$$

where y is the error between real numbers of cattle and carrying capacity and t is time in years. Determine if the differential equation of the system is homogeneous or non homogeneous.

Solution: The reader is requested to find the first and second derivatives and rewrite the differential equation in the proper manner, as it was done in previous examples. Hint, make

$$A = e^{-0.069t}\cos(2.09t)$$
$$B = e^{-0.69t}\sin(2.09t)$$

Then, the following is the second order differential equation of the system:

$$\frac{d^2y}{dt^2} + 1.39\frac{dy}{dt} + 4.85y = 0$$

This is a homogeneous differential equation with constant coefficients.

Example 3.5.9 The rumination pattern of a group of steers can be described by the following state equation[6]:

$$y = 239e^{-1.34t}\,t^{1.18}$$

where y is percent of animals ruminating and t is time in hours after feeding. Determine if the differential equation of the system is homogeneous or non homogeneous.

Solution: The following is the first derivative of the state equation:

---

[6]Vohnout, K., Unpublished data.

$$\frac{dy}{dt} = 239e^{1.34t}\, t^{1.18} \left(\frac{1.18}{t} - 1.34 \right)$$

This equation can also be written as

$$\frac{dy}{dt} + \left(1.34 - \frac{1.18}{t} \right) y = 0$$

which is homogeneous.

The Time Variable

As with difference equations, time is always the independent variable of a system and is the only independent variable of the system, unless otherwise specified. The same definitions applied to difference equations also apply to differential equations:

Definition 3.5.7 When one or more terms depend explicitly on the time variable, the differential equation is called *time variant*.

Definition 3.5.8 If none of the terms depends explicitly on the variable time, the differential equation is called *time invariant*.

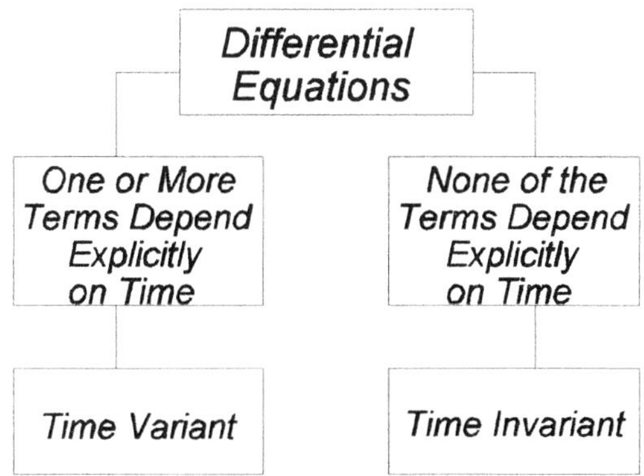

Figure 3.5.4

From the above definitions, any differential equation with constant coefficients, such as equation in Example 3.5.8, is clearly always time invariant:

$$\frac{d^2y}{dt^2} + 1.39\frac{dy}{dt} + 4.85y = 0$$

Conversely, the differential equation in Example 3.5.7 is time variant:

$$\frac{dy}{dt} + \left(1.34 - \frac{1.18}{t}\right)y = 0$$

An additional example follows.

Example 3.5.9 The following is a fitted equation for the feeding pattern of a group of feedlot steers[7]:

$$y = 29.5e^{-0.476(t-6.6)^2} + 47.5e^{-0.286(t-18.1)^2}$$

where y is percent of steers eating and t is the time of the day in hours. Determine if the corresponding differential equation is time variant.

Solution: The set of derivatives of the system is

$$\frac{dy}{dt} = -39.88\,(t-6.6)e^{-0.676(t-6.6)^2} - 27.17(t-18.1)e^{-0.286(t-18.1)^2}$$

$$\frac{d^2y}{dt^2} = -53.92\,(t-6.6)^2 e^{-0.676(t-6.6)^2} - 15.54(t-18.1)^2 e^{-0.286(t-18.1)^2}$$

By using the procedure outlined in Example 3.5.6, the differential equation of the system was determined as follows:

---

[7]Computed from Ray, D.E. and R.E. Roubicek

$$\frac{d^2y}{dt^2} - 2[(0.476 + 0.286)t - 6.6(0.476) - 18.1(0.286)]\frac{dy}{dt}$$
$$+ 4(0.476)(0.286)[t + (6.6 + 18.1)t + 6.6(18.1)]y = 0$$

After simplification, the above equation becomes

$$\frac{d^2y}{dt^2} - (1.524t - 16.636)\frac{dy}{dt} + (0.545t^2 + 13.450t + 65.051)y = 0$$

This is a time variant system.

Summary

Differential equations are expressions involving derivatives or differentials. The derivative of highest order determines the order of the equation and the degree is determined by the value of the largest exponent affecting the largest order differential term. When the degree of the dependent variable and any of its derivatives are not greater than one, the equation is called linear, otherwise it is non linear. When the dependent variable appears only once in each term of the equation, the equation is called homogeneous, otherwise it is non homogeneous. When one or more terms of the equation depend explicitly on the time variable, the equation is called time variant. Otherwise, it is called time invariant. Time is always the independent variable of a system. Systems are usually named according to the type of differential equations of the model assigned to the system.

4

SELECTED TRANSFORM PROCEDURES

Several procedures, used in finding solutions to differential and difference equations, are based on the replacement of functions of a real variable by functions of a complex variable. Two important methods for solving linear differential and difference equations with constant coefficients are introduced in this chapter, the *Laplace transform* and the *Z transform*. A good knowledge of partial fractions and complex numbers is necessary for the manipulation of these transform procedures.

4.1 PARTIAL FRACTION EXPANSIONS

A *rational function* is a fraction of which the numerator and the denominator are polynomials. If the denominator can be broken into its real prime factors, then complex rational functions can be expressed into simpler forms called *partial fractions*. This process of representing rational functions is called *partial fraction expansion*. Thus, a partial fraction expansion may be defined as follows:

Definition 4.1.1 A partial fraction expansion is the process of representing a rational function as the sum of partial fractions, each one of which has a real prime factor as denominator

Two situations will be discussed here:

• The denominator of a fraction contains only first degree factors of the form $(ax+b)^n$

• The denominator of a fraction contains second degree factors of the form $(ax^2+bx+c)^n$

The denominator is of the form $(ax+b)^n$. An expression of the form

$$\frac{c_1 x^{n-1} + c_2 x^{n-2} + \cdots + c_n}{(a_1 x + b_1)(a_2 x + b_2) \ \cdots \ (a_n x + b_n)}$$

in which the denominator is the product of n first degree linear factors, can be expanded as a sum of n simpler terms, such as

$$\frac{A_1}{a_1x+b_1} + \frac{A_2}{a_2x+b_2} + \cdots + \frac{A_n}{a_nx+b_n}$$

A procedure for solving the above identity for coefficients A_1, A_2, $\cdots$, A_n is shown in the following examples.

Example 4.1.1 Find partial fractions for the following equation:

$$\frac{2x^4-5x^3-5x^2+5x+3}{2x^3+x^2-2x-1}$$

Solution: The degree of the numerator of this fraction is greater than the degree of the denominator. Therefore, by dividing the numerator by the denominator, this fraction may be reduced to a mixed expression:

$$x-3 + \frac{2x^2 + 4x}{(2x+1)(x+1)(x-1)} = x-3 + \frac{A}{2x+1} + \frac{B}{x+1} + \frac{C}{x-1}$$

The last term is a fraction having the degree of the numerator less than the degree of the denominator. This new fraction can be written as

$$\frac{2x^2 + 4x}{(2x+1)(x+1)(x-1)} = \frac{A}{2x+1} + \frac{B}{x+1} + \frac{C}{x-1}$$

By taking a common denominator, the following expression is obtained:

$$\begin{aligned} 2x^2 + 4x &= A(x+1)(x-1) + B(2x+1)(x-1) + C(2x+1)(x+1) \\ &= (A+2B+2C)x^2 + (3C-B)x - (A+B-C) \end{aligned}$$

This identity is true if

$$A + 2B + 2C = 2$$
$$- B + 3C = 4$$
$$A + B - C = 0$$

Then, the solution for this system of equations is $A = 2$, $B = -1$, $C = 1$. Thus

$$\frac{2x^4 - 5x^3 - 5x^2 + 5x + 3}{2x^3 + x^2 - 2x - 1} = x - 3 + \frac{2}{2x+1} - \frac{1}{x+1} + \frac{1}{x-1}$$

Example 4.1.2 Find partial fractions for

$$\frac{s^2 + 0.845s + 0.149}{s^2 + 1.190s + 0.127}$$

Solution: The degree of the numerator and denominator are the same. Therefore this expression can be expanded as follows:

$$\frac{s^2 + 0.845s + 0.149}{s^2 + 1.190s + 0.127} = 1 + \frac{B}{s + 1.071} + \frac{C}{s + 0.119}$$

where 1.071 and 0.119 are the roots of the quadratic expression in the denominator. After taking a common denominator, the equation becomes

$$s^2 + 0.845s + 0.119 = s^2 + 1.190s + 0.127 + (B+C)s + 0.119B + 1.071C$$

Then $-0.345s + 0.22 = (B+C)s + 0.119B + 1.071C$. This equality is true if

$$\begin{bmatrix} 1 & 1 \\ 0.119 & 1.071 \end{bmatrix} \begin{bmatrix} B \\ C \end{bmatrix} = \begin{bmatrix} -0.345 \\ 0.022 \end{bmatrix}$$

The solution of this system is $B = -0.4112$, $C = 0.0662$. Thus

$$\frac{s^2+0.845s+0.149}{s^2+1.190s+0.127} = 1 - \frac{0.4112}{s+1.071} + \frac{0.0662}{s+0.119}$$

Example 4.1.3 Find partial fractions for the following equation:

$$\frac{3x^2+4x-1}{(x+2)^2(2x+1)}$$

Solution: For every n linear factors in the denominator, there must be the sum of n partial fractions. The denominator of the above fraction is of the third degree. Therefore, this expression can be expanded to three partial fractions:

$$\frac{3x^2+4x-1}{(x+2)^2(2x+1)} = \frac{A}{x+2} + \frac{B}{(x+2)^2} + \frac{C}{2x+1}$$

Then $3x^2+4x-1 = (2A+C)x^2 + (5A+2B+4C)x + 2A+B+4C$ and

$$\begin{bmatrix} 2 & 0 & 1 \\ 5 & 2 & 4 \\ 2 & 1 & 4 \end{bmatrix} \begin{bmatrix} A \\ B \\ C \end{bmatrix} = \begin{bmatrix} 3 \\ 4 \\ -1 \end{bmatrix}$$

The solution of the above system is $A = 2$, $B = -1$, $C = -1$. Thus

$$\frac{3x^2+4x-1}{(x+2)^2(2x+1)} = \frac{2}{x+2} - \frac{1}{(x+2)^2} - \frac{1}{2x+1}$$

The denominator is of the form $(ax^2+bx+c)^n$. When the denominator contains quadratic factors of the form $(ax^2+bx+c)^n$, where n is a positive integer, for every such factor there will be a corresponding sum of n partial fractions of the form

$$\frac{A_1x+B_1}{ax^2+bx+c} + \frac{A_2x+B_2}{(ax^2+bx+c)^2} + \cdots + \frac{A_nx+B_n}{(ax^2+bx+c)^n}$$

Non quadratic factors are dealt with as before.

Example 4.1.4 Find partial fractions for the following equation:

$$\frac{4}{(x^2+1)^2(x-1)^2}$$

Solution: The second degree factor is here $(x^2+1)^2$ for $n=2$. Factor $(x-1)^2$ is first degree and should be treated accordingly. This equation can be written as

$$\frac{4}{(x^2+1)^2(x-1)^2} = \frac{Ax+B}{x^2+1} + \frac{Cx+D}{(x^2+1)^2} + \frac{E}{x-1} + \frac{F}{(x-1)^2}$$

By taking a common denominator, the following expression is obtained:

$$4 = (Ax+B)(x^2+1)(x-1)^2 + (Cx+D)(x-1)^2 + E(x-1)(x^2+1)^2 + F(x^2+1)^2$$

For $x = 1$, the above equation becomes $4 = F(1+1)^2$. Then, $F = 1$. Substituting this value in the equation and dividing throughout by $(x-1)$, it is found that

$$-(x^3+x^2+3x+3) = (Ax+B)(x^2+1)(x-1) + (Cx+D)(x-1) + E(x^2+1)^2$$

As before, for $x=1$, $E=-2$. Substituting this value in the equation and dividing again throughout by $(x-1)$, the following expression is obtained:

$$2x^3+x^2+4x+1 = (Ax+B)(x^2+1) + Cx + D$$
$$= Ax^3 + Bx^2 + (A+C)x + B + D$$

Finally, the following results are obtained after equating the coefficients:

$$A = 2, \ B = 1, \ A + C = 4, \ C = 2, \ B + D = 1, \ D = 0$$

Thus

$$\frac{4}{(x^2+1)^2(x-1)^2} = \frac{2x+1}{x^2+1} + \frac{2x}{(x^2+1)^2} - \frac{2}{x-1} + \frac{1}{(x-1)^2}$$

Summary

Partial fractions are expressions derived from more complex rational fractions, provided that the denominator can be broken into its real prime factors. Two cases were discussed: when the denominator contains only first degree factors of the form $(ax + b)^n$ and when the denominator contains second degree factors of the form $(ax^2 + bx + c)^n$, where n is a positive integer.

4.2 COMPLEX NUMBERS

Complex numbers may be defined as follows:

Definition 4.2.1 A complex number is an expression having the form $\alpha + i\beta$, where α and β are real numbers and $i = \sqrt{-1}$.

The α value is called the *real part*, β is called the *imaginary part* and i is the *imaginary unit*. Operations with complex numbers are the same as in the algebra of real numbers, replacing i^2 by -1 when it occurs. Inequalities for complex numbers are not defined.

The Complex Plane

A complex number $\alpha + i\beta$ can be represented as a point in an XY plane, called the *complex plane,* with the α value plotted along the X axis and the β value plotted along the Y axis.

Example 4.2.1 Locate the following points in the complex plane:

$$\rho_1 = 4 + 3i, \ \rho_2 = 4 - 3i, \ \rho_3 = -4 + 2i, \ \rho_4 = -3 - 2i \ \ \rho_5 = 2 + i$$

Solution: The requested points are shown in Fig. 4.2.1

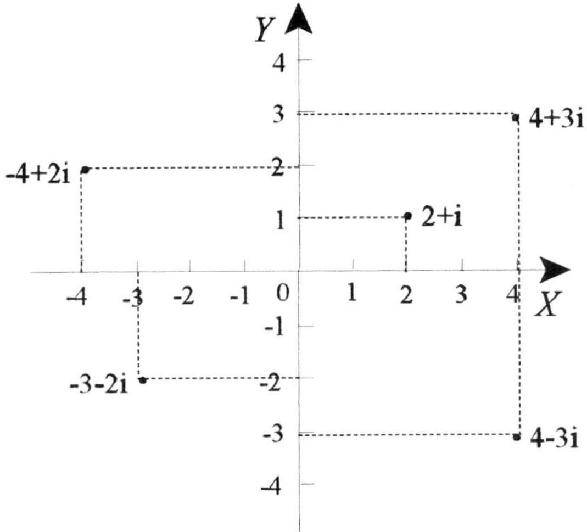

Figure 4.2.1

Example 4.2.2 The difference between the carrying capacity of a pasture land and the real number of cattle is given by the state equation

$$y = e^{-0.69t}[10.00\cos(2.09t) + 14.43\sin(2.09t)]$$

where y is % difference and t is time in years. The following is the corresponding second order linear differential equation:

$$\frac{d^2y}{dt^2} + 1.386\frac{dy}{dt} + 4.848y = 0$$

Define the characteristic equation[1] of the system and the roots of the equation as complex numbers.

Solution: The following is the characteristic equation of the system:

---

[1]The concept of characteristic equations was defined in Chapter 2.

$$\lambda^2 + 1.386\lambda + 4.848 = 0$$

where λ is a characteristic root. Then

$$\alpha = \frac{-1.386}{2} = -0.69$$

$$i\beta = \frac{\sqrt{1.386^2 - 4(4.848)}}{2} = 2.09\,i$$

$$\lambda = -0.69 \pm 2.09\,i$$

Note that α and β are coefficients of the state equation. This is a cyclical or periodic function and can be defined on a *polar coordinate system.*

Polar Form of Complex Numbers

As shown in Fig. 4.2.2, the x and y values of a complex number $\alpha + i\beta$ are

$$x = r\cos\theta$$
$$y = r\sin\theta$$

where $r = \sqrt{x^2 + y^2}$ is the distance between $\rho(x, y)$ and the origin 0 and $\theta = \tan^{-1}(y/x)$ is the angle, in radians, between r and the abscissa.

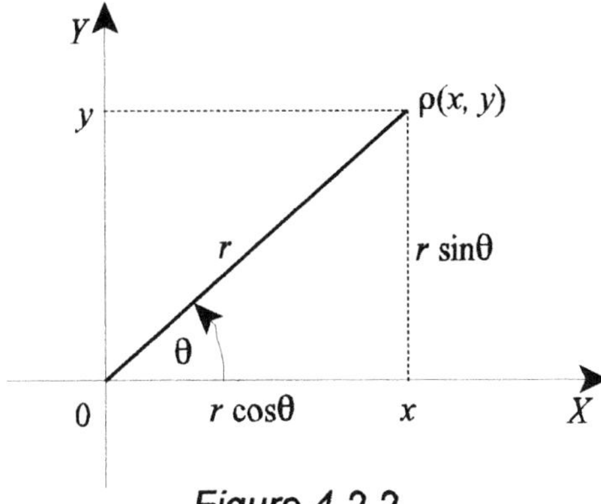

Figure 4.2.2

Then, if $y = \sin\theta$, $x = \cos\theta$ and $r = 1$, it follows that $x + iy = r(\cos\theta + i\sin\theta)$ and

$x - iy = r(\cos\theta - i\sin\theta)$. These expressions are called *polar forms* of a complex number and r and θ are called *polar coordinates*. If x is measured in radians, then

$$\cos\theta + i\sin\theta = e^{i\theta}$$
$$\cos\theta - i\sin\theta = e^{-i\theta}$$

These expressions are known as the *Euler's formula*. As a test, note that

$$(\cos\theta + i\sin\theta)(\cos\theta - i\sin\theta) = \cos^2\theta + \sin^2\theta = 1 \text{ and } e^{i\theta}e^{-i\theta} = 1.$$

Thus

$$x + iy = re^{i\theta}$$
$$x - iy = re^{-i\theta}$$

The relation between Cartesian and polar coordinates is shown in Fig. 4.2.3.

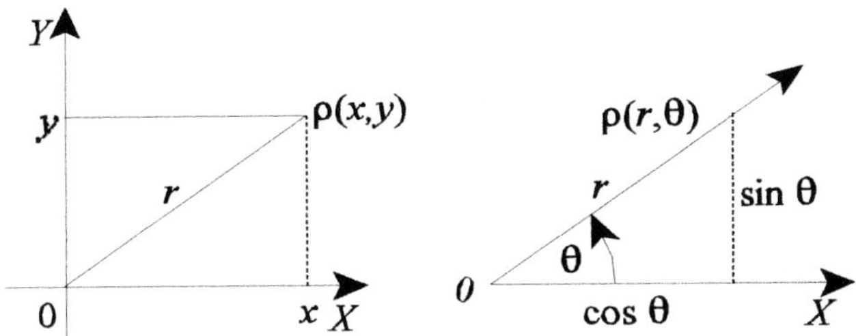

Figure 4.2.3

The radius r modulates the *amplitude* of a cyclical function and the angle θ modulates the *frequency*. This is illustrated in the following example.

Example 4.2.3 Show the graphs for functions $y = r\sin at$ and $y = r\cos at$, where a is the angle between the radius r and the polar axis and t is value of the independent variable.

Solution: As shown in Fig. 4.2, a full cycle or period of the function has the value $2\pi/a$. Thus, by increasing the value of a the cycle is decreased. Note also that the amplitude is increased by increasing the value of r.

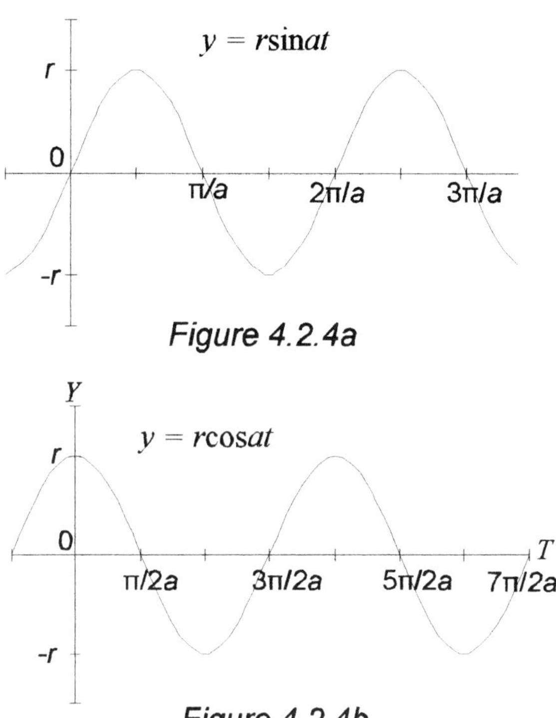

Figure 4.2.4a

Figure 4.2.4b

Example 4.2.4 Show the graph of equation

$$y = e^{-0.69t}[10.00\cos(2.09t) + 14.43\sin(2.09t)]$$

as defined in Example 4.2.2.

Solution: As shown in Fig. 4.2.5, the cycle of this function is $2\pi/a = 6.28/2.09 = 3$ years. Note also that the amplitude decreases with time according to the exponential expression $e^{-0.69t}$. As was shown in Example 4.2.2, the roots of the characteristic equation of this system are $\lambda = -0.69 \pm 2.09i$.

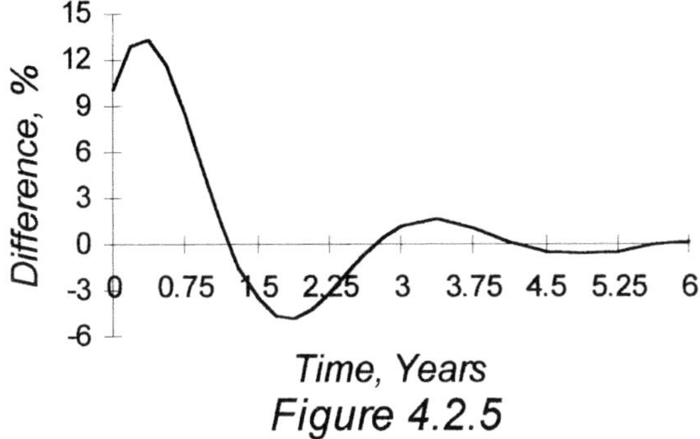

Figure 4.2.5

Summary

Complex numbers are expressions having the form $x + iy$ where x and y are real numbers and $i = \sqrt{-1}$. Complex numbers can be represented as a point in the XY plane, called the complex plane or in a polar form such that $x = r\cos\theta$ and $y = r\sin\theta$, where $r = \sqrt{x^2 + y^2}$ and $\theta = \tan^{-1}(y/x)$.

4.3 THE LAPLACE TRANSFORM

The process of solving linear differential equations by Laplace transforms is outlined in Fig. 4.3.1.

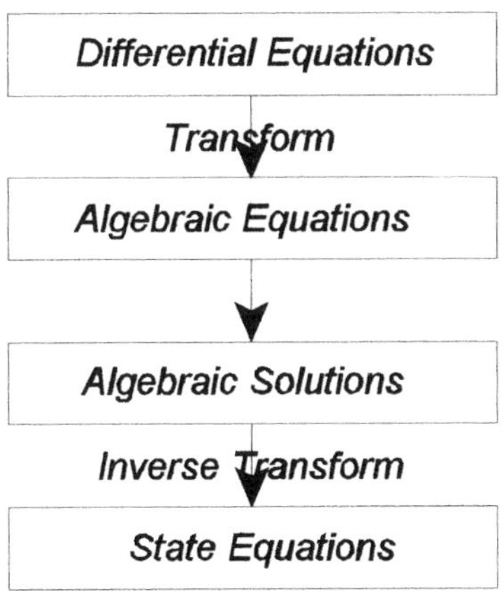

Figure 4.3.1

The Laplace transform allows complicated differential equations to be reduced to simple algebraic expressions. Then, the *inverse transform* of algebraic solutions become solutions of the differential equations. When the independent variable is time, the Laplace procedure transforms differentiation and integration operations in the time domain into multiplication and division operations in the frequency domain.

The relationship between a Laplace transform and its inverse is similar to the relationship between a logarithm and its antilogarithm or between the derivative and the anti-derivative. The Laplace transform is the technique of choice for solving linear differential equations with constant coefficients.

Definition of the Laplace transform

The following notation is often used to indicate that a function $F(s)$ is the Laplace transform of $f(t)$. Herein, the symbols f, g and h are used for defining an input function, a state function and an output function, respectively. The symbol L is called *the Laplace operator* and indicates the Laplace transformation.

$$F(s) = L[f(t)]$$

The following notation signifies that $f(t)$ is the function whose transform is $F(s)$:

$$f(t) = L^{-1}[F(s)]$$

where the symbol L^{-1} is called *the inverse Laplace operator.* The new variable s is the complex variable

$$s = \sigma + i\omega$$

where σ and ω are real variables and $i = \sqrt{-1}$.

Then, the following is the formal definition of the Laplace transform:

$$L[f(t)] = F(s) = \int_{0^+}^{\infty} f(t)e^{-st}\, dt$$

where $t > 0$ is a real variable, $s = \sigma + i\omega$ and $i = \sqrt{-1}$. This process transforms a function $f(t)$ to $F(s)$ by multiplying the function by e^{-st} and then integrating over t, between zero and infinity. The symbol 0^+ is usually used to deal with functions that are discontinuous at $t = 0$. Whereas t could be any variable, it is used here only to denote time.

Example 4.3.1 Find the Laplace transform of e^{-t}.

Solution: $L[e^{-t}] = \int_0^\infty e^{-st} e^{-t} dt = \dfrac{-1}{s+1} e^{-(s+1)t} \Big|_0^\infty = \dfrac{1}{s+1}$

    Laplace transforms, for the most frequently used equations, can be located in tables. However, when a transform of a particular equation is not found, determining it by the above process is possible.

Example 4.3.2 Find the transform of function $x = t^{1/2}$.

Solution: $F(s) = \int_0^\infty t^{1/2} e^{-st} dt$. Let $t = u^2$ and $dt = 2u\,du$. Then

$$F(s) = \int_0^\infty u e^{-su^2} (2u)\,du = 2\int_0^\infty u^2 e^{-su^2}\,du$$

The solution of this integral can be found directly in a table of definite integrals:

$$F(s) = \frac{\sqrt{\pi}}{2s^{3/2}}$$

    Laplace transforms of linear differential equations with constant coefficients are rational fractions. When the inverse transform is not found in tables, the inversion process is greatly simplified by partial fraction expansions, as shown in the following example.

Example 4.3.3 Find the inverse of the following transform:

$$F(s) = \frac{s^2 + 2}{s(s+2)(s+3)}$$

Solution:

$$F(s) = \frac{s^2 + 2}{s(s+2)(s+3)} = \frac{A}{s} + \frac{b}{s+2} + \frac{C}{s+3}$$
$$= \frac{s^2(A+B+C) + s(5A+3B+2C) + 6A}{s(s+2)(s+3)}$$

Clearly

$$
\begin{array}{rcrcrcl}
A & + & B & + & C & = & 1 \\
5A & + & 3B & + & 2C & = & 0 \\
6A & & & & & = & 2
\end{array}
$$

Then, $A = 1/3$, $B = -3$ and $C = 11/3$. The new transform is

$$F(s) \; = \; \frac{1/3}{s} \; - \; \frac{3}{s+2} \; + \; \frac{11/3}{s+3}$$

The inverse of the above fractions are easily found in tables. Thus

$$f(t) \; = \; \frac{1}{3} \; - \; 3\,e^{-2t} \; + \; \frac{11\,e^{-3t}}{3}$$

Selected Properties

The following properties of the Laplace transform will be used extensively in this book.

Property 1. The Laplace transform and its inverse are linear transformations between functions defined in the domain of the real variable $t > 0$ and functions defined in the domain of the complex variable s, that is

$$k_1 F_1(s) + k_2 F_2(s) \text{ is the Laplace transform of } k_1 f_1(t) + k_2 f_2(t)$$

and

$$k_1 f_1(t) + k_2 f_2(t) \text{ is the inverse of } k_1 F_1(s) + k_2 F_2(s)$$

Example 4.3.4 Find the Laplace transform of $3e^{-t} + e^{-2t}$.

Solution: From a table of Laplace transforms, $L[e^{-t}] = 1/(s+1)$ and $L[e^{-2t}] = 1/(s+2)$. Then

$$L[3e^{-t} + e^{-2t}] \; = \; 3L[e^{-t}] \; + \; L[e^{-2t}] \; = \; \frac{3}{s+1} \; + \; \frac{1}{s+2}$$

Property 2. The Laplace transforms of the derivatives of a function $f(t)$ whose transform is $f(s)$ are

$$L\left[\frac{d}{dt}f(t)\right] = sF(s) - f(0)$$

$$L\left[\frac{d^2}{dt^2}f(t)\right] = s^2F(s) - sf(0) - \frac{d}{dt}f(0)$$

$$\vdots \qquad \vdots$$

$$L\left[\frac{d^n}{dt^n}f(t)\right] = s^nF(s) - s^{n-1}f(0) - s^{n-2}\frac{d}{dt}f(0) - s^{n-3}\frac{d^2}{dt^2}f(0) - \ldots - \frac{d^{n-1}}{dt^{n-1}}f(0)$$

where $\dfrac{d^n}{dt^n}f(0)$ is the limit of $\dfrac{d^n}{dt^n}f(t)$ as $t \to 0^+$.

Example 4.3.5 The digestion of the cell walls of a forage is represented by the following differential equation[2]:

$$\frac{dy}{dt} + 0.082y = 0$$

where y is percent digestion and t is time in hours. Find the state equation for an initial value of 54.

Solution:

$$L\left[\frac{dy}{dt}\right] + 0.082\,L[y] = sG(s) - y_0 + 0.082\,G(s) = 0$$

$$(s + 0.082)G(s) = y_0$$

$$G(s) = \frac{y_0}{s + 0.082}$$

where y_0 is an initial value. Then

$$y = L^{-1}\left[\frac{y_0}{s + 0.082}\right]$$

----

[2] Computed from Van Soest, P.J.

Given $y_0 = 54$, the inverse transform, as found in tables, is the state equation

$$y = 54e^{-0.082t}$$

Note that the notation G was used here to symbolize a state related function.

Example 4.3.6 The concentration of ammonia in the rumen of sheep, after eating a food containing urea, is given by the following first order differential equation[2]:

$$\frac{dy}{dt} + 0.5y = 45e^{-0.5t}$$

where y is NH_3 in mM/liter and t is time in hours after eating. Find the state equation.

Solution: The following is the transformed differential equation:

$$sG(s) - y_0 + 0.5G(s) = \frac{45}{s+0.5}$$

$$G(s) = \frac{y_0}{s+0.5} + \frac{45}{(s+0.5)^2}$$

If $g(0) = 11$, the state equation is the inverse transform of the above, that is

$$y = 11e^{-0.5t} + 45te^{-0.5t} = e^{-0.5t}(45t + 11)$$

Example 4.3.7 The growth of a group of steers is represented by the following differential equation:

$$\frac{d^2y}{dt^2} + 1.98\frac{dy}{dt} + 0.789y = 616$$

-----

[2]Computed from Streeter, C.L. et.al.

where y is body weight in Kg and t is years. Find the state equation.

Solution:

$$L\left[\frac{d^2y}{dt^2}\right] + L\left[1.98\frac{dy}{dt}\right] + L[0.789y] = L[616]$$

$$s^2 G(s) - sy_0 - y_0{}' + 1.98[sG(s) - y_0] + 0.789G(s) = \frac{616}{s}$$

$$G(s)(s^2 + 1.98s + 0.789) - y_0(s + 1.98) - y_0{}' = \frac{616}{s}$$

Then

$$G(s) = \frac{y_0(s + 1.98) + y_0{}'}{(s + 1.427)(s + 0.553)} + \frac{616}{s(s + 1.427)(s + 0.553)}$$

For initial values of $y_0 = 30$ and $y_0{}' = 183$, the state equation is the following inverse:

$$y = L^{-1}\left[\frac{30(s + 8.08)}{(s + 1.427)(s + 0.553)}\right] + L^{-1}\left[\frac{616}{s(s + 1.427)(s + 0.553)}\right]$$

By looking at transform tables, it is found that

$$y = 30\left[\frac{(8.08 - 1.427)e^{-1.427t} - (8.08 - 0.553)e^{-0.553t}}{0.553 - 1.427}\right]$$
$$+ \frac{616}{1.427(0.553)}\left[1 - \frac{0.553\,e^{-1.427t} + 1.427e^{-0.553t}}{0.553 - 1.427}\right]$$

Thus

$$y = 780 + 265\,e^{-1.427t} - 1016\,e^{-0.553t}$$

Example 4.3.8 Body weight and efficiency of milk production in a group of Holstein cows were related by the following set of differential equations[3]:

$$\frac{dY}{dt} = \begin{bmatrix} -0.399526 & 0 \\ -0.007892 & 0 \end{bmatrix} Y + \begin{bmatrix} 238.890 \\ 4.569 \end{bmatrix} + \begin{bmatrix} 4.17028 \\ 0.07011 \end{bmatrix} t$$

for $Y = (y_1, y_2)$, where y_1 is body weight in kilograms, y_2 is kilograms of milk per Mcal of metabolizable energy, t is months after calving and matrix A defines the relations between the state variables. Determine the state equations.

Solution: This is a multidimensional linear model reducible to the form

$$\frac{dY}{dt} = AY + X$$

The following is the Laplace transform of the above equation:

$$sG(s) - Y_0 = AG(s) + F(s) \quad \text{or}$$
$$(sI - A)G(s) = Y_0 + F(s)$$

where $(sI\text{-}A)$ is the characteristic equation of the system, $G(s)$ is the set of Laplace transforms corresponding to the set of state variables, Y_0 is the set of initial conditions and $F(s)$ is the set of Laplace transforms of the input functions represented by X. Then, the following is the Laplace transform of the system equations:

$$\begin{bmatrix} s+0.3995 & 0 \\ 0.007892 & s \end{bmatrix} G(s) = Y_0 + \begin{bmatrix} 238.890 \\ 4.569 \end{bmatrix} \frac{1}{s} + \begin{bmatrix} 4.170 \\ 0.07011 \end{bmatrix} \frac{1}{s^2}$$

where the characteristic equation of the system is as follows:

---

[3]Computed from Miller, R.H. and N.W. Hooven, Jr.

$$|sI - A| = \begin{vmatrix} s+0.3995 & 0 \\ 0.007892 & s \end{vmatrix} = s(s+0.3995)$$

The following is the Laplace transform for body weight:

$$G_1(s) = \frac{1}{s(s+0.3995)} \begin{vmatrix} y_{01} + \dfrac{238.890}{s} + \dfrac{4.170}{s^2} & 0 \\ y_{02} + \dfrac{4.569}{s} + \dfrac{0.07011}{s^2} & s \end{vmatrix}$$

$$= \frac{y_{01}}{s+0.3995} + \frac{238.890}{s(s+0.3995)} + \frac{4.170}{s^2(s+0.3995)}$$

where $y_{01} = 607$ and $y_{02} = 1.25$ are initial values. After finding the inverse for the above transform and rearranging terms, the state equation for body weight is

$$y_1 = 572 + 10.4t + 35.1e^{-0.400t}$$

The following is the Laplace transform of efficiency:

$$G_2(s) = \frac{1}{s(s+0.3995)} \begin{vmatrix} s+0.3995 & y_{01} + \dfrac{238.890}{s} + \dfrac{4.170}{s^2} \\ 0.007892 & y_{02} + \dfrac{4.569}{s} + \dfrac{0.07011}{s^2} \end{vmatrix}$$

$$= \frac{y_{02}}{s} + \frac{4.569}{s^2} + \frac{0.07011}{s^3} - \frac{0.007892}{s(s+0.3995)}\left[y_{01} + \frac{238.890}{s} + \frac{4.170}{s^2} \right]$$

The inverse of the above transform is the state equation of efficiency :

$$y_2 = 0.5577 + 0.05634t + 0.02892t^2 + 0.6923e^{-0.400t}$$

The reader is encouraged to check the above solutions.

Property 3. The Laplace transform of the integral $\int_0^t f(\tau)\,d\tau$ of a function $f(t)$ whose Laplace transform is $f(s)$ is

$$L\left[\int_0^t f(\tau)\,d\tau\right] = \frac{f(s)}{s}$$

Example 4.3.9 The lactation curve of a group of dairy cows is given by equation

$$g(t) = 972e^{-0.387t} - 722e^{-1.178t}$$

where $g(t)$ is milk production in Kg/month and t is time in months. Find the cumulative curve $h(t)$ for milk production.

Solution:

$$h(t) = \int_0^t g(t)\,dt = 972\int_0^t e^{-0.387t}\,dt - 722\int_0^t e^{-1.178t}\,dt$$

The corresponding Laplace transform is here

$$H(s) = \frac{972}{s(s+0.387)} - \frac{722}{s(s+1.178)}$$

Then

$$h(t) = \frac{972}{0.387}(1 - e^{-0.387t}) - \frac{722}{1.178}(1 - e^{-1.178t})$$

Thus, the cumulative milk production curve is as follows:

$$h(t) = 1899 + 613e^{-1.178t} - 2512e^{-0.387t}$$

Property 4. If $F_1(s)$ and $F_2(s)$ are the Laplace transforms of $f_1(t)$ and $f_2(t)$, then the product of two functions is

$$L^{-1}\left[F_1(s)F_2(s)\right] = \int_0^t f_1(\tau)f_2(t-\tau)\,d\tau = \int_0^t f_2(\tau)f_1(t-\tau)\,d\tau$$

The above integrals are called *convolution integrals*.

Example 4.3.10 Find the inverse of the following transform:

$$F(s) = \frac{3s+2}{(s+3)(s+2)^2}$$

Solution:

$$F(s) = [F_1(s)][F_2(s)] = \left[\frac{1}{s+3}\right]\left[\frac{3s+2}{(s+2)^2}\right]$$

$$f_1(t) = L^{-1}\left[f_1(s)\right] = e^{-3t}$$

$$f_2(t) = L^{-1}\left[f_2(s)\right] = 3e^{-2t} - 4te^{-2t}$$

Then, by Property 4

$$L^{-1}[F(s)] = L^{-1}[F_1(s)F_2(s)] = \int_0^t f_1(t-\tau)f_2(\tau)\,d\tau$$

$$= \int_0^t e^{-3(t-\tau)}(3e^{-2\tau} - 4\tau e^{-2\tau})\,d\tau$$

$$= e^{-3t}\left[3\int_0^t e^{\tau}\,d\tau - 4\int_0^t \tau e^{\tau}\,d\tau\right]$$

$$= e^{-3t}\left[3(e^t - 1) - 4(1 - e^t + te^t)\right]$$

$$= 7e^{-2t} - 7e^{-3t} - 4te^{-2t}$$

Property 5. The following is the Laplace transform of a function of the form $e^{-at}f(t)$:

$$L[e^{-at}f(t)] = F(s+a)$$

where $F(s) = L\big[f(t)\big]$.

Example 4.3.11 Find the Laplace transform of $e^{-2t}\cos t$.

Solution: The Laplace transform of $\cos t$ is $s/(s^2+1)$. Thus

$$L[e^{-2t}\cos t] = \frac{s+2}{(s+2)^2+1} = \frac{s+2}{s^2+4s+5}$$

In conclusion, the following are selected properties of Laplace transforms, to be applied in further chapters:

- $k_1F_1(s) + k_2F_2(s)$ is the transform of $k_1f_1(t) + k_2f_2(t)$

- $s^nF(s) - s^{n-1}f(0) - s^{n-2}\dfrac{d}{dt}f(0) - s^{n-3}\dfrac{d^2}{dt^2}f(0) - \ldots - \dfrac{d^{n-1}}{dt^{n-1}}f(0)$

  is the transform of $\dfrac{d^n}{dt^n}f(t)$

- $\dfrac{F(s)}{s}$ is the transform of $\displaystyle\int_0^t f(\tau)d\tau$

- $F_1(s)F_2(s)$ is the inverse transform of $\displaystyle\int_0^t f_1(\tau)f_2(t-\tau)\,d\tau = \int_0^t f_2(\tau)f_1(t-\tau)\,d\tau$

- $F(s+a)$ is the transform of $e^{-at}f(t)$

Summary

Laplace procedures transform differentiation and integration operations in the time domain of variable t, into multiplication and division operations in the frequency domain of the complex variable s. Thus, differential equations are reduced to algebraic forms that, by inverse transformations, become the solution of the differential equation.

4.4 THE Z TRANSFORM

In the same way the Laplace transform is used to solve linear differential equations with constant coefficients, the Z transformation is used to solve linear difference equations with constant coefficients. The process of solving linear difference equations by Z transform procedures is outlined in Fig. 4.4.1.

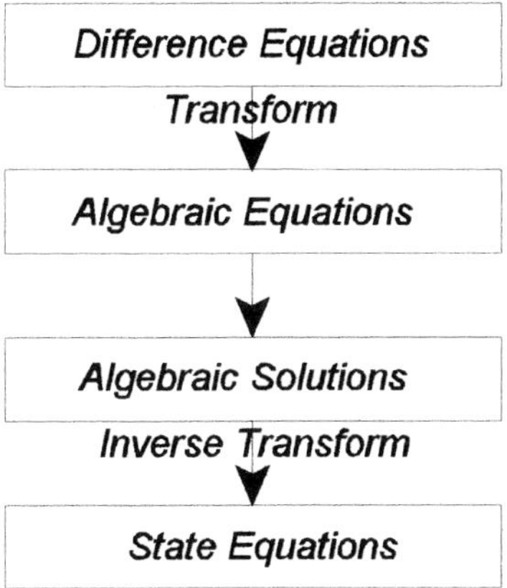

Figure 4.4.1

Definition of the Z Transform

The following notation is frequently used to denote that a function $F(z)$ is the Z transform of a sequence $f(n)$ of real values, where n is a positive integer :

$$F(z) = Z[f(n)]$$

where Z is called *the Z operator*. The following indicates that $f(n)$ is a sequence whose transform is $F(z)$:

$$f(n) = Z^{-1}[F(z)]$$

where Z^{-1} is called *the inverse Z operator*. The new variable z is the complex variable

$$z = u + iv$$

where u and v are real variables and $i = \sqrt{-1}$. Then, the formal definition of the Z transformation is as follows:

$$Z[f(n)] = \sum_{0}^{\infty} f(n)z^{-n} = F(z)$$

where n is a positive integer and $z = u + iv$. This process transforms the sequence $f(n)$ into $F(z)$ by multiplying $f(n)$ by z^{-n} and then summing over n from zero to infinity. When the resulting series is convergent, the sequence is transformable. Then, there exists a real finite number r such that $F(z)$ converges for $r < |z|$. Number r is called the *radius of convergence* of the series.

Example 4.4.1 Find the radius of convergence of $F(z)$, where $f(n) = a^n$, n is a positive integer and a is any finite complex number.

Solution:

$$Z[a^n] = \sum_0^\infty a^n z^{-n} = 1 + az^{-1} + a^2 z^{-2} + \ldots + a^n z^{-n}$$

Then

$$S_n = 1 + az^{-1} + (az^{-1})^2 + \ldots + (az^{-1})^{n-1}$$
$$a^n z^{-n} S_n = az^{-1} + (az^{-1})^2 + (az^{-1})^3 + \ldots + (az^{-1})^n$$
$$= S_n - 1 + a^n z^{-n}$$
$$S_n = \frac{1 - (az^{-1})^n}{1 - az^{-1}}$$

Thus

$$F(z) = \lim_{n \to \infty} S_n = \frac{1}{1 - az^{-1}} = \frac{z}{z - a}$$

If $|az^{-1}| < 1$ or $r = |a| < |z|$, this series is convergent. This relationship is shown in Fig. 4.4.2.

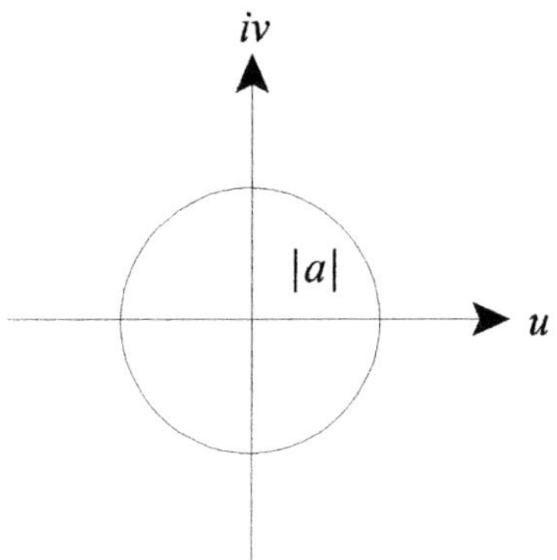

Figure 4.4.2

Selected Properties

The following properties of the Z transformation will be used extensively in this treatise.

Property 1. The Z transform and its inverse are linear transformations between sequences defined in the domain of the real variable $n > 0$ and functions defined in the domain of the complex variable z, that is

$$k_1 F_1(z) + k_2 F_2(z) \text{ is the Z transform of } k_1 f_1(n) + k_2 f_2(n)$$

and

$$k_1 f_1(n) + k_2 f_2(n) \text{ is the inverse of } k_1 F_2(z) + k_2 F_2(z)$$

Example 4.4.2 Find the Z transform of

$$L\left[2\left(-\frac{1}{2}\right)^n - 4\left(\frac{1}{4}\right)^n\right]$$

Solution: The following is obtained from a table of Z transforms:

$$L\left[2\left(-\frac{1}{2}\right)^{n}\right] = \frac{2z}{z+1/2}$$

$$L\left[4\left(\frac{1}{4}\right)^{n}\right] = \frac{4z}{z-1/4}$$

Then

$$L\left[2\left(-\frac{1}{2}\right)^{n} - 4\left(\frac{1}{4}\right)^{n}\right] = \frac{2z}{z+1/2} - \frac{4z}{z-1/4}$$

Property 2. If $F(z)$ is the Z transformation of a sequence $f(n)$ for $n > 1$, then

$$\begin{aligned}
L[f(n+1)] &= zF(z) - zf(0) \\
L[f(n+2)] &= z^{2}F(z) - z^{2}f(0) - zf(1) \\
\vdots\ &=\ \vdots \\
L[f(n+k)] &= z^{n}F(z) - z^{n}f(0) - z^{n-1}f(1) - \dots - zf(n-1)
\end{aligned}$$

where $f(0)$ is the initial value of function f.

Example 4.4.3 The growth of colonies of a bacteria is given by the following difference equation:

$$y_{n+1} - 1.021y_{n} = 0$$

where y is the number of colonies and n is time in hours. Find the state equation.

Solution:

$$\begin{aligned}
L[y_{n+1}] - 1.021L[y_{n}] &= zG(z) - zy_{0} - 1.021G(z) \\
(z - 1.021)G(z) &= zy_{0} \\
G(z) &= y_{0}\frac{z}{z - 1.021}
\end{aligned}$$

Then

$$y_n = L^{-1}\left[y_0 \frac{z}{z-1.021}\right]$$

Given an initial value of $y_0 = 0.21$, the inverse transform, as found in tables, is the state equation

$$y_n = 0.21(1.021)^n$$

Example 4.4.4 A rancher sells every month 3.6% of his feedlot steers and buys 90 new animals. Find the state equation assuming that he started the business with 460 animals.

Solution: The system is represented by the following difference equation:

$$y_{n+1} - y_n = -0.036 y_n + 90$$

Upon rearrangement, the difference equation becomes

$$y_{n+1} - (1-0.036)y_n = 90$$

The transformed equation is

$$z G(z) - 460 z - (1-0.036)G(z) = 90 \frac{z}{z-1}$$

Then

$$G(z) = 90 \frac{z}{(z-1)(z-0.964)} + 460 \frac{z}{z-0.964}$$

By looking at transform tables, it is found that

$$y_n = \frac{90}{1-0.964}[1 - (0.964)^n] + 460(0.964)^n$$
$$= 2500 - 2040(0.964)^n$$

Example 4.4.5 The changes in an insect population are represented by the following difference equation:

$$y_{n+2} - 1.212y_{n+1} + 0.367y_n = 0$$

where y is the number of insects and n is months. Find the solution when $y_0 = 250$ and $y_1 = 424$.

Solution: The transformed equation is

$$z^2 G(z) - z^2 y_0 - zy_1 - 1.212[zG(z) - zy_0] + 0.367G(z) = 0$$

Then

$$G(z) = \frac{y_0(z^2 - 1.212z) + y_1 z}{z^2 - 1.212z + 0.367}$$

The denominator of the above transform is the characteristic equation of the system[7]. Note that the coefficients of the characteristic equation are the same as the coefficients of the difference equation. The characteristic equation has two equal roots. Then, upon rearranging and replacing the y_0 and y_1 values, $G(z)$ becomes

$$G(z) = \frac{250z^2 + 121z}{(z - 0.606)^2}$$

This equation can be written as

---

[7]Characteristic equations were defined in Chapter 2.

$$\frac{G(z)}{z^2} = \frac{250z + 121}{z(z - 0.606)^2} = \frac{A}{z(z - 0.606)} + \frac{B}{(z - 0.606)^2}$$

where $A = -200$ and $B = 450$. Then

$$F(z) = \frac{-200z}{z - 0.606} + \frac{450z^2}{(z - 0.606)^2}$$

The inverse transform of $F(z)$ is the state equation

$$y_n = -200(0.606)^n + 450(n + 1)(0.606)^n$$
$$= (0.606)^n(450n + 250)$$

Example 4.4.6 The population of a type of bird doubles every year. The introduction of predators reduces the number of birds by ten times the number of predators. The number of predators also doubles every year. Some 200 new birds move into the ecosystem each year and some 30 predators are hunted down. Determine the state equations of the system, assuming 1000 initial birds and 50 initial predators.

Solution: The difference equation of the system is as follows:

$$Y_{n+1} = \begin{bmatrix} 2 & -10 \\ 0 & 2 \end{bmatrix} Y_n + \begin{bmatrix} 200 \\ -30 \end{bmatrix}$$

for   where   is birds,   is predators and n is years. This is a first order multidimensional linear model represented by difference equations reducible to the form

$$Y_{n+1} = AY_n + X$$

The Z transform of the above equation is written as follows:

$$zG(z) - zY_0 = AG(z) + F(z)$$
$$(zI - A)G(z) = zY_0 + F(z)$$

where $(zI - A)$ is the characteristic equation of the system, $G(z)$ is the set of Z transforms of the state variables, is a set of initial conditions and $F(z)$ is the set of transforms of X. Then, the following is the Z transform of the system equation:

$$\begin{bmatrix} z-2 & 10 \\ 0 & z-2 \end{bmatrix} G(z) = \begin{bmatrix} 1000 \\ 50 \end{bmatrix} z + \begin{bmatrix} 200 \\ -30 \end{bmatrix} \frac{z}{z-1}$$

The characteristic equation of the system is here

$$|zI - A| = \begin{vmatrix} z-2 & 10 \\ 0 & z-2 \end{vmatrix} = (z-2)^2$$

The following is the Z transform of the birds:

$$G_b(z) = \frac{1}{(z-2)^2} \begin{vmatrix} 1000z & 10 \\ 50z & z-2 \end{vmatrix} + \frac{1}{(z-2)^2} \begin{vmatrix} \dfrac{200z}{z-1} & 10 \\ \dfrac{-30z}{z-1} & z-2 \end{vmatrix}$$

$$= \frac{z(1000z - 2500)}{(z-2)^2} + \frac{200z}{(z-2)(z-1)} + \frac{300z}{(z-2)^2(z-1)}$$

$$= \frac{Az}{z-2} + \frac{Bz^2}{(z-2)^2} + \frac{200z}{(z-2)(z-1)} + 300z\left(\frac{C}{(z-2)^2} + \frac{D}{z-1} + \frac{E}{z-2} \right)$$

where $A = 1250$, $B = -250$, $C = 1$, $D = 1$ and $E = -1$.

The following is the Z transform of the predators:

$$G_p(z) = \frac{1}{(z-2)^2} \begin{vmatrix} z-2 & 1000z \\ 0 & 50z \end{vmatrix} + \frac{1}{(z-2)^2} \begin{vmatrix} z-2 & \dfrac{200z}{z-1} \\ 0 & \dfrac{-30z}{z-1} \end{vmatrix}$$

$$= \frac{50z}{z-2} - \frac{30z}{(z-2)(z-1)}$$

After taking the inverse transforms and rearranging terms, the following is the set of state equations of the system:

$$Y_n = \begin{bmatrix} 100 \\ 30 \end{bmatrix} + \begin{bmatrix} 100(9-n) \\ 20 \end{bmatrix} 2^n$$

Testing that the solution is correct is possible by equating the system solution with the system difference equation, such that

$$Y_n = \begin{bmatrix} 100 \\ 30 \end{bmatrix} + \begin{bmatrix} 100(9-n) \\ 20 \end{bmatrix} 2^n = \begin{bmatrix} 2 & -10 \\ 0 & 2 \end{bmatrix}^{-1} \begin{bmatrix} Y_{n+1} - \begin{bmatrix} 200 \\ -3 \end{bmatrix} \end{bmatrix}$$

The reader may wish to check that the above holds true.

In conclusion, the following are selected properties of the Z transforms, to be applied in further chapters:

- $k_1 F_1(z) + k_2 F_2(z)$ is the Z transform of $k_1 f_1(n) + k_2 f_2(n)$

- $z^n F(z) - z^n f(0) - z^{n-1} f(1) - \ldots - z f(n-1)$ is the transform of $f(n+k)$

Summary

The Z transformation procedure is used to convert difference and sum operations in the time domain of variable t, into multiplication and division operations in the frequency domain of the complex variable z. Difference equations with constant coefficients are transformed to algebraic forms. Then, by inverse transformations, the algebraic expressions become the solution of the difference equation.

5

CURVE FITTING AND EVALUATION

In a system analysis problem, the data is defined as a time series and the name of the game is developing equations for predicting the system behavior over time. As such, the most important parameters in the system evaluation process are the constant coefficients of the equations representing the hypotheses. Thus, the purpose of this chapter is to present appropriate procedures for determining and for evaluating the state and the output functions of the system.

5.1 THEORETICAL BASIS OF NONLINEAR CURVE FITTING

Several methods are available for a non linear curve fitting problem and many were developed exclusively for specific mathematical models. This section is related only to a general non linear regression procedure. A major difficulty of this procedure is guessing the initial values of the equation parameters. If the guesses are not correct, the process may not converge to the least sum of squared errors. Moreover, it is not always possible to know if the process converged to the best estimate of the least sum of squared errors. Thus, the real problem becomes getting too many answers to the curve fitting problem.

The Least Squares Concept

Curve fitting is the process of finding numerical values for the constant coefficients of the mathematical model representing the system. The least squares regression is the procedure for finding the best possible curve fitting for the data. The simplest model, the straight line $\hat{y} = a + bt$, will be used to illustrate this criterion.

As indicated, the problem is here finding numerical values to coefficients a and b, for the best possible fitting equation to the data. Then

$$
\begin{aligned}
y_1 - \hat{y}_1 &= y_1 - (a + bt_1) = d_1 \\
y_2 - \hat{y}_2 &= y_2 - (a + bt_2) = d_2 \\
&\ \ \vdots \qquad\qquad \vdots \qquad\quad \vdots \\
y_n - \hat{y}_n &= y_n - (a + bt_n) = d_n
\end{aligned}
$$

are the differences between each data value $c_i = (t_i, y_i)$ and the corresponding point $\hat{y}$ in equation $\hat{y} = a + bt$. The following is the sum of squares of those differences:

$$SS = \sum_1^n \left[y_i - (a + bt_i) \right]^2$$

To find numerical values for coefficients a and b, the partial derivatives of SS, defined over a and b, are required. Note that, for minimizing SS, the values of the partial derivatives are zero. Thus

$$\frac{\partial SS}{\partial a} = -2 \sum_1^n \left[y_i - (a + bt_i) \right] = 0$$

$$\frac{\partial SS}{\partial b} = -2 \sum_1^n t_i \left[y_i - (a + bt_i) \right] = 0$$

The above system of simultaneous *normal linear equations* may be written as

$$an + b \sum_1^n t_i = \sum_1^n y_i$$

$$a \sum_1^n t_i + b \sum_1^n t_i^2 = \sum_1^n t_i y_i$$

This system of two equations and two unknowns is solved easily. The following are the solutions for coefficients a and b, corresponding to the minimum SS, written in an abbreviated form:

$$a = \frac{\sum y \sum t^2 - \sum t \sum ty}{n \left[\sum t^2 - \left(\sum t \right)^2 / n \right]}$$

$$b = \frac{\sum ty - \left(\sum t \sum y \right)/n}{\sum t^2 - \left(\sum t \right)^2 / n}$$

The expression for the sum of squares of the deviations from regression is obtained by replacing the a and b values in SS. Then

$$SS = \sum y^2 - \left(\sum y\right)^2/n - \frac{\left[\sum ty - \left(\sum t\sum y\right)/n\right]^2}{\sum t^2 - \left(\sum t\right)^2/n}$$

where *SS* is a minimum. The sum of squares of the deviations from regression is known as $\sum d_{y.t}^2$.

Example 5.1.1 The following are the average heights of soybean plants, sampled at random each week:

Age, weeks	1	2	3	4	5	6	7
Height, cm	5	13	16	23	33	38	40

Compute the regression line, corresponding to the least sum of squared errors, for the above data.

Solution: The squares and products of the data are given in the following table:

Table 5.1.1

Age	Height	Squares		Products
t	y	t^2	y^2	ty
1	5	1	25	5
2	13	4	169	26
3	16	9	256	48
4	23	16	529	92
5	33	25	1089	165
6	38	36	1444	228
7	40	49	1600	280
$\sum t = 28$ $\bar{t} = 4$	$\sum y = 168$ $\bar{y} = 24$	$\sum t^2 = 140$	$\sum y^2 = 5112$	$\sum ty = 844$

The following are the equation coefficients:

$$b = \frac{844 - 28(168)/7}{140 - 28(28)/7} = 6.143$$

$$a = \bar{y} - b\bar{t} = 24 - 6.143(4) = -0.572$$

Thus, the system equation is $\hat{y} = 6.143\,t - 0.572$ and the following is the sum of squares of the deviations from regression:

$$\sum d_{y.t}^2 = 5112 - 168^2/7 - \frac{844 - 28(168)/7}{140 - 28^2/7} = 23.4286$$

The General Method for Nonlinear Curve Fitting

As in linear least squares fitting, the goal in nonlinear least squares fitting is to minimize the sum of squares of the deviations from regression. In linear least squares fitting, the simultaneous normal equations are linear. In nonlinear curve fitting, the normal equations are not expected to be linear. Therefore, the sum of squares is not obtained by direct calculations but by iterative procedures. One of the simplest non linear expression, the exponential curve $y = a(1 - e^{-bt})$, will be used to illustrate the procedure.

The following is the sum of squares of the differences between each data value and the corresponding point $\hat{y}$ in equation $\hat{y} = a(1 - e^{-bt})$:

$$SS = \sum_{1}^{n} \left[y_i - a\left(1 - e^{-bt_i}\right) \right]^2$$

The corresponding partial derivatives are here

$$\frac{\partial S}{\partial a} = -2\sum_{1}^{n} \left[y_i - a\left(1 - e^{-bt_i}\right) \right]\left(1 - e^{-bt_i}\right) = 0$$

$$\frac{\partial S}{\partial b} = -2\sum_{1}^{n} \left[y_i - a\left(1 - e^{-bt_i}\right) \right]\left(at_i e^{-bt_i}\right) = 0$$

Note that the above system of equations is not linear and that there is not a direct easy solution for the unknowns. An alternative procedure follows.

The following are the residuals R_i between each data value and the matching point $\hat{y}$ in equation $\hat{y} = a(1 - e^{-bt})$:

$$
\begin{aligned}
y_1 - \hat{y}_1 &= y_1 - \alpha(1 - e^{\beta t_1}) = R_1 \\
y_2 - \hat{y}_2 &= y_2 - \alpha(1 - e^{\beta t_2}) = R_2 \\
&\;\;\vdots \\
y_n - \hat{y}_n &= y_n - \alpha(1 - e^{\beta t_n}) = R_n
\end{aligned}
$$

where α and β are initial guesses assigned to coefficients a and b of the mathematical model. Manipulation of α and β should result in a progressive reduction of the R_i residuals, such that α and β should progressively approach the asymptotic values represented by coefficients a and b.

The changes in the R_i residuals may be defined as the sum of the changes due to changes in α and in β. Thus

$$
\begin{aligned}
\Delta R_1 &= \frac{\partial R_1}{\partial \alpha}\Delta\alpha + \frac{\partial R_1}{\partial \beta}\Delta\beta \\
\Delta R_2 &= \frac{\partial R_2}{\partial \alpha}\Delta\alpha + \frac{\partial R_2}{\partial \beta}\Delta\beta \\
&\;\;\vdots \\
\Delta R_n &= \frac{\partial R_n}{\partial \alpha}\Delta\alpha + \frac{\partial R_n}{\partial \beta}\Delta\beta
\end{aligned}
$$

The $\partial R_i/\partial \alpha$ and $\partial R_i/\partial \beta$ values may be obtained from the partial derivatives of R_i with respect to α and β, such that

$$
\begin{aligned}
\frac{\partial R_i}{\partial \alpha} &= -\left(1 - e^{-\beta t_i}\right) \\
\frac{\partial R_i}{\partial \beta} &= -\alpha\, t_i e^{-\beta t_i}
\end{aligned}
$$

After replacing these values in ΔR_i, the following expression is obtained:

$$
\Delta R_i = -\left[\Delta\alpha\left(1 - e^{-\beta t_i}\right) + \Delta\beta\,\alpha\, t_i e^{-\beta t_i}\right]
$$

The goal here is having $\Delta R_i = -R_i$. Then, $E_i = R_i + \Delta R_i$ is an error term, as was the difference $d_i = y_i - \hat{y}_i$ in linear regression. Therefore, the objective of the procedure should be minimizing the sum of squares of these newly defined errors. The following is the sum of squares of E_i:

$$SS = \sum_{1}^{n} \left\{ R_i - \left[\Delta\alpha\left(1 - e^{-\beta t_i}\right) + \Delta\beta\,\alpha\,t_i e^{-\beta t_i} \right] \right\}^2$$

where $\Delta\alpha$ and $\Delta\beta$ are now the unknowns. Then

$$\frac{\partial SS}{\partial\Delta\alpha} = -2\sum_{1}^{n} \left\{ R_i - \left[\Delta\alpha\left(1 - e^{-\beta t_i}\right) + \Delta\beta\,\alpha\,t_i e^{-\beta t_i} \right] \right\}\left(1 - e^{-\beta t_i}\right) = 0$$

$$\frac{\partial SS}{\partial\Delta\beta} = -2\sum_{1}^{n} \left\{ R_i - \left[\Delta\alpha\left(1 - e^{-\beta t_i}\right) + \Delta\beta\,\alpha\,t_i e^{-\beta t_i} \right] \right\}\left(\alpha\,t_i e^{-\beta t_i}\right) = 0$$

Note that α and β are not the variables here, but $\Delta\alpha$ and $\Delta\beta$. Therefore, the system of simultaneous normal equations is now linear and may be written as

$$\Delta\alpha\sum_{1}^{n} A^2 + \Delta\beta\sum_{1}^{n} AB = \sum_{1}^{n} R_i A$$

$$\Delta\alpha\sum_{1}^{n} AB + \Delta\beta\sum_{1}^{n} B^2 = \sum_{1}^{n} R_i B$$

where $A = 1 - e^{-\beta t_i}$, $B = \alpha\,t_i e^{-\beta t_i}$ and $R_i = y - \alpha(1 - e^{-\beta t_i})$. Then the following are the solutions for $\Delta\alpha$ and $\Delta\beta$, written in a condensed form:

$$\Delta\alpha = \frac{\sum RA \sum B^2 - \sum RB \sum AB}{\sum A^2 \sum B^2 - \left(\sum AB\right)^2}$$

$$\Delta\beta = \frac{\sum RB \sum A^2 - \sum RB \sum AB}{\sum A^2 \sum B^2 - \left(\sum AB\right)^2}$$

The sum of squares for the error is obtained by replacing $\Delta\alpha$ and $\Delta\beta$ in *SS*:

$$SS = \sum \left[R - (\Delta\alpha\,A + \Delta\beta\,B) \right]^2$$

This sum of squares may also be written as $\sum E^2$. The cycle is completed after determining new values for α and β:

$$\alpha' = \alpha + \Delta\alpha$$
$$\beta' = \beta + \Delta\beta$$

The process is repeated until the changes in α and β and the changes in SS fulfill some given convergence criteria. Convergence is defined here as the approximation of α and β and of $\sum E^2$ to their corresponding asymptotic values a, b and $\sum d_{y.t}^2$.

Note that, except for very simple mathematical models, the process is not feasible without a computer. Note also that, if the initial guesses of the system parameters are not appropriate, the process may not converge to the correct solution or may not converge at all.

Example 5.1.2 The following are the microbial digestibility values of cell walls of a pasture sample:

Hours	6	12	18	24	30	36	42	48
% Digestion	19	35	43	47	49	51	52	53

Compute the constant coefficients for equation $\hat{y} = a(1 - e^{-bt})$.

Solution: The initial guesses may be obtained from a graph, as the one in Fig. 5.1.1. For the above mathematical model, coefficient a is an asymptotic value and b is a slope. As seen in the graph, coefficient a is around 50. If a is guessed as 50, coefficient b may be obtained by plotting $y - a$ on a semi-log paper or simply by solving expression

$$b = -\frac{\ln}{t}\left(\frac{a - y}{a}\right)$$

where t and y may be selected from the data. From the above, coefficient b is found to be roughly 0.1. If the initial guesses are $\alpha=50$ and $\beta=0.1$, then

$$A = 1 - e^{-0.1t}$$
$$B = 50te^{-0.1t}$$
$$R = y - 50(1 - e^{-0.1t})$$

where A, B and R were previously defined.

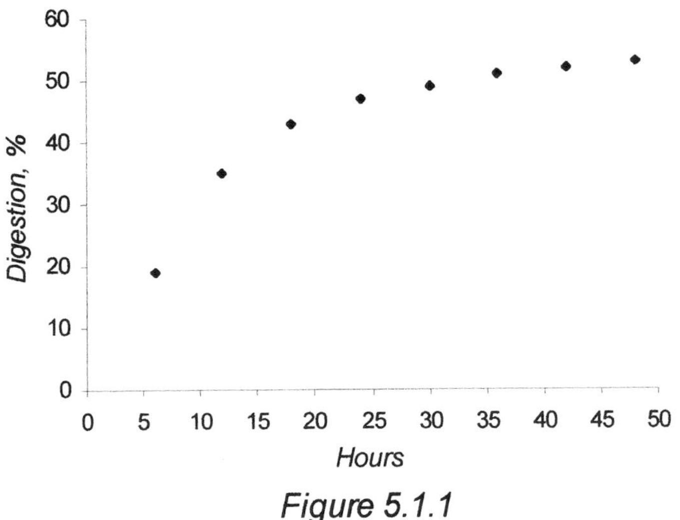

Figure 5.1.1

Table 5.1.2 shows the values required for computing $\Delta\alpha$ and $\Delta\beta$. For simplicity, only four data points will be used.

Table 5.1.2

T	Y	R	A	B	RA	RB	A^2	B^2	AB
6	19	-3.56	0.451	164.6	-1.65	-586.0	0.203	27107	74.3
18	43	1.27	0.835	148.8	1.06	188.2	0.697	22132	124.2
30	49	1.49	0.950	74.7	1.42	112.2	0.903	5577	70.9
42	52	2.75	0.985	31.5	2.71	86.6	0.970	992	31.0
$\sum$		1.95			3.58	-199.0	2.273	55808	300.4

Then

$$\Delta\alpha = \frac{\sum RA \sum B^2 - \sum RB \sum AB}{\sum A^2 \sum B^2 - \left(\sum AB\right)^2} = 4.02$$

$$\Delta\beta = \frac{\sum RB \sum A^2 - \sum RB \sum AB}{\sum A^2 \sum B^2 - \left(\sum AB\right)^2} = -0.0252$$

New values for α and β can now be computed:

$$\alpha' = \alpha + \Delta\alpha = 50 + 4.02 \approx 54$$
$$\beta' = \beta + \Delta\beta = 0.1 - 0.0252 \approx 0.075$$

As shown in the next table, the R_i', residuals computed with the new α' and β' values, are smaller that the R_i residuals computed with α and β. Note that the sum of squares of the new residuals was also reduced:

Table 5.1.3

T	Y	R	R^2	R'	$(R')^2$
6	19	-3.56	12.666	-0.534	0.285
18	43	1.27	1.600	3.035	9.209
30	49	1.49	2.217	0.708	0.501
42	52	2.75	7.565	0.314	0.099
$\sum$			24.046		10.046

A full cycle has now been completed. The iterative process is repeated until the convergence criteria are fulfilled.

Several nonlinear computational methods have been developed. They differ mainly in how they compute the change of the equation parameters. For a particular problem one method may perform better that others. Several statistical packages, like SPSS for Windows, SAS/STAT for Windows, S-PLUS for Windows, to mention a few, provide computation programs for nonlinear regression.

Summary

Curve fitting is the process of finding numerical values for the constant coefficients of the mathematical model representing a set of data. In linear regression, the curve fitting error is the difference $d_i = y - \hat{y}$, where y is a data point and $\hat{y}$ is the corresponding estimate. The simultaneous normal equations, derived from $\sum d_i^2$, are linear and the system is easily solved for its unknowns. In nonlinear regression, the system of normal equations is not linear and may not have an easy solution. A general method of nonlinear least squares curve fitting defines the error as the difference $E_i = R_i - \Delta R_i$. R_i is the difference between a data point and the corresponding conditional estimate and ΔR_i is the change in the value of R_i, corresponding to a change in value of the estimate. The set of normal equations is derived from $\sum E_i^2$ and the unknowns are the changes in the numerical value of the equation parameters. The iterative process is repeated until the changes in the equation parameters and the changes in $\sum E_i^2$ fulfil some given convergence criteria.

5.2 COMPUTATION OF THE MODEL PARAMETERS

As indicated in the previous section, the nonlinear regression process starts with the initial guesses of the model parameters. If these initial values are not appropriate, the process may not converge to the least sum of squares of the error terms. In addition, there is always some uncertainty whether the process converged to the least sum of squared errors or to a trap. The outcome is a trap, when different regression methods or different initial guesses yield different results. Lack of convergence and traps may be caused by incorrect initial guesses or highly correlated parameters, by the size of the change across successive iterations, by an inappropriate mathematical model and even by the quality of the data.

Models with some exponential terms and powers may cause *underflow* or *overflow* convergence problems. A number that is too small for the computer to handle may cause underflow. An overflow is caused if the number is too large. A transformation of the time scale may often correct the problem, for example using years instead of months or subtracting the smallest time value from all the other time values.

Limiting the size of the parameter changes across iterations may help solve some convergence and trap problems, but it would slow down the process. Deleting some parameters may correct over parameterization. A model with fewer parameters that fit the data does not necessarily mean that the original model was inappropriate. It may mean that the data was not sufficient to estimate all the parameters.

If the culprit of computational problems is the quality of data, imposing bounds on parameters may prevent jumps of the iterations in the wrong direction, forcing the function trough the expected path. Main sources for poor quality data are, too large experimental errors and lack of data points where the function is expected to have critical and extreme values.

Guessing the values of the initial parameters is often a combination of technique and artistry. There are no fixed rules and only the experience of the research team may determine the best pathway for defining initial values for the constant coefficients.

As disclosed in the first chapter, the system difference or differential equations may be determined, by linear regression, from difference tables of the data. Then, the numerical values of the constant coefficients of the resulting state equations may be used as the initial values for nonlinear regression. As shown in the next examples, this pathway is often the simplest approach for determining the initial values of the model parameters.

Example 5.2.1 The following is the energy content of milk of a group of cows[1]:

Days after calving	10	20	40	60	100	120	140	150
Mjoules/Kg	3.45	2.97	2.90	2.88	2.90	2.74	2.84	2.82

Determine the equation representing the above data.

Solution: The first step is defining a difference table for the data:

Table 5.2.1

t	y	Δy	$\Delta y/\Delta t$
10	3.45	-0.20	-0.02
20	3.25	-0.28	-0.014
40	2.97	-0.07	0.0035
60	2.90	-0.02	-0.001
80	2.88	0.02	0.001
100	2.90	-0.16	-0.008
120	2.74	0.10	0.005
140	2.84	-0.02	-0.002
150	2.82		

The following differential equation was fitted by linear regression to the data from the difference table:

$$\frac{\Delta y}{\Delta t} + 0.03312y = 0.09382$$

where y is energy content in Mjoules/Kg and t is days after calving. As indicated in

---

[1]Computed from Lowman, B.G. et.al.

Chapter 3, $\Delta y \neq dy$, but it may be a good approximation, provided that Δt is small enough. The corresponding Laplace transform of the above equation is here

$$G(s) = \frac{y_0}{s+0.03312} + \frac{0.09382}{s(s+0.03312)}$$

where the initial energy content of $y_0 = 4.0$ Mjoules/Kg was estimated from a graph of the data. The following is the resulting response equation:

$$y = 2.8 + 1.2e^{-0.0331t}$$

The constant coefficients of the above equation may be used as the initial values for the non linear process. This equation has the form $y = a + be^{-ct}$. Then, the following are the partial derivatives for the unknowns of the model, required by the nonlinear procedure:

$$\frac{\partial y}{\partial a} = 1$$

$$\frac{\partial y}{\partial b} = e^{-ct}$$

$$\frac{\partial y}{\partial c} = -bte^{-ct}$$

Note that some statistical packages, like SPSS for Windows, do not always require the partial derivatives from the user. The following results were obtained from nonlinear regression[2]:

Table 5.2.2

Source	DF	Sum of Squares	Mean Square
Regression	3	79.91208	26.63736
Residual	6	0.01382	2.303615E-03
Uncorrected Total	9	79.92590	
(Corrected Total)	8	0.41896	

R squared = 1 - Residual SS / Corrected SS = 0.96701

---

[2]SPSS Professional Statistics 7.5.

| | | Asymptotic | |
Parameter	Estimate	Std. Error	"t"
a	2.821334360	0.025511076	110.64
b	0.964619071	0.115501347	8.32
c	0.042261182	0.008612802	4.91

Asymptotic Correlation Matrix of the Parameter Estimates

	a	b	c
a	1.0000	0.4077	0.6386
b	0.4077	1.0000	0.8582
c	0.6386	0.8582	1.0000

Thus, the following is the resulting equation for energy content:

$$y = 2.821 + 0.965\,e^{-0.0423\,t}$$

The graph of this function is shown in Fig. 5.2.1.

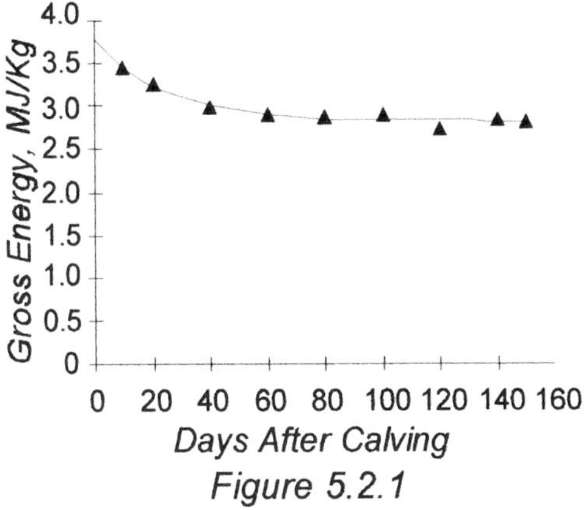

Figure 5.2.1

Iterations stopped after 10 model evaluations and 5 derivative evaluations, because the relative reduction between successive residual sums of squares and the relative difference between successive parameter estimates matched a given criteria.

Note that the coefficient of determination and the equation parameters are statistically significant. These statistics are the main criteria for the evaluation of the process and the mathematical model of the system. Note also that the correlation between coefficients *a* and *b* are not significant, confirming again that the mathematical model is appropriate for the data.

Determining the difference or differential equations of the system may also require sometimes a nonlinear regression approach. This assertion is explained in the following example.

Example 5.2.2 The following mathematical model is proposed for the lactation curve of a group of dairy cows:

$$y = (a + bt)e^{-ct}$$

where y is Kg/month and t is months. Determine the corresponding differential and state equations from the following data:

Months	0.5	1	2	3	4	5	6	7	8	9	10
Milk, Kg/month	400	430	425	360	290	205	150	110	75	45	30

Solution: The following is the differential equation of the model:

$$\frac{dy}{dt} + ay = be^{-at}$$

Clearly, nonlinear regression is required for determining the numerical values of the coefficients of the above equation. The following are the results of a first attempt to obtain provisional estimates of the parameters using linear regression and data from a difference table:

Table 5.2.3

Source	DF	Sum of Squares	Mean Square
Regression	2	4083.06985	2041.53492
Residual	7	6476.93015	925.27574

Variable	b	SE b	"t"	Sig t
y	-0.693281	0.337930	-2.052	0.0793
t	-36.201751	17.235579	-2.100	0.0738
(Constant)	300.344925	162.033905	1.854	0.1062

The resulting equation is $\Delta y/\Delta dt = 300 - 36t - 0.69y$. This equation may be modified to $\Delta y/\Delta t = 300e^{-0t} - 36t - 0.7y$ to yield the initial guessing of the parameter values. Note that an exponential term with a 0 exponent was added to the new equation, because any number to the zero power equals one. The following are the nonlinear regression results of a second regression round:

Table 5.2.4

Source	DF	Sum of Squares	Mean Square
Regression	4	23918.32779	5979.58195
Residual	6	331.67221	55.27870
Uncorrected Total	10	24250.00000	
(Corrected Total)	9	10560.00000	

R squared = 1 - Residual SS / Corrected SS = 0.96859

Parameter	Estimate	Asymptotic Std. Error	"t"
a	241.19590320	109.19316215	2.21
b	0.410718065	0.403085491	1.02
c	0.710669594	1.530578335	0.46
d	0.400183173	0.314375479	1.27

The new equation is $\Delta y/\Delta t = 241e^{-0.41t} - 0.71t - 0.40y$. Note that this equation is now over parameterized. The best candidate for removal is coefficient c. Note also that coefficients b and d are virtually the same. The results of the next nonlinear regression round are shown in Table 5.2.5. These results are now acceptable and the following is the final adopted equation of the system:

Table 5.2.5

Source	DF	Sum of Squares	Mean Square
Regression	2	23839.95913	11919.97956
Residual	8	410.04087	51.25511
Uncorrected Total	10	24250.00000	
(Corrected Total)	9	10560.00000	

R squared = 1 - Residual SS / Corrected SS = 0.96117

Parameter	Estimate	Asymptotic Std. Error	"t"
a	251.68452603	12.243564107	21.37
b	0.416681056	0.010715303	38.94

The following is the Laplace transform of the above equation:

$$G(s) = \frac{y_0}{s+0.417} + \frac{239}{(s+0.417)^2}$$

where $y_0 = 400$ is an estimated initial value. Thus, the resulting state equation is

$$y = (400 + 239t)e^{-0.417t}$$

The constant coefficients of this equation may now be the initial guesses for a final nonlinear regression round. Iterations stopped after 12 model evaluations and 6 derivative evaluations. The results are shown in Table 5.2.6.

Table 5.2.6

Source	DF	Sum of Squares	Mean Square
Regression	3	824140.12783	274713.37594
Residual	8	259.87217	32.48402
Uncorrected Total	11	824400.00000	
(Corrected Total)	10	247090.90909	

R squared = 1 - Residual SS / Corrected SS = 0.99895

Parameter	Estimate	Asymptotic Std. Error	"t"
a	298.01044072	11.670179081	25.54
b	411.19839087	16.848179856	24.4
c	0.483780363	0.007414271	65.28

Asymptotic Correlation Matrix of the Parameter Estimates

	a	b	c
a	1.0000	-0.8655	-0.7321
b	-0.8655	1.0000	0.9399
c	-0.7321	0.9399	1.0000

Thus, the following is the resulting lactation curve equation:

$$y = (298 + 411t)e^{-0.484t}$$

The corresponding graph is shown in Fig. 5.2.2.

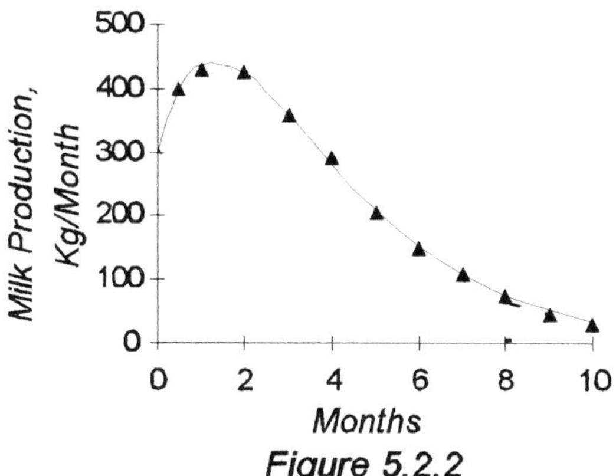

Figure 5.2.2

Example 5.2.3 Determine the mathematical expression for the following data of ammonia
and protein nitrogen of the rumen of steers fed a soybean meal diet[3]:

t	1	2	3	4	5	6	7	8	9	10	11	12	13
y_a	15.0	24.5	26.0	26.5	24.3	21.5	20.0	19.1	18.2	17.1	16.6	15.5	15.0
y_p	127	154	152	142	144	140	139	141	142	142	141	141	140

where y_a is ammonia nitrogen y_p, is protein nitrogen, in Mg/100ml of ruminal fluid and
t is hours after feeding.

Solution: The following set of equations was fitted to the above data:

$$\frac{\Delta Y}{\Delta dt} = \begin{bmatrix} -0.7132 & 0.05305 \\ -1.1440 & -0.6978 \end{bmatrix} Y + \begin{bmatrix} 13.2910 & 0 \\ 0 & 132.7176 \end{bmatrix} \begin{bmatrix} e^{-0.1381t} \\ e^{-0.01346t} \end{bmatrix}$$

Data of a difference table and procedures outlined in the previous example were used here.
The following table shows the statistics for the ammonia differential equation:

---

[3]Computed from Davis G.V. and O.T. Stallcup

Table 5.2.7

Source	DF	Sum of Squares	Mean Square
Regression	4	109.81357	27.45339
Residual	8	2.40643	0.30080
Uncorrected Total	12	112.22000	
(Corrected Total)	11	112.22000	

R squared = 1 - Residual SS / Corrected SS = 0.97856

Parameter	Estimate	Asymptotic Std. Error	"t"
K1	-0.713185655	0.066077062	10.79
K2	0.053045980	0.012721538	4.17
K3	13.290957161	2.239732720	5.93
K4	0.138065737	0.046170577	2.99

The following is the summary of statistics for the protein differential equation:

Table 5.2.8

Source	DF	Sum of Squares	Mean Square
Regression	4	813.79945	203.44986
Residual	8	45.70055	5.71257
Uncorrected Total	12	859.50000	
(Corrected Total)	11	845.41667	

R squared = 1 - Residual SS / Corrected SS = 0.94594

Parameter	Estimate	Asymptotic Std. Error	"t"
K1	-1.144027890	0.315665445	3.62
K2	-0.697849486	0.164338043	4.25
K3	132.71754638	18.514228294	7.17
K4	0.013458092	0.003497463	3.85

The Laplace transform of the system has the following expression:

$$\begin{bmatrix} s+0.7132 & -0.05305 \\ 1.1440 & s+0.6978 \end{bmatrix} G(s) = Y_0 + \begin{bmatrix} \dfrac{13.2910}{s+0.1381} \\ \dfrac{132.7176}{s+0.01346} \end{bmatrix}$$

where Y_0 is the set of initial values, for $y_{a0} = 15$ and $y_{b0} = 127$ are data values for ammonia and for protein. Then, the characteristic equation of the system is

$$|sI - A| = s^2 + 1.4110s + 0.5584 = (s + \lambda_1)(s + \lambda_2)$$
$$= [s + (0.7055 + 0.2462i)][s + (0.7055 - 0.2462i)]$$

where $\lambda = \alpha \mp \beta i = 0.7055 \pm 0.2462i$. Thus, the state equations have a periodic form.

The above system may be first solved symbolically. For such, the set of differential equations may be written as

$$\frac{dY}{dt} = \begin{bmatrix} b_{11} & b_{12} \\ b_{21} & b_{22} \end{bmatrix} Y + \begin{bmatrix} c_1 & 0 \\ 0 & c_2 \end{bmatrix} \begin{bmatrix} \exp^{-d_1 t} \\ \exp^{-d_2 t} \end{bmatrix}$$

The corresponding solutions have the form

$$Y = \begin{bmatrix} -\dfrac{c_1(b_{22}+d_1)}{k_1} & \dfrac{c_2 b_{12}}{k_2} \\[2ex] \dfrac{c_1 b_{21}}{k_1} & -\dfrac{c_2(b_{11}+d_2)}{k_2} \end{bmatrix} \begin{bmatrix} e^{-d_1 t} \\ e^{-d_2 t} \end{bmatrix}$$

$$+ e^{-\alpha t} \begin{bmatrix} \dfrac{c_1(b_{22}+d_1)}{k_1} - \dfrac{c_2 b_{12}}{k_2} + y_{p0} & \dfrac{c_1 k_3}{k_1 \beta} + \dfrac{c_1 b_{12}(d_2 - \alpha)}{k_2 \beta} + \dfrac{k_5}{\beta} \\[2ex] -\dfrac{c_1 b_{21}}{k_1} + \dfrac{c_2(b_{11}+d_2)}{k_2} + y_{a0} & \dfrac{c_1 b_{21}(d_1 - \alpha)}{k_1 \beta} + \dfrac{c_2 k_4}{k_2 \beta} + \dfrac{k_6}{\beta} \end{bmatrix} \begin{bmatrix} \cos \beta t \\ \sin \beta t \end{bmatrix}$$

where

$$k_1 = \alpha^2 + \beta^2 - 2\alpha d_1 + d_1^2 \qquad\qquad k_2 = \alpha^2 + \beta^2 - 2\alpha d_2 + d_2^2$$
$$k_3 = \alpha^2 + \beta^2 + \alpha b_{22} - d_1(\alpha + b_{22}) \qquad k_4 = \alpha^2 + \beta^2 + \alpha b_{11} - d_2(\alpha + b_{11})$$
$$k_5 = -y_{p0}(\alpha + b_{22}) + y_{a0} b_{12} \qquad k_6 = y_{p0} b_{21} - y_{a0}(\alpha + b_{11})$$

The state equations may now be obtained by replacing symbols by their numerical values:

$$Y = \begin{bmatrix} -39.70 & 172.13 \\ 19.42 & 13.05 \end{bmatrix} \begin{bmatrix} e^{-0.1381t} \\ e^{-0.01346t} \end{bmatrix}$$

$$+ \; e^{-0.7055t} \begin{bmatrix} -32.47 + y_{p0} & -27.53 - 0.03128 y_{p0} + 0.2155 y_{a0} \\ -132.43 + y_{a0} & 146.68 - 4.647 y_{p0} + 0.03128 y_{a0} \end{bmatrix} \begin{bmatrix} \cos 0.2462\,t \\ \sin 0.2462\,t \end{bmatrix}$$

The above state equations provide excellent initial guesses for the model parameters. However, the reader is reminded that the relationship between the state variables is not absolute. If all these coefficients are allowed to change independently for each state equation, the relationships between the state variables, established by the set of differential equations, could be disrupted. If preserving the relationships between the state variables is wanted, then only the initial values $g_1(0) = 15$ and $g_2(0) = 127$ should be allowed to change, because these are data values that include an experimental error. By adopting this criterion, the following state equations were obtained by non linear regression:

$$Y = \begin{bmatrix} -39.70 & 172.13 \\ 19.42 & 13.05 \end{bmatrix} \begin{bmatrix} e^{-0.1381t} \\ e^{-0.01346t} \end{bmatrix}$$

$$+ \; e^{-0.7055t} \begin{bmatrix} -32.47 + 15.06 & -27.53 + 51.32 \\ -132.43 + 129.30 & 146.68 - 25.79 \end{bmatrix} \begin{bmatrix} \cos 0.2462\,t \\ \sin 0.2462\,t \end{bmatrix}$$

The resulting initial values are $y_{a0} = 15.06$ and $y_{p0} = 129.30$. The nonlinear regression statistics for dependent variable y_a are shown in Table 5.2.9. Iterations stopped after 4 model evaluations and 2 derivative evaluations.

Table 5.2.9

Source	DF	Sum of Squares	Mean Square
Regression	2	5363.29904	2681.64952
Residual	11	2.65096	0.24100
Uncorrected Total	13	5365.95000	
(Corrected Total)	12	217.82000	

R squared = 1 - Residual SS / Corrected SS = 0.98783

Parameter	Estimate	Asymptotic Std. Error	"t"
y_{a0}	15.057922231	0.476088061	31.63
y_{p0}	240.32223677	12.841811683	18.77

The value 240.32 is significantly different from 127.

The following are the statistical results for variable y_p:

Table 5.2.10

Source	DF	Sum of Squares	Mean Square
Regression	2	261787.93725	130893.96862
Residual	11	129.81275	1 1.80116
Uncorrected Total	13	261917.75000	
(Corrected Total)	12	495.26923	

R squared = 1 - Residual SS / Corrected SS = 0.73789

Parameter	Estimate	Asymptotic Std. Error	"t"
y_{a0}	6.419969406	4.185956705	1.53
y_{p0}	129.34073046	3.331537927	38.83

Iterations stopped after 4 model evaluations and 2 derivative evaluations. The value 6.42 is statistically different from 15, $p \le 0.10$.

The reader may wish to experiment with additional parameter changes to improve the coefficient of determination of the protein equation. However, if more than four or five coefficients are allowed to change simultaneously in each equation, the non linear regression procedure may find the model over parameterized.

The graph of the above equations is shown in Fig. 5.2.3.

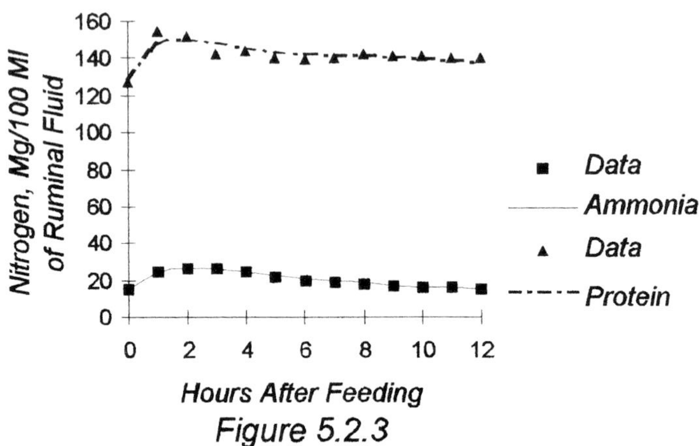

Figure 5.2.3

A graphic approach may often help determining the initial parameter values. This approach is particularly helpful in periodic functions, because sometimes determining the state equations from difference or differential equations may be a long process prone to errors.

Example 5.2.4 The following are the production data of a Kikuyo pasture field[4]:

Months	S	O	N	D	J	F	M	A	M	J	J	A
Time	1	2	3	4	5	6	7	8	9	10	11	12
Dry Matter, Kg/Ha/day	15	16	20	51	47	78	67	44	25	22	16	20

Determine the numerical values for the constant coefficients of the following model:

$$y = a + be^{\alpha t}\cos[\beta(t - c)]$$

where y is pasture production in Kg/Ha/day and t is months.

Solution: The geometrical meaning of the model parameters is shown in Fig. 5.2.4. This graph provides the following parameter estimates:

$a = 40$ is the distance between the abscissa and the axes of the response curve
$be^{\alpha t} = 40$ modulates the amplitude
$\beta = 2\pi/12$ modulates the frequency response, for a $2\pi/\beta$ cycle
$c = 6$ is the out-of-phase coefficient, may be also a negative number

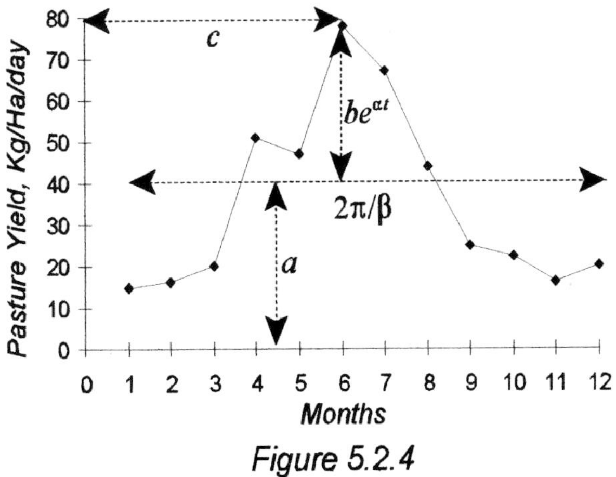

Figure 5.2.4

[4]Computed from Murthagh, G.J. et.al.

The following are the results of a first nonlinear regression round, excluding the α coefficient:

Table 5.2.11

Source	DF	Sum of Squares	Mean Square
Regression	4	19498.07116	4874.51779
Residual	8	466.92884	58.36610
Uncorrected Total	12	19965.00000	
(Corrected Total)	11	5194.91667	

R squared = 1 - Residual SS / Corrected SS = 0.91012

Parameter	Estimate	Asymptotic Std. Error	"t"
a	40.686033337	2.395914989	16.98
b	27.735169241	3.142103289	8.83
β	0.690857972	0.040456179	17.06
c	6.164223027	0.175905181	35.02

The exponent α will now be added, using the above parameters as initial values and assuming an initial value of zero. The following parameter values were obtained:

Table 5.2.12

Parameter	Estimate	Asymptotic Std. Error	"t"
a	40.660655707	2.573407459	15.80
b	28.362500936	6.381951019	4.44
β	0.690155050	0.044064108	15.65
c	6.165792283	0.188149562	32.80
α	-0.003640961	0.031355789	-0.12

Clearly, parameter α is not significant and the function is now over parameterized. Thus, the following is the resulting equation when the exponential term α is deleted:

$$y = 40.7 + 27.7\cos\left[0.691(t - 6.16)\right]$$

The graph related to the system equation is shown in Fig. 5.2.5.

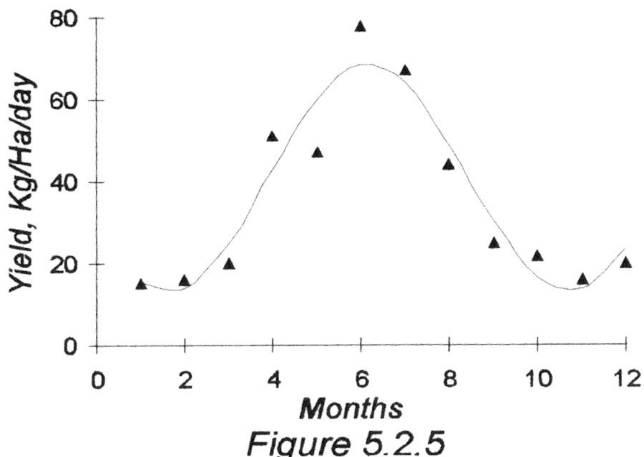

Figure 5.2.5

As shown in this example, sometimes a graphic approach, for determining parameter initial values, may save a substantial amount of work.

The problem of determining the model parameters becomes more complex as more input variables are added to the system. The following example illustrates a case with one input variable.

Example 5.2.5 The following mathematical model was defined for the *in situ* digestibility of the cell walls of sugarcane leafs:

$$y = \frac{c}{b} + \left(y_0 - \frac{c}{b} e^{-bt} \right)$$

where y is percent digestibility, y_0 is an initial value, t is weeks, c/b is an asymptotic value and b is a relative rate. Determine how supplementation of green bananas to experimental steers affects the system. Table 5.2.13 shows the available data[5].

Solution: The following is the proposed model of the differential equation representing the system:

$$\frac{\partial y}{\partial t} + by = f(x)$$

where $f(x)$ is the system input and x is percent of green bananas. Note that this is a partial

---

[5]Computed from San-Martin, F.A.

differential equation because the banana inputs are fixed in each treatment.

Table 5.2.13

Time, Hours	Digestibility, %					
	Green Bananas, % in the Diet					
	0	**21.6**	**35.9**	**55.1**	**60.1**	**70.7**
6	0.00	3.51	2.00	1.20	0.00	0.00
12	16.51	19.36	4.97	4.28	0.00	0.53
18	21.73	25.96	13.33	5.04	0.08	0.61
24	26.01	35.87	22.23	12.29	8.25	3.48
48	40.44	47.02	32.79	37.19	28.71	7.06
72	46.91	50.14	39.38	41.54	33.25	16.86
96	51.05	53.63	49.52	52.40	35.60	28.13
120	52.88	60.67	52.46	54.72	41.39	33.37

The solution of the differential equation has the form

$$y = \frac{f(x)}{b} + \left(y_0 - \frac{f(x)}{b} e^{-bt} \right)$$

Thus, the first step in determining the state equation of the system is finding $f(x)$. The following results were obtained by linear regression using a difference table:

Table 5.2.14

Source	DF	Sum of Squares	Mean Square
Regression	2	7.11301	3.55650
Residual	39	10.20353	0.26163

Variable	b	SE b	"t"	Sig t
Y	-0.021561	0.004729	-4.559	0.0000
X	-0.014729	0.003553	-4.146	0.0002
(Constant)	1.657543	0.222851	7.438	0.0000

Thus, the resulting equation is $\Delta y/\Delta t + 0.0216y = 1.658 - 0.0147x$, where the input is $f(x) = 1.658 - 0.0147x$. Then, the following is the solution of the differential equation:

$$y = \frac{f(x)}{0.0216} + \left(y_0 - \frac{f(x)}{0.0216}\right) e^{-0.0216t}$$

where y_0 is an initial state value, estimated as $y_0 = 8$ from a graphic approach. This equation provides the initial guesses for the final regression round shown below:

Table 5.2.15

Source	DF	Sum of Squares	Mean Square
Regression	4	46887.43088	11721.85772
Residual	44	2211.08952	50.25203
Uncorrected Total	48	49098.52040	
(Corrected Total)	47	18372.17637	

R squared = 1 - Residual SS / Corrected SS = 0.87965

Parameter	Estimate	Asymptotic Std. Error	"t"
K1	1.699820480	0.264108342	6.44
K2	0.011083298	0.002260906	4.90
K3	0.025055015	0.005426163	4.62
Y0	-8.139136574	3.687234943	2.21

All the above parameters are statistically significant. Thus, the following is the final expression for the state equation of the system:

$$y = \frac{f(x)}{0.0251} - \left(8.139 + \frac{f(x)}{0.0251}\right) e^{-0.0251t}$$

$$f(x) = (1.700 - 0.0111x)$$

where $f(x)$ is the input of green bananas. The reader is invited to apply the procedure to determine the more accurate equation shown below:

$$y = \frac{f(x)}{0.0258} - \left(8.566 + \frac{f(x)}{0.0258}\right) e^{-0.0258t}$$

$$f(x) = (1.515 + 0.0800x) e^{-0.0303x}$$

Summary

The first step in nonlinear regression is guessing initial values of the model parameters. If the initial values are not appropriate, the process may not converge to the least sum of squares of the error terms. Wrong initial parameters, highly correlated parameters, too large parameter changes across interactions, overflows or underflows, inappropriate mathematical models or poor quality data, may impair convergence. A solution to these problems may require, among other tactics, defining smaller parameter changes across interactions, imposing bounds on parameters, redefining the time scale of the system or even making changes in the mathematical model of the system. In linear systems, the simplest approach for guessing the initial parameter values is often determining the difference or differential equations of the system by linear regression. The initial values are then obtained from the resulting state equations. When possible, a graphic approach for determining the initial parameters may save some work in mathematical manipulations. No single rule is valid for all cases for determining the initial guesses of the model parameters.

5.3 EVALUATION OF THE MATHEMATICAL MODELS AND SYSTEM BEHAVIOR

Several statistical tests of the outcomes of the regression analysis are required before the mathematical model of the system can be accepted or rejected. These tests include evaluations of the constant coefficients of the mathematical model and evaluations of the predictive value and accuracy of the model.

Evaluation of the Constant Coefficients

Evaluation of the constant coefficients of the mathematical model should be done at two levels of resolution:

- Within components
- Between components

A "t" test should be used to evaluate each coefficient of the mathematical model within components, such that

$$t = \frac{k - k_0}{S_k}$$

where k represents a constant coefficient, k_0 is the corresponding hypothetical value, S_k is the standard error of the k coefficient and $k - k_0 = 0$ is the null hypothesis. Usually, k_0 is zero. Several mathematical models were displayed in the previous section with the complete statistical outcomes. When the null hypothesis for a parameter was accepted, the mathematical model was considered over-parameterized and the corresponding coefficient was deleted. The reader may review those examples and the criteria used to accept, modify or reject a model.

If the system has more than one component, the "t" test expression for comparing pairs of parameters is as follows:

$$t = \frac{k_i - k_j}{\sqrt{S_{k_i}^2 + S_{k_j}^2}}$$

where k_i represents a coefficient of the i component, k_j represents the coefficient of the j component and S_{k_i} and S_{k_j} are their related errors. The null hypothesis is $k_i - k_j = 0$. The expression for degrees of freedom for the above test is $DF = n_i - m + n_j - m = n_i + n_j - 2m$, where n_i and n_j are the number of observations in the i and j components and m is the number of constant coefficients in the regression equation.

Example 5.3.1 The following is the energy consumption of Jersey calves grown in a heat chamber at 10 °C and at 27 °C [6]:

Age, Months		1	3	4	5	6	7	8	9	10	11	12
NDT, %	10 °C	2.2	3.0	2.9	2.7	1.9	-	1.9	1.7	1.6	-	1.5
Body Weight*	27 °C	2.0	2.4	2.4	2.5	2.1	1.8	1.9	1.8	1.6	1.6	1.4

*1 Kg of NDT = 4.4 Mcal of digestible energy

Determine if the environmental temperature affects the energy consumption of the two groups of calves.

Solution: The following is the mathematical model used for this system:

$$y = a + bt^c e^{-dt}$$

The results for the first 10 °C group are shown below:

---

[6]Computed from Johnson, H.D.

Table 5.3.1

Source	DF	Sum of Squares	Mean Square
Regression	4	44.31358	11.07840
Residual	5	0.14642	0.02928
Uncorrected Total	9	44.46000	
(Corrected Total)	8	2.64222	

R squared = 1 - Residual SS / Corrected SS = 0.94459

Parameter	Estimate	Asymptotic Std. Error	"t"
a	1.512640404	0.153029984	9.89
b	1.463399174	0.326449846	4.49
c	2.186617924	0.701647784	3.12
d	0.780566265	0.244944768	3.19

The results for the 27 $^{\circ}$C group are as follows:

Table 5.3.2

Source	DF	Sum of Squares	Mean Square
Regression	4	43.24803	10.81201
Residual	7	0.10197	0.01457
Uncorrected Total	11	43.35000	
(Corrected Total)	10	1.32727	

R squared = 1 - Residual SS / Corrected SS = 0.92317

Parameter	Estimate	Asymptotic Std. Error	"t"
a	1.201402929	0.360151004	3.34
b	1.098037273	0.374336172	2.94
c	1.046353355	0.573048232	1.83
d	0.337156859	0.182164352	1.85

Thus, the resulting equations for the 10 $^{\circ}$C and the 27 $^{\circ}$C groups are

$$y_1 = 1.52 + 1.46\, t^{2.19}\, e^{-0.781t}$$
$$y_2 = 1.20 + 1.10\, t^{1.05}\, e^{-0.337t}$$

The graph of the above equations is shown in Fig. 5.3.1:

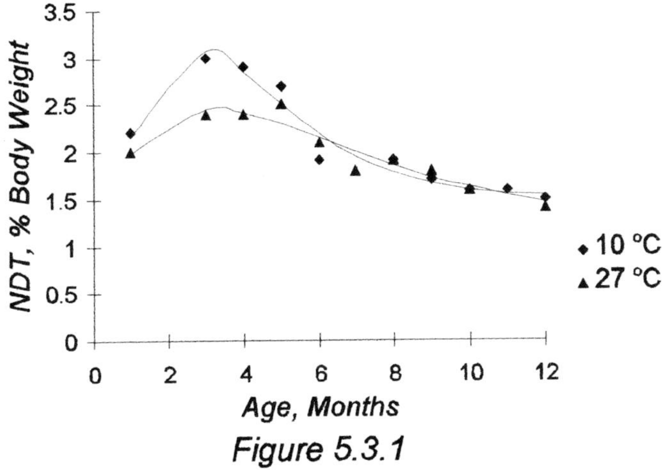

Figure 5.3.1

The two energy consumption curves in Fig 5.3.1 may look different. However, as shown by the "t" tests, the only significant difference between the curves relates to coefficient *d*, with a mild statistical significance:

$$t_d = \frac{0.7806 - 0.3372}{\sqrt{(0.2449)^2 + (0.1882)^2}} = 1.436 \quad ; \quad P<0.20$$

The degrees of freedom are here 9+11-8=12. Coefficient *d* is related to the rate of energy consumption.

Example 5.3.2 The following is the energy consumption of a group of Jersey, Holstein and Brown Swiss calves, grown in a heat chamber at 27 °C of environmental temperature[7].

Age, Months		1	3	4	5	6	7	8	9	10	11	12
NDT, %	**Jersey**	2.0	2.4	2.4	2.5	2.1	1.8	1.9	1.8	1.6	1.6	1.4
Body Weight*	**Holstein**	1.8	2.1	2.2	2.0	-	1.7	1.7	1.6	1.4	1.4	1.4
	Swiss	-	2.1	2.3	2.3	2.0	1.8	1.8	1.7	1.5	-	1.4

*1 Kg of NDT = 4.4 Mcal of digestible energy

---

[7]Computed from Johnson, H.D.

Determine if the breed of the calves affects energy consumption.

Solution: The following mathematical model was used in the required evaluations:

$$y = a + bt^{c}e^{-dt}$$

The results for the Jersey group are shown below:

Table 5.3.3

Source	DF	Sum of Squares	Mean Square
Regression	4	43.24803	10.81201
Residual	7	0.10197	0.01457
Uncorrected Total	11	43.35000	
(Corrected Total)	10	1.32727	

R squared = 1 - Residual SS / Corrected SS = 0.92317

Parameter	Estimate	Asymptotic Std. Error	"t"
a	1.201405070	0.360131236	3.34
b	1.098034810	0.374320474	2.94
c	1.046357260	0.573059581	1.83
d	0.337158043	0.182166716	1.85

The following are the results for the Holstein group:

Table 5.3.4

Source	DF	Sum of Squares	Mean Square
Regression	4	30.68319	7.67080
Residual	6	0.02681	4.468262E-03
Uncorrected Total	10	30.71000	
(Corrected Total)	9	0.78100	

R squared = 1 - Residual SS / Corrected SS = 0.96567

Parameter	Estimate	Asymptotic Std. Error	"t"
a	1.219996492	0.156572078	7.77
b	0.836027400	0.164482133	5.10
c	1.166526980	0.403027490	2.90
d	0.386706189	0.132456924	2.93

The following results were obtained for the Brown Swiss group:

Table 5.3.5

Source	DF	Sum of Squares	Mean Square
Regression	4	32.53502	8.13376
Residual	5	0.03498	6.995549E-03
Uncorrected Total	9	32.57000	
(Corrected Total)	8	0.83556	

R squared = 1 - Residual SS / Corrected SS = 0.95814

Parameter	Estimate	Asymptotic Std. Error	"t"
a	1.392388259	0.113864207	12.21
b	0.172633010	0.151988348	1.14
c	3.748846679	1.531365765	2.45
d	0.890486073	0.336431886	2.65

Thus, the following set of equations represents the three component systems:

$$y_1 = 1.20 + 1.10t^{1.05}e^{-0.337t}$$
$$y_2 = 1.22 + 0.836t^{1.17}e^{-0.387t}$$
$$y_3 = 1.39 + 0.172t^{3.75}e^{-0.890t}$$

As shown by the "t" tests, no significant differences were found between the Jersey and the Holstein calves. By conventional criteria, some significant differences were found between Brown Swiss and the other two groups:

$$t_{(b_1-b_3)} = 2.291 \; ; \; P \leq 0.05$$
$$t_{(b_2-b_3)} = 2.962 \; ; \; P \leq 0.025$$

Thus, $b_3 < b_2 = b_1$. Some mild statistical differences between Brown Swiss and the other two groups, also exist in the c and d coefficients at the $P \leq 0.20$ tolerance level.

Predictive Value and Accuracy of the Mathematical Model

The "t" test for the constant coefficients is valid only as a criterion for accepting or rejecting parameters of the mathematical model of the system. This "t" test provides no information regarding the predictive value and accuracy of the mathematical model. For such, the *coefficient of determination* and the *standard deviation from regression* are required.

The coefficient of determination R^2 is the square of the correlation coefficient and tells how much the mathematical model affects the total variability, that is

$$R^2 = \frac{\sum \hat{y}^2}{\sum y^2}$$

where $\sum \hat{y}^2$ is the sum of squares attributable to regression and $\sum y^2$ is sum of squares of the total. Thus, as with the constant coefficients, the mathematical model of the system also affects the R^2 coefficient.

The coefficient of determination is tested by means of the multiple correlation coefficient. When highly inter-correlated variables are included in the equation, the R^2 value may be significant, while the constant coefficients may not. High correlations between independent variables inflate the variances of the estimates, making individual coefficients unreliable.

A question arises on whether a change in R^2 resulting from a change in the mathematical model is significant. The change in the coefficient of determination is defined as follows:

$$R^2_{change} = R^2 - {}^*R^2$$

where ${}^*R^2$ is the coefficient of determination when one or more variables are excluded from the equation. Then, the null hypothesis that $R^2_{change} = 0$ is verified by the following F test:

$$F_{change} = \frac{(n-p-1)R^2_{change}}{q(1-R^2)}$$

where n is the total number of cases in the equation, p is the number of variables related to R^2 and q number of variables related to ${}^*R^2$. Then, n-p-1 is degrees of freedom for the residual related to R^2. Note that, in nonlinear regression, the residual degrees of freedom related to R^2 is written as n-p, where p is the number of parameters in the equation. The statistical significance of F_{change} is obtained from the F distribution, with q and n-p-1 degrees of freedom.

Example 5.3.3 The following are the differential equations proposed for the lactation curve of a group of cows:

$$\frac{dy}{dt} = \left(\frac{a}{t} - b\right)y + c$$

$$\frac{dy}{dt} = ae^{-bt} - by$$

where y is milk production, Kg/month and t is months. The statistical results for the first equation are as follows:

Table 5.3.6

Source	DF	Sum of Squares	Mean Square
Regression	3	22928.34409	7642.78136
Residual	7	1321.65591	188.80799
Uncorrected Total	10	24250.00000	
(Corrected Total)	9	10560.00000	

R squared = 1 - Residual SS / Corrected SS = 0.87484

Parameter	Estimate	Asymptotic Std. Error	"t"
a	0.174882315	0.025002471	6.99
a	0.206780539	0.042671018	4.84
c	-15.78533600	9.278828992	1.70

The following are the statistics of the second equation:

Table 5.3.7

Source	DF	Sum of Squares	Mean Square
Regression	2	23839.95913	11919.97956
Residual	8	410.04087	51.25511
Uncorrected Total	10	24250.00000	
(Corrected Total)	9	10560.00000	

R squared = 1 - Residual SS / Corrected SS = 0.96117

Parameter	Estimate	Asymptotic Std. Error	"t"
a	251.68452603	12.243564107	21.37
b	0.416681056	0.010715303	38.94

Determine if the coefficients of determination of the two equations are statistically different.

Solution: The following is the F_{change} value for the above coefficients of determination:

$$F_{change} = \frac{R^2_{change}(n-p)}{q(1-R^2)} = \frac{(0.96117-0.87484)(10-2-1)}{3(1-0.96117)} = 5.188 \qquad P \leq 0.05$$

The hypothesis that $R^2_{change} = 0$ is equivalent to the hypothesis that non significant parameters are also zero. Then, deleting non significant parameters should not affect significantly the coefficient of determination. This statement is shown in the following example:

Example 5.3.4 The following differential equation is proposed for the lactation curve of the group of cows in Example 5.2.2:

$$\frac{dy}{dt} = ae^{-bt} + ct + dy$$

where y is milk production and t is months. Determine the effect of deleting non significant parameters on the R^2 value.

Solution: The following statistical results fit the data:

Table 5.3.8

Source	DF	Sum of Squares	Mean Square
Regression	4	26736.10290	6684.02572
Residual	6	213.89710	35.64952
Uncorrected Total	10	26950.00000	
(Corrected Total)	9	15390.00000	

R squared = 1 - Residual SS / Corrected SS = 0.98610

Parameter	Estimate	Asymptotic Std. Error	"t"
a	327.86977097	49.692096538	6.60
b	0.330155226	0.083324925	3.96
c	-1.067232252	1.416257767	0.75
d	-0.547996260	0.125794564	4.35

Clearly, parameter c is not significant. A new round with parameter c deleted shows the following results:

Table 5.3.9

Source	DF	Sum of Squares	Mean Square
Regression	3	26705.71069	8901.90356
Residual	7	244.28931	34.89847
Uncorrected Total	10	26950.00000	
(Corrected Total)	9	15390.00000	

R squared = 1 - Residual SS / Corrected SS = 0.98413

Parameter	Estimate	Asymptotic Std. Error	"t"
a	287.76427522	59.496146589	4.84
b	0.440587310	0.193108177	2.28
d	-0.432546460	0.170836477	2.53

Note that the coefficients of determination in the two statistical evaluations are almost identical. Note also that coefficients b and d are also similar. Thus, the final equation is

$$\frac{dy}{dt} + 0.433\hat{y} = 288e^{-0.441t}$$

The R^2 coefficient was defined as a measure of goodness of the mathematical model for fitting the data. The opposite criterion to the goodness of fit is the sum of squares of the deviations from regression $\sum d^2$, defined by the following expression:

$$\sum d^2 = \sum y^2 - \sum \hat{y}^2$$

that is, the sum of squares of the deviations from regression is the sum of squares of the total $\sum y^2$ minus de sum of squares due to regression $\sum \hat{y}^2$. The sum of squares of the deviations from regression is the basis for estimating the standard deviation from regression. The standard deviation from regression is defined as follows:

$$S_{y.t} = \sqrt{\frac{\sum d^2}{n-p-1}}$$

The $S_{y.t}$ value is an estimate of the failure of the mathematical model in fitting the data.

Example 5.3.5 Equation $y = 40.7 + 27.7\cos[0.691(t-6.16)]$ was fitted to the pasture production data of Example 5.2.4. The predictive value of this equation is $R^2 = 0.910$. The lack of fit, for a standard deviation of $S_{y.t} = 7.64$, is shown in Fig. 5.3.2:

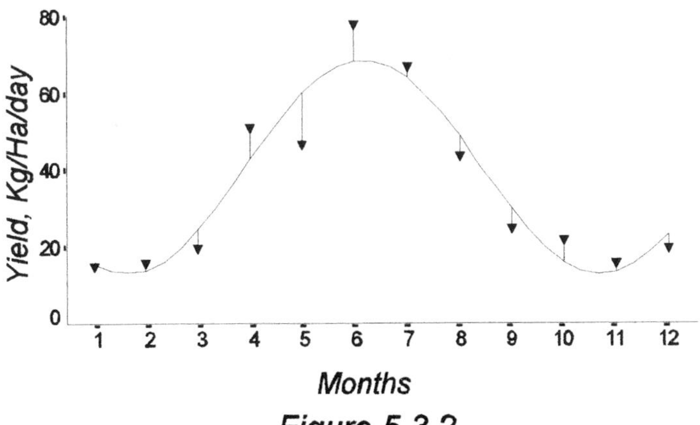

Figure 5.3.2

The standard deviation estimates de accuracy of the mathematical model for fitting the data, whereas R^2 estimates the predictive value of the model. It is expected that a smaller $S_{y.t}$ corresponds to a larger R^2, if the non significant coefficients are deleted. This statement is illustrated in the next example.

Example 5.3.6 The following mathematical models were fitted to the rumen concentration of ammonia in lambs fed a diet containing urea[8]:

$$y = =(a + bt)e^{-ct}$$
$$y = a + bte^{-ct}$$

where y is rumen ammonia, mMoles/liter and t is hours after feeding. Determine the numerical values of the constant coefficients and the statistical parameters for the two mathematical models.

Solution: The following are the results for the first model:

Table 5.3.10

Source	DF	Sum of Squares	Mean Square
Regression	3	5824.71512	1941.57171
Residual	6	105.07488	17.51248
Uncorrected Total	9	5929.79000	
(Corrected Total)	8	1043.78000	

R squared = 1 - Residual SS / Corrected SS = 0.89933

Parameter	Estimate	Asymptotic Std. Error	"t"
a	7.829388620	4.020972876	1.958
b	34.543303266	5.458689171	6.326
c	0.395602703	0.037130787	10.663

Asymptotic Correlation Matrix of the Parameter Estimates

	a	b	c
a	1.0000	-0.6144	-0.3656
b	-0.6144	1.0000	0.8650
c	-0.3656	0.8650	1.0000

The corresponding state equation for model one is here:

---

[8]Computed from Streeter, C.L. et.al.

$$y = (7.83 + 34.54t)e^{-0.397t}$$

The results for the second model are as follows:

Table 5.3.11

Source	DF	Sum of Squares	Mean Square
Regression	3	5855.48225	1951.82742
Residual	6	74.30775	12.38462
Uncorrected Total	9	5929.79000	
(Corrected Total)	8	1043.78000	

R squared = 1 - Residual SS / Corrected SS = 0.92881

Parameter	Estimate	Asymptotic Std. Error	"t"
a	7.469339752	2.550686920	2.929
b	37.522345821	4.972439824	7.549
c	0.497918668	0.053916065	9.237

Asymptotic Correlation Matrix of the Parameter Estimates

	a	b	c
a	1.0000	-0.3140	0.5444
b	-0.3140	1.0000	0.5221
c	0.5444	0.5221	1.0000

The state equation for model two is

$$y = 7.47 + 37.52te^{-0.498t}$$

A summary for the above statistics is shown in the following table:

Table 5.3.12

State Equation	R^2	$S_{y.t}$
$y = (7.83 + 34.54t)e^{-0.396t}$	0.899	4.18
$y = 7.47 + 37.5te^{-0.498t}$	0.929	3.52

Note that the second equation has a larger R^2, a smaller standard deviation and more reliable constant coefficients. Note also that the correlations between the parameter estimates of the second model are smaller that the correlations in the first model.

In conclusion, in selecting a mathematical model, the research team should look at the following statistics. It is expected also that results should agree with the experimental hypothesis.

- Reliability of the constant coefficients
- The coefficient of determination
- The standard deviation from regression
- The correlation matrix of the parameter estimates

Summary

The main criteria for accepting or rejecting the experimental hypothesis are the "t" tests for the constant coefficients of the mathematical model of the system. However, the predictive value and accuracy of the model are estimated from the coefficient of determination and the standard deviation from regression. Sometimes, the correlation matrix of the parameter estimates may also be included in the evaluation of the mathematical model.

6

FRAMEWORK FOR MODELING AGRICULTURAL SYSTEMS

A model is a characterization of a real system. It may take the form of a drawing, a simple written verbal description or may be a complicated set of equations to be used in the simulation of the system.

This chapter is an extension of Chapter 1. Is a conceptual overview of the modeling process, as is further developed for specific applications in chapters 7, 8 and 9. For such, simple examples are introduced and developed for explaining general modeling principles.

6.1 THE SYSTEM VARIABLES

As disclosed previously, the following variables are required for defining the mathematical model of a system:

- The time scale adopted for the system
- Input variables
- State variables
- Output variables

The Time Scale

Time is a continuous process. However, when sampling takes place at fixed intervals of time, then a discrete signal is generated. Depending on the time scale adopted for the model, system models are grouped into two categories:

- Continuous systems
- Discrete systems

Actually, any system is neither continuous nor discrete, they are simply systems. It is in the modeling process, according to the human interpretation of the system, that systems are given specific definitions and features.

For continuous systems, the time scale is the set of all nonnegative real numbers. Continuous systems are frequently called differentiable systems when they are represented by differential equations and their solutions.

When the state variables can be accepted as discrete, adopting a discrete model for such system may be appropriate. This may be the case of state variables defined as number of individuals or as qualitative traits. For discrete systems the time scale is the set of all nonnegative integers. Discrete systems are not differentiable, because the state variables are discrete. Discrete systems are sometimes represented by difference equations and their solutions.

Input Variables

An *input* is anything admitted into the system, either in physical terms or as information. Any agricultural system is bombarded by different kinds of inputs, most of them not explicitly related to the research problem. Some inputs can be manipulated, for example the application of fertilizers. In agricultural research, manipulation of inputs may determine experimental treatments and designs. Most inputs, however, are not subject to manipulation, like the weather factors and may add uncertainties for the modeler with respect to the response of the system.

An input variable is named here x_i. A set of input variables determines a Cartesian product X, such that

$$X = X_1 \times ... \times X_l = \left\{ x = (x_1 ... x_l); x_i \in X_i \right\}$$

where the l-tuple x is an input, X_i is the range of input variable x_i, and $i = 1,2,...,l$ is the identification of the system input variables. Each X_i is also called an *input port*[1].

Example 6.1.1 The following levels of fertilizer are applied to a pasture experimental field to test the pasture response to sodium nitrate and to superphosphate:

$$X_1 = \{0, 300, 600\}; X_2 = \{0, 200\}$$

where X_1 is sodium nitrate and X_2 is superphosphate, in kilograms per hectare. Define the set of inputs of the system.

Solution: The set of inputs of the system is determined by the following Cartesian product:

$$X = X_1 \times X_2 = \{(0,0),(0,300),(0,600),(200,0),(200,300),(200,600)\}$$

---

[1] Waymore, A.W.

where each of the six ordered pairs in the above product is an input.

Inputs organized over time are called *input trajectories*. Thus, an input trajectory f has the following form:

$$f = \left\{(t,x): t \in \tau; x \in X; X = X_1 \times ... \times X_I; x = (x_1,...,x_I); x_i \in X_i\right\}$$

for continuous systems and

$$f = \left\{(n,x): n \in N; x \in X; X = X_1 \times ... \times X_I; x = (x_1,...,x_I); x_i \in X_i\right\}$$

for discrete systems, where τ and N are the continuous and the discrete time scales.

Example 6.1.2 Define an input trajectory for treatment $x_2 = (0,3000)$ for the experiment in Example 6.1.1, assuming applications of the fertilizers every three months, during a full year. Define also a fertilization program, where no fertilizer is applied in winter, that is treatment $x_1 = (0,0)$ and maximum levels are applied during summer, that is treatment $x_6 = (200,600)$.

Solution: The requested trajectories are shown in the following table:

Table 6.1.1

Months	Trajectories	
	f_1	f_2
0	x_2	x_1
3	x_2	x_2
6	x_2	x_6
9	x_2	x_5
12	x_2	x_1

Note that in agricultural experiments inputs subject to manipulation are usually held at constant values over time, as is the case of trajectory f_1 in the above table. Note also that the input trajectories in the table correspond to inputs that can be manipulated.

The following example corresponds to inputs that cannot be manipulated.

Example 6.1.3 The following are average temperatures and rainfall data for a country region in Panama:

Month	J	F	M	A	M	J	J	A	S	O	N	D
Temperature*	26	27	26	25	24	24	23	23	23	24	24	25
Rainfall, mm	18	50	70	27	8	2	1	0	0	0	0	3

*Centigrade

Each ordered pair $x =$ (temperature, rainfall) in the above table is an input of the system and the table specifies an input trajectory.

Note that the Cartesian product X holds true for both, discrete and for continuos models, because all combinations of values within the range of the input variables are possible. However, what actually defines the system response is the input trajectories accepted by the system.

State Variables

The notion of a state is related to what is going on inside the boundaries of the system. The state of the system is a static condition that can be determined by many variables. Often many of these variables may not be even related to the research problem. State variables may be either quantitative, like the weight of a cow or qualitative, like the color of a cow.

A state variable is named here y_i. Then, the state of the system is represented by the Cartesian product Y, such that

$$Y = Y_1 \times \cdots \times Y_m = \left\{ y = (y_1 \ldots y_m) : y_i \in Y_i \right\}$$

where the n-tuple y is a state, Y_i is the range of state variable y_i and $i = 1,2,\ldots,m$ are labels of the state variables.

Due to the randomness of most inputs, the operation of all agricultural systems must be considered subject to some kind of uncertainty. Depending on wether uncertainties are taken into consideration or are ignored, models of systems are assigned to two categories:

- Stochastic models
- Deterministic models

In stochastic models, the states of the system are defined as probability distributions. In deterministic models, the states are defined as the expected value of the outcomes.

Example 6.1.4 The trees of a citrus plantation are classified by size as small (s), medium (m) and large (l) and by health as healthy (h) or diseased (d). Define the set of states of the system.

Solution: The state of this system is represented by two variables, namely size s and health h. Thus, the set of states of the system is represented by the following product:

$$Y = \{(s,h), (s,d), (m,h), (m,d), (l,h), (l,d)\}$$

States organized over time are the *state trajectories* of the system. Thus, a state trajectory is any function defined over the time scale with values in the set of all states of the system. Thus, a state trajectory g is defined in the following form:

$$g = \left\{(t,y) : t \in \tau; y \in Y; Y = Y_1 \times ... \times Y_m; y = (y_1, ... , y_m); y_i \in Y_i\right\}$$

for continuous systems and

$$g = \left\{(n,y) : n \in N; y \in Y; Y = Y_1 \times ... \times Y_m; y = (y_1, ... , y_m); y_i \in Y_i\right\}$$

for discrete systems, where τ and N are the continuous and discrete time scales and y_i is a state variable.

Example 6.1.5 The following fitted equation represents the growth curve of a group of steers[2]. This equation represents the body weight trajectory and is shown in Fig. 6.1.1.

$$y = 780 + 265 e^{-1.427t} - 1015 e^{0.553t}$$

Here y is the state variable of the system as the body weight of the steers in Kg and t is the age of the steers in years.

---

[2]Vohnout. K., Unpublished

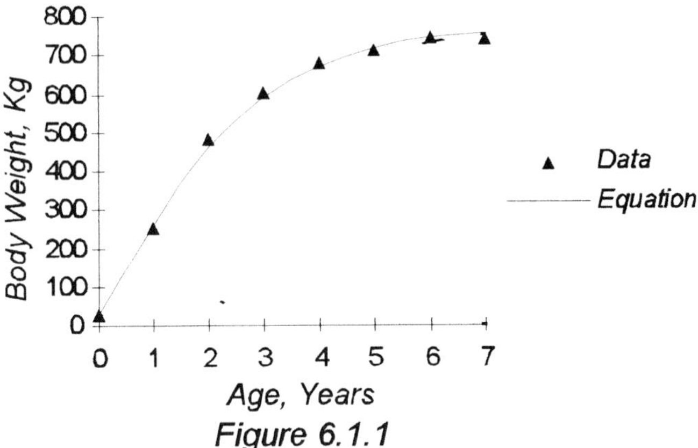

Figure 6.1.1

Output Variables

An output is anything produced by the system across its boundaries. As with inputs, outputs can be either in a physical form or as information packages. Oranges produced by the citrus plantation in Example 6.1.4 are a physical output of the system, while knowledge of the state of the trees is an information output. Many outputs of a system may not be related to the research problem. Therefore, defining such outputs is not necessary. This is the case in Example 6.1.4, where the problem is only related to the size and health of the trees. Then, defining oranges as an output of the system is not necessary, unless the problem is defined in terms of production of oranges in relation to the size and health of the trees. Furthermore, being explicit in defining the size and health of the trees is an information output of the system. In this case, the state is also the output of the system.

An output variable is named here z_i. Then, the output of the system is denoted here by the Cartesian product Z, where the n-tuple z is an output, Z_i is the range of variable z_i and $i = 1,2,...,n$ are labels of the output variables. Each Z_i is also called an *output port*[3]:

$$Z = Z_1 \times ... \times Z_n = \left\{ z = (z_1,...,z_n); z_i \in Z_i \right\}$$

Example 6.1.6 Production of a pasture field is determined in terms of grass and in terms of milk. Define the output of the system.

---

[3]Waymore, A.W.

Solution: The output of the system is represented by two output variables, where z_1 and z_2 stands for grass and for milk, such that

$$Z = Z_1 \times Z_2 = \left\{ z = (z_1, z_2) ; z_1 \in Z_1 ; z_2 \in Z_2 \right\}$$

Outputs organized over time are the *output trajectories* of the system. Then, an output trajectory *h* is defined as follows. For continuous systems

$$h = \left\{ (t,z) : t \in \tau ; z \in Z ; Z = Z_1 \times ... \times Z_n ; z = (z_1, ..., z_n) ; z_i \in Z_i \right\}$$

and for discrete systems, where τ and N are the continuous and the discrete time scales.

$$h = \left\{ (n,z) : n \in N ; z \in Z ; Z = Z_1 \times ... \times Z_n ; z = (z_1, ..., z_n) ; z_i \in Z_i \right\}$$

Example 6.1.7 The following is the equation fitted to the lactation curve of a group of dairy cows[4]:

$$z = e^{-0.484t}(298 + 411t)$$

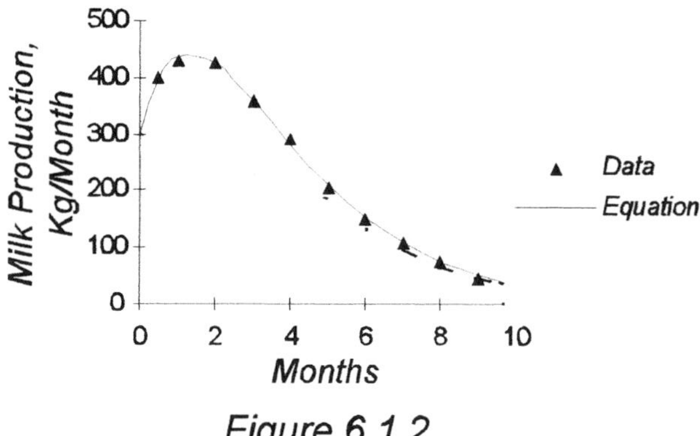

Figure 6.1.2

---

[4]Vohnout, K., Unpublished

This lactation curve is also shown in Fig. 6.1.2, where the output variable of the system is z, as milk production in kilograms/month and t is time in months.

The following is an outline of the system related variables:

• Time scale for continuous systems

$$\tau = \{t: t \text{ is a real number}; 0 \leq t < \infty\}$$

• Time scale for discrete systems

$$N = \{n:n \text{ is an integer}; 0 \leq n < \infty\}$$

• Set of inputs

$$X = X_1 \times ... \times X_l = \left\{x = (x_1 .... x_l): x_i \in X_i\right\}$$

• Set of states

$$Y = Y_1 \times ... \times Y_m = \left\{y = (y_1, ..., y_m); y_i \in Y_i\right\}$$

• Set of outputs

$$Z = Z_1 \times ... \times Z_n = \left\{(z = z_1, ..., z_n); z_i \in Z_i\right\}$$

where X_i and Z_i are input and output ports.

Summary

An input is anything admitted to the system, either as physical objects or as information packages. Depending on the time scale adopted, systems are classified as continuous or discrete. Depending on whether uncertainties in the admission of inputs are considered or ignored, systems are classified as stochastic or deterministic. States are traits that characterize the system and outputs are anything produced by the system, either in a physical form or as information packages, as a function of the state.

6.2 SYSTEM DYNAMICS

The notion of a state is related only to an instantaneous or static condition of the system. As disclosed before, the dynamic condition is represented by the state transition function. The state transition function represents the changes in the state of the system over time, as determined by the initial state and by inputs. It was also disclosed that the output depends only on the state of the system. These statements are discussed in more detail in this section.

The State Transition Function

The state transition of a continuous system is determined usually by a derivative function μ that depends only on a state $y = g(t)$ and an input $x = f(t)$, such that

$$\frac{dg(t)}{d\tau} = \mu(g(t), f(t))$$

A state trajectory of the system is the solution μ of the above differential equation for a given initial state $y_0 = g(0)$. Thus, if the system is started at a state y_0, is supplied by an input trajectory f and is run to some time t, then

$$y = u(y_0, f, t)$$

Clearly, given the initial conditions, a continuous system is completely determined by a differential equation or a set of interconnected differential equations.

The next state function v of discrete systems is equivalent to the derivative function μ of continuous systems. Thus, a state y_{n+1} at the discrete time $n+1$ is completely determined by the state y_n and the input x_n at time n. Then

$$y_{n+1} = v(y_n, x_n)$$

Therefore, given the initial conditions, a discrete system is completely determined by a difference equation or a set of interconnected difference equations.

From the above, it is clear that the state transition function represents the dynamic behavior of the system. The state transition function may be defined by a graph, a table, or by mathematical expressions.

Example 6.2.1 The movement of DDT from plant to soil is 25% per month, from soil to plant is 2% and carried out with ground water is 5%. Define the state transition function and the set of state trajectories of the system.

Solution: The dynamics of the system is shown in Fig. 6.2.1. This is a two-compartment[5] open system and Fig. 6.2.1 symbolizes a state transition function. It is a continuous system because there is a continuous flow of DDT within the system and between the system and the outside environment.

---

[5]The term "compartment" is widely used in tracer kinetics and was accepted and adopted for this book

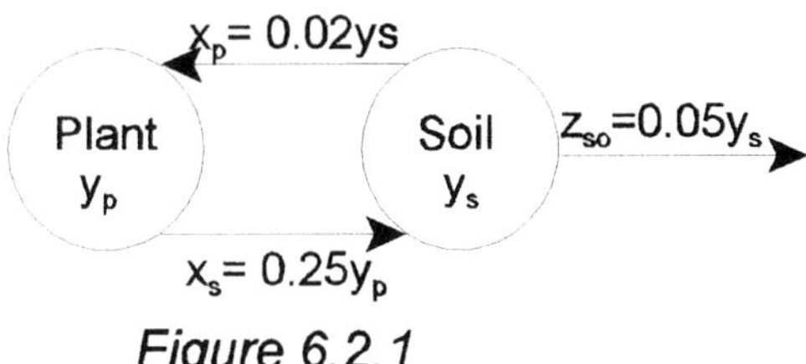

Figure 6.2.1

The system is completely determined by a set of initial conditions $Y_0 = (0.6, 0.4)$ and by the following set of differential equations:

$$\frac{dY}{dt} = AY = \begin{bmatrix} -0.25 & 0.02 \\ 0.25 & -(0.02+0.05) \end{bmatrix} Y$$

for $Y = (y_p, y_s)$, where y_p is the plant concentration of DDT, y_s is the soil concentration of DDT, t is months, $x_{sp} = 0.02y_s$ is the input to the plant compartment, $x_{ps} = 0.25y_p$ is the input to the soil compartment, $z_{ps} = 0.25y_p$ is the output of the plant compartment to the soil compartment, $z_{sp} = 0.02y_s$ is the output of the soil compartment to the plant compartment and $z_{so} = 0.05y_s$ is the output of the soil compartment to the outside. Note that coefficients with a positive sign are inputs and coefficients with negative signs are outputs. The Laplace transform of this set of differential equations is given by the expression $(sI - A)G(s) = Y_0$, where

$$|sI - A| = \begin{vmatrix} s+0.25 & -0.02 \\ -0.25 & s+0.07 \end{vmatrix} = (s + 0.0455)(s + 0.2745)$$

is the characteristic equation of the system. Then

$$G_1(s) = \frac{1}{(s+0.0455)(s+0.2745)} \begin{vmatrix} 0.6 & -0.02 \\ 0.4 & s+0.07 \end{vmatrix}$$

$$G_2(s) = \frac{1}{(s+0.0455)(s+0.2745)} \begin{vmatrix} s+0.25 & 0.6 \\ -0.25 & 0.4 \end{vmatrix}$$

The reader is encouraged to check that the solution of the above transforms is the following set of state trajectories:

$$Y = \begin{bmatrix} 0.0990 & 0.5010 \\ 1.0122 & -0.6122 \end{bmatrix} \begin{bmatrix} e^{-0.0455t} \\ e^{-0.2745t} \end{bmatrix}$$

These trajectories are shown in Fig. 6.2.2. Since this system has an output to the outside environment but does not have inputs from outside, DDT values will approach zero as the time variable gets very large.

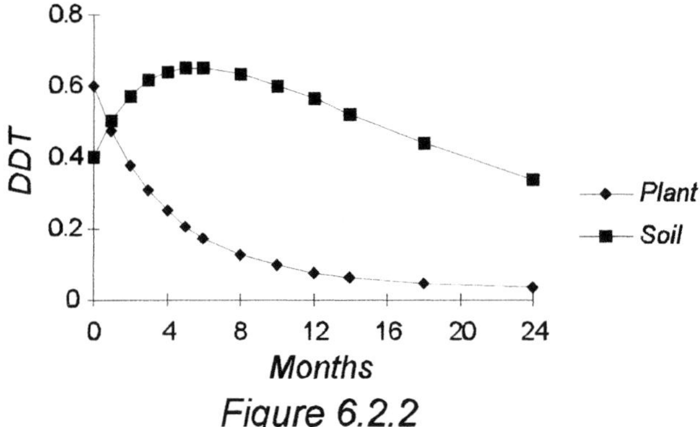

Figure 6.2.2

Example 6.2.2 A rancher sells each month 3.6% of his feedlot steers and buys 90 new animals. The initial number of steers is 460. Define the next state function and the state trajectory of the system.

Solution: This system is discrete because the state variable steers is discrete. The system is depicted in Fig 6.2.3:

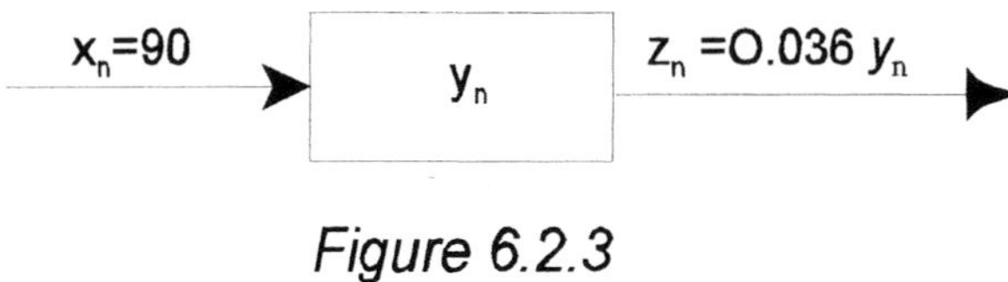

Figure 6.2.3

The following is the corresponding difference equation, where y_n is the present state of the system, y_{n+1} is the next state, $x = 90$ is the input and $z_n = 0.036y_n$ is the output:

$$y_{n+1} - y_n = 90 - 0.036y_n$$

The above equation may also be written in the next state form $y_{n+1} = 90 + 0.964y_n$. Then, the Z transform of this difference equation is

$$G(z) = \frac{90z}{(z-1)(z-0.964)} + \frac{460z}{z-0.964}$$

The following state trajectory is the corresponding inverse

$$y_n = 2500 - 2040(0.964)^n$$

also shown in Fig. 6.2.4:

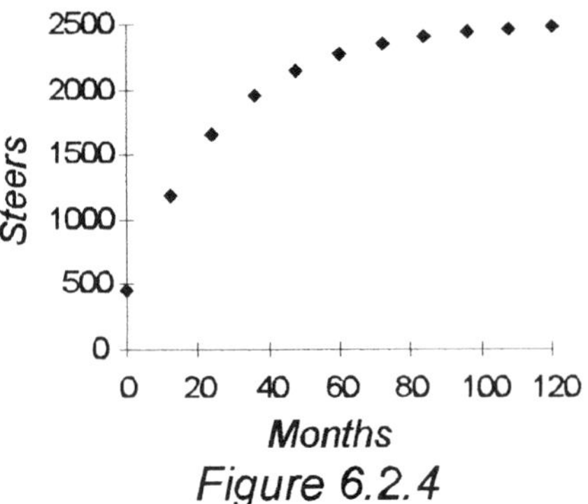

Figure 6.2.4

Example 6.2.3 A forest area is chopped down and burned. After the first year, 20% of the burned area is regrown by trees and 30% is colonized by grasses. The remaining area stays as bare soil. Mortality of trees is 15% and mortality of grasses is 25%. Define the next state function and the state trajectories of the system.

Solution: This forest area may be defined as a finite discrete system with three state variables, the bare soil state variable, the grass state variable and the trees state variable. The next state function of the system is shown in Fig. 6.2.5:

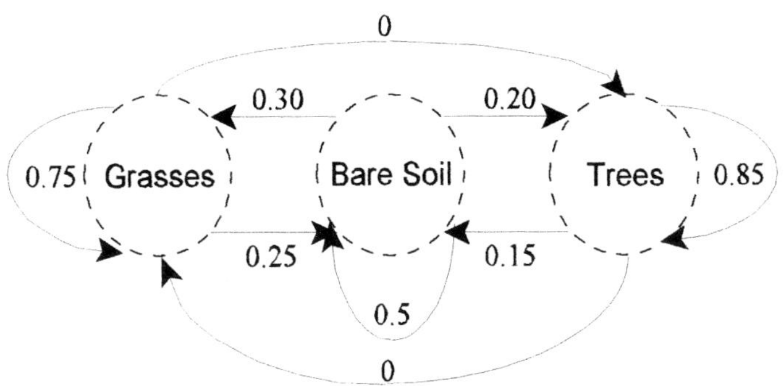

Figure 6.2.5

The corresponding state transition matrix is shown in the following table:

Table 6.2.1

Present State	Next State		
	Bare Soil	**Grasses**	**Trees**
Bare Soil	0.50	0.30	0.20
Grasses	0.25	0.75	0
Trees	0.15	0	0.85

The above table shows that 50% of bare soil may remain as bare soil in the next state, 30% may become grasses and 20% may become regrowth of trees. It also shows that 25% of the grasses may die out, reverting to bare soil and that 75% may remain as grasses. Fifteen percent of the trees may die and revert to bare soil and 85% may stay alive.

The system is represented by the following set of next state equations:

$$Y_{n+1} = \begin{bmatrix} 0.50 & 0.30 & 0.20 \\ 0.25 & 0.75 & 0 \\ 0.15 & 0 & 0.85 \end{bmatrix} Y_n$$

Then, the corresponding Z transform and its inverse are

$$G(z) \;=\; G(0)\frac{z}{z-Q}$$

$$Y_n \;=\; G(0)Q^n$$

where $G(0)$ is the set of initial states of the system and Q is the state transition matrix. The set of initial states is here

$$G(0) \;=\; \begin{bmatrix} 1 & 0 & 0 \\ 0 & 1 & 0 \\ 0 & 0 & 1 \end{bmatrix}$$

Note that the first row of the above matrix shows that the initial state is bare soil. Then, the following is the set of state trajectories of the system:

$$Y_n \;=\; \begin{bmatrix} 1 & 0 & 0 \\ 0 & 1 & 0 \\ 0 & 0 & 1 \end{bmatrix}\begin{bmatrix} 0.50 & 0.30 & 0.20 \\ 0.25 & 0.75 & 0 \\ 0.15 & 0 & 0.85 \end{bmatrix}^n$$

After solving matrix Q^n, the following are the state trajectories when the initial state is bare soil[6]:

$$y_{1n} \;=\; 0.2830 + 0.6937(0.285)^n + 0.0233(0.815)^n$$

$$y_{2n} \;=\; 0.3396 - 0.4479(0.285)^n + 0.1083(0.815)^n$$

$$y_{3n} \;=\; 0.3773 - 0.2457(0.285)^n - 0.1316(0.815)^n$$

Fig. 6.2.6 shows the above state trajectories. Since this is a closed system, states will approach an asymptotic value as time gets larger.

---

[6]The procedure for determining the powers of a matrix will be discussed in the next chapter

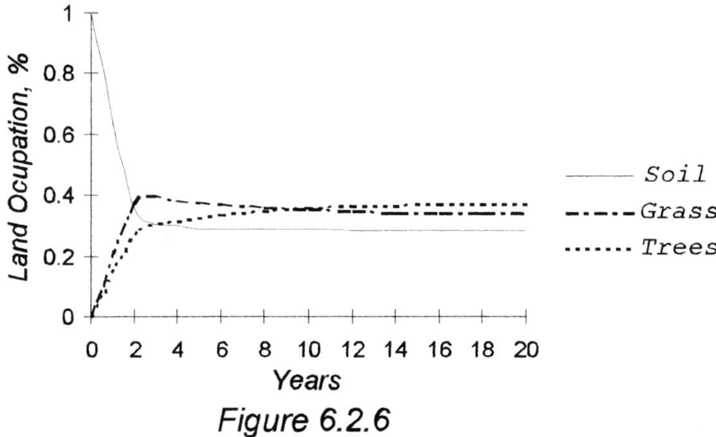

Figure 6.2.6

The Output Function

As specified in the first section of this chapter, the notion of an output function is related to the production of outputs as a response to the state of the system, such that

$$z = w(y)$$

where w is an output function. Then, an output z of the system is completely determined by the state y.

Defining output functions is not always necessary. In addition, when dealing with empirical models, an output function may often have only an abstract meaning.

Example 6.2.4 Define the output function of the DDT system in Example 6.2.1.

Solution: This system was represented by the following set of differential equations:

$$\frac{dY}{dt} = \begin{bmatrix} -0.25 & 0.02 \\ 0.25 & -(0.02+0.05) \end{bmatrix} Y$$

where y_p is DDT concentration in the plant compartment and y_s is the concentration of the insecticide in the soil compartment. Because there are no external inputs defined for the system, the above set of equations is also the output of the system, such that

$$\frac{dZ}{dt} = \begin{bmatrix} -0.25 & 0.02 \\ 0.25 & -0.07 \end{bmatrix} \frac{dY}{dt}$$

Since the state trajectories are known, the following is the set of output trajectories of the system:

$$Z = \begin{bmatrix} -0.25 & 0.02 \\ 0.25 & -0.07 \end{bmatrix} \begin{bmatrix} 0.0990 & 0.5010 \\ 1.0122 & -0.6122 \end{bmatrix} \begin{bmatrix} e^{-0.0455t} \\ e^{-0.2745t} \end{bmatrix}$$

These trajectories may have only a theoretical meaning. More important are here the input-output relationships between compartments. As shown in the graph of Fig. 6.3.7, the outputs of each individual compartment are determined by the coefficients with negative signs, that is $z_{ps} = 0.25y_p$ and $z_s = z_{sp} + z_{so} = 0.07y_s$, where z_{ps} is the output from the plant compartment to the soil compartment, z_{sp} is the output from the soil compartment to the plant compartment, z_{so} is the output from the soil compartment to the outside and $z_s = z_{sp} + z_{so}$ is the total output from the soil compartment. Fig. 6.2.7 represents the output function of the system.

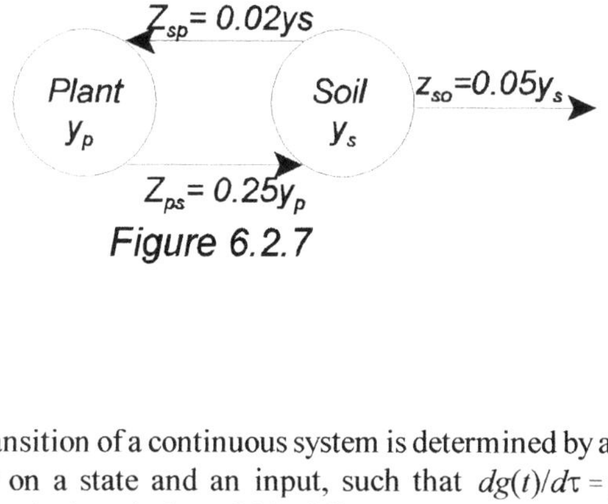

Figure 6.2.7

Summary

The state transition of a continuous system is determined by a derivative function μ that depends only on a state and an input, such that $dg(t)/d\tau = \mu(g(t), f(t))$. A state trajectory of the system is the solution of this differential equation for a given initial state. Thus, if the system is started at a state y_0, is supplied by an input trajectory f and is run to some time t, then $y = u(y_0, f, t)$. The next state function v of discrete systems is equivalent to the derivative function μ of continuous systems. Thus, a state at the discrete time $n+1$ is completely determined by the state and the input at time n such that $y_{n+1} = v(y_n, x_n)$. Given the initial conditions, a system is completely determined by a differential or a difference equation or by a set of interconnected differential or difference equations. An output function *w* relates outputs and states, such that $z = w(y)$, where *z* is the output.

6.3 RESPONSE FUNCTIONS

A continuous linear system was represented, in the first chapter, as a tank with devises for water admission and for water discharge. The change of the water level was defined as the difference between admission and discharge, such that

$$\frac{dy}{dt} = x - by$$

where *x* is the water input, *y* is the height of water, *t* is time and *by* is the water output. Two processes are taking place in the tank. One process is the filling and the other is the emptying of the tank. The emptying process may take place even if the water input is turned off, independently of the water input. In this case, the emptying of the tank is due exclusively to y_0, the height of water at time zero. Conversely, the filling process determines a system response that is due exclusively to the input *x*, independently of the initial conditions. These two processes are called the *free response* and the *forced response*. The portrait of this system for an input of $x = c$ is shown in Fig. 6.3.1.

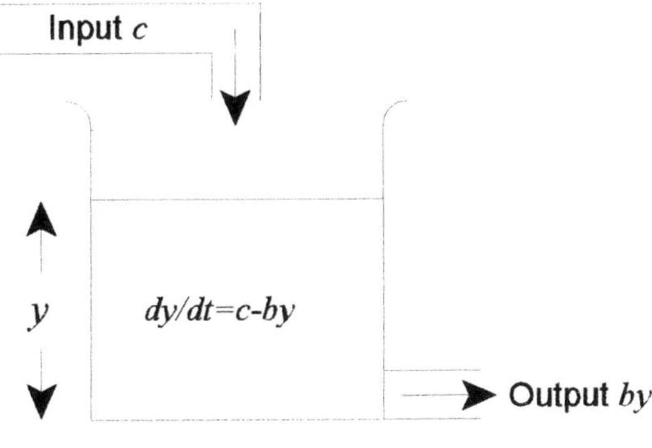

Figure 6.3.1

From the above, the following definition applies for the free response:

Definition 6.3.1 The free response is the system response due only to the initial conditions in the absence of inputs.

The following definition applies for the forced response:

Definition 6.3.2 The forced response is the system response due only to inputs, regardless of the initial condition.

Thus, if the input c is zero, the free response of the system is a homogeneous differential equation:

$$\frac{dy}{dt} + by = 0$$

The free response is always represented by a homogeneous differential equation.

The two responses are clearly seen in the following Laplace transform of the differential equation of the system, for an input $x = c$:

$$G(s) = \frac{y_0}{s+b} + \frac{c}{s(s+b)}$$

The first fraction of the above expression corresponds to the free response and the second to the forced response. Thus, the inverse of the first fraction is the state trajectory of the free response y_A and the inverse of the second fraction is the state trajectory of the forced response y_B:

$$y_A = y_0 e^{-bt}$$
$$y_B = \frac{x}{b}\left(1 - e^{-bt}\right)$$

The sum of the two responses is the *total response* of the system. Thus, the following definition applies:

Definition 6.3.3 The total response of the system is the sum of the free response and the forced response.

Then, the state trajectory of the total response is the sum

$$y = y_A + y_B = y_0 e^{-bt} + \frac{c}{b}\left(1 - e^{-bt}\right)$$
$$= \frac{c}{b} + \left(y_0 - \frac{c}{b}\right)e^{-bt}$$

The *steady state response* and the *transient response* are two other quantities whose sum equals to the total response. Note that, when the output of the system is equal to the input, the system is at a steady state, that is

$$\frac{dy}{dt} = c - by = 0$$

Then, at steady state, the response of the system approaches the constant c as time approaches infinity. Clearly, changing c also changes the steady state of the system. Thus, the following definition applies for steady state:

Definition 6.3.4 A steady state is the response of the system when the input and the output are equal.

The transient response is defined as follows:

Definition 6.3.5 A transient response is the system response when the input and the output are not equal.

Thus

$$\frac{dy}{dt} = x - by \neq 0$$

Example 6.3.1 An individual with an immunodeficiency problem was dosed with 9.9 grams of gamma globulin intravenously. The blood concentration of the patient gamma globulin is described by the following fitted state equation[7]:

$$y = 218 + 245\,e^{-0.0386t}$$

where y is gamma globulin concentration in mg/dl and t is time in days. Define the response functions of the system.

Solution: The following is the differential equation related to the state equation:

$$\frac{dy}{dt} + 0.0386y = 180$$

---

[7]Vohnout, K., Unpublished

The corresponding Laplace transform of the above differential equation is

$$G(s) = \frac{463}{s+0.0386} + \frac{180}{s(s+0.0386)}$$

where 463mg/ml is the blood gamma globulin at zero time and 180mg/day is the patient gamma globulin input. The first fraction represents the free response due to the dose of gamma globulin given to the patient. The second fraction represents the forced response and is attributable to the patient's own gamma globulin contribution. The following state trajectories are the free response y_A and the forced response y_B:

$$y_A = 463e^{-0.0386t}$$
$$y_B = 218\left(1 - e^{-0.0386t}\right)$$

Then, the total response is the sum

$$y = 218 + 245e^{-0.0386t}$$

The steady state response is the asymptotic value 218. The system responses are shown in Fig. 6.3.2.Note that the gamma globulin dosed to the patient approaches zero as time increases. Conversely, the patient's own gamma globulin contribution approaches the asymptotic value of 218 mg/ml.

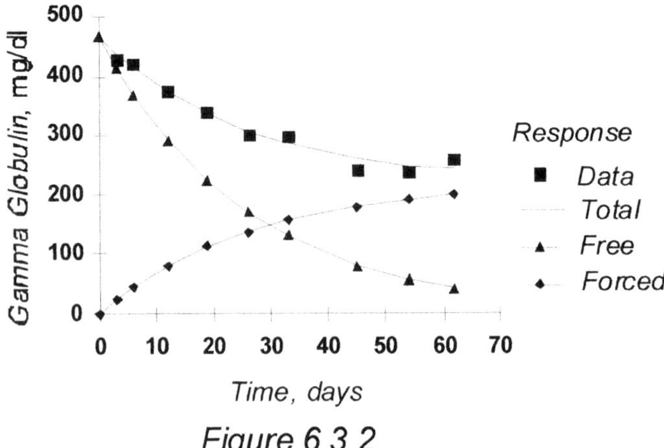

Figure 6.3.2

Example 6.3.2 The following equation was fitted to the microbial digestion of the cell walls of a forage sample[8]:

$$y = 54.0e^{-0.0820t}$$

where y is percent of residual cell walls and t is time in hours. Define the system response functions.

Solution: The following is the differential equation of the system:

$$\frac{dy}{dt} + 0.0820y = 0$$

This is a homogeneous equation and the change of state of the system is determined only by the output $0.0820y$. Therefore, the fitted equation is a free response function depending only on the initial condition 54.0%.

The same principles described for continuous systems apply also for discrete systems. A simple first order system is represented by the following difference equation:

$$\Delta y_n = x_n - by_n$$
$$y_{n+1} = x_n + (1 - b)y_n$$

where x_n is the input and by_n is the output. This difference equation has the following Z transform for a constant input of $x_n = c$:

$$G(z) = \frac{g(0)z}{z-(1-b)} + \frac{cz}{(z-1)[z-(1-b)]}$$

where $g(0)$ represents the initial condition of the system. If the input c is zero, then the following homogeneous difference equation represents the free response of the system:

$$y_{n+1} - (1 - b)y_n = 0$$

The Z transform of this difference equation is

---

[8]Computed from Van Soest, P.J.

$$G_A(z) = \frac{g(0)z}{z-(1-b)}$$

which is the first fraction of the total response transform. When the initial condition $g(0)$ is zero, then the transform of the forced response is

$$G_B(z) = \frac{cz}{(z-1)[z-(1-b)]}$$

which is the second fraction of the transform of the total response. Thus, the following sequences are the free response y_{An} and the forced response y_{Bn} of the system:

$$y_{An} = g(0)(1-b)^n$$

$$y_{Bn} = \frac{c}{b}\left[1 - (1-b)^n\right]$$

The total response is the sum

$$y_n = y_{An} + y_{Bn} = g(0)(1-b)^n + \frac{c}{b}\left[1 - (1-b)^n\right]$$

$$= \frac{c}{b} + \left(f(0) - \frac{c}{b}\right)(1-b)^n$$

Example 6.3.3 Each month 3.6% of farm workers of a county are laid-off or quit and are replaced by 90 newcomers. Define the response functions of the system if the initial number of workers is 460.

Solution: The following is the difference equation representing this system:

$$y_{n+1} - y_n = 90 - 0.036y_n$$

Thus

$$y_{n+1} - 0.964y_n = 90$$

The following is the Z transform of the above equation:

$$G(z) = \frac{460z}{z - 0.964} + \frac{90z}{(z-1)(z-0.964)}$$

where 460 is the initial number of workers and 90 is the monthly input of newcomers. Then, the following solutions are the free response y_{An}, the forced response y_{Bn} and the total response y_n:

$$y_{An} = 460(0.964)^n$$
$$y_{Bn} = 2500\left[1 - (0.964)^n\right]$$
$$y_n = 2500 - 2040(0.964)^n$$

These responses are shown in Fig. 6.3.3. Note that, as time increases, the number of old workers approaches zero and the number of newcomers approaches 2500.

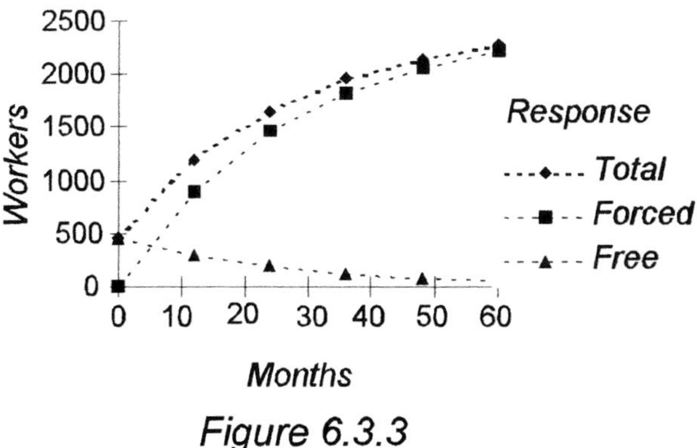

Figure 6.3.3

Summary

The response of a linear system is represented by two types of functions, the free response and the forced response. The free response is the reaction of the system to initial conditions in the absence of inputs and the forced response is the system reaction due exclusively to inputs. The sum of the free and the forced response of the system is called the total response. A steady state is the response of the system when inputs and outputs are equal.

6.4 TRANSFER FUNCTIONS

In a broad sense, a transfer function relates the response function of the system with an input function. Consider the following example:

Example 6.4.1 Define the transfer function of the system:

$$b_1\frac{dy}{dt} + by = c_1\frac{dx}{dt} + cx$$

Solution: The following is the Laplace transform of the above differential equation:

$$b_1[sG(s) - g(0)] + bG(s) = c_1[sF(s) - f(0)] + cF(s)$$

Then

$$G(s) = \frac{c_1 s + c}{b_1 s + b}F(s) - \frac{c_1 f(0) - b_1 g(0)}{b_1 s + b}$$

The above transform may also be written as

$$G(s) = P(s)F(s) - \frac{c_1 f(0) - b_1 g(0)}{b_1 s + b}$$

where $P(s)$ is called the transfer function of the system. The transform of the input function is $F(s)$, the transform of the response function is $G(s)$ and $f(0)$ and $g(0)$ are initial values for the input and for the state variables. When all initial values are zero, the transfer function relates the response function of the system and the input by the expression $G(s) = P(s)F(s)$. This relation is shown in Fig. 6.4.1.

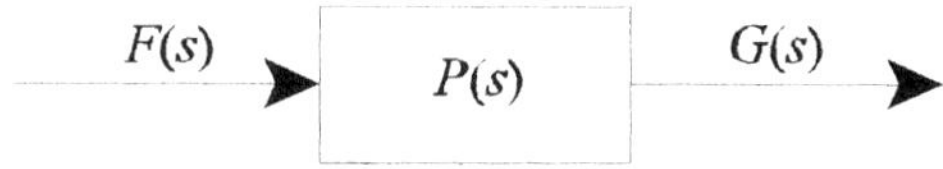

Figure 6.4.1

Thus, the following definition applies for the transfer function:

Definition 6.4.1 In a linear system, a transfer function $P(s)$ is the ratio between the transform of the response function of the system and the transform of the input function, when all the initial values are zero.

By using the Laplace or the Z transforms, this definition is valid for continuous and for discrete systems.

The general expression for the Laplace transform of the response function can be written as

$$G(s) \; = \; \frac{\displaystyle\sum_{0}^{m} c_i s^{i}}{\displaystyle\sum_{0}^{n} b_i s^{i}} F(s) \; + \; (\text{all terms for initial conditions})$$

Then, the general expression for the transfer function is

$$P(s) \; = \; \frac{\displaystyle\sum_{0}^{m} c_i s^{i}}{\displaystyle\sum_{0}^{n} b_i s^{i}} \; = \; \frac{c_m s^{m} + c_{m-1} s^{m-1} + \ldots + c}{b_n s^{n} + b_{n-1} s^{n-1} + \ldots + b}$$

Note that the denominator of the transfer function is the characteristic polynomial of the system.

Example 6.4.2 Define the transfer function for the following system:

$$y_{n+2} + b_1 y_{n+1} + b_2 y_n \; = \; x_{n+2} + c_1 x_{n+1} + c_2 x_n$$

Solution: The following is the Z transform of the system:

$$(z^2 G(z) - g(0)z^2 - g(1)z + b_1[zG(z) - g(0)z] + b_2 G(z)$$

$$= z^2 F(z) - f(0)z^2 - f(1)z + c_1[zF(z) - f(0)z] + c_2 zF(z)$$

where $g(0)$, $g(1)$, $f(0)$ and $f(1)$ are the initial conditions. Then, the transform of the response may be written as

$$G(z) = \frac{z^2 + c_1 z + c_2}{z^2 + b_1 z + b_2} F(z) + \frac{g(0)z(z + b_1) + g(1)z}{z^2 + b_1 z + b_2} - \frac{f(0)z(z + c_1) + f(1)z}{z^2 + b_1 z + b_2}$$

The transfer function is here

$$P(z) = \frac{z^2 + c_1 z + c_2}{z^2 + b_1 z + b_2}$$

where $z^2 + b_1 z + b_2$ is the characteristic equation of the system.

As will be shown in the next example, when the input depends on time, defining the response function of the system requires the appropriate handling of the transfer function.

Example 6.4.3 The following is the differential equation representing the yield of a Kikuyu grass field, in response to rainfall[9]:

$$\frac{dy}{dt} + 0.4916y = 0.1049\frac{dx}{dt} + 0.1090x$$

where y is pasture yield, as kg/ha/day of dry green leaves and x is rainfall in mm/month, as defined by the following corresponding rainfall equation:

$$x = 206 - 152.6\cos 0.809t - 43.5\sin 0.809t$$

Determine the response functions of the system.

Solution: The above differential equation may be expressed symbolically as

$$\frac{dy}{dt} + by = c_1\frac{dx}{dt} + c_2 x$$

---

[9]Computed from Murtagh, G.J. et.al.

and the symbolic expression for rainfall has the form

$$f(t) = k_0 + k_1 \cos\theta t + k_2 \sin\theta t$$

Then, the Laplace transform of the system differential equation is

$$G(s) = \frac{c_1 s + c_2}{s + b} F(s) - \frac{c_1 f(0) - g(0)}{s + b}$$

As disclosed in the previous section, the response of a system can be separated into two components, the free response depending only on initial conditions of the system and the forced response depending only on the input.

Free Response. The Laplace transform of the free response is here

$$G_A(s) = \frac{g(0)}{s+b}$$

where $g(0)$ is the initial pasture yield. Then, the free response of the system is simply the inverse transform of the above equation:

$$g_A(t) = g(0)e^{-bt}$$

Forced Response. The Laplace transform of the forced response is

$$G_B(s) = \frac{c_1 s + c_2}{s + b} F(s) - \frac{c_1 f(0)}{s + b}$$

where $f(0)$ is the initial condition for rainfall. Note that the forced response includes the initial conditions of the input. The initial conditions of the input should not be confused with the initial conditions of the system.

The first term of the forced response may be written as $G_b(s) = P(s)F(s)$, where $P(s)$ is the transfer function of the system and $F(s)$ is the Laplace transform of the rainfall input. For practical purposes, this first term will be solved first.

As disclosed in Chapter 4 and Property 4, the inverse Laplace transform of the product of functions $P(s)$ and $F(s)$ is given by the following convolution integral:

$$g_b(t) = L^{-1}[P(s)F(s)] = \int_0^t p(t-\tau)f(\tau)\,d\tau = \int_0^t p(\tau)f(t-\tau)\,d\tau$$

where

$$P(s) = \frac{c_1 s + c_2}{s+b} = c_1\left[1 - \frac{b}{s+b}\right] + \frac{c_2}{s+b}$$

The inverse of the above transfer function is

$$p(t) = c_1\left(\delta(t) - be^{-bt}\right) + c_2 e^{-bt} = c_1\delta(t) - \left(c_1 b - c_2\right)e^{-bt}$$

The term $\delta(t)$ is the inverse transform of integer 1 and is called *a unit impulse function,* or *delta function*. The delta function represents a spike whose ordinate approaches infinity and the width of the independent variable approaches zero. The area under the curve is equal to one, that is

$$\int_{-\infty}^{\infty} \delta(t)\,dt = 1$$

meaning that of a unit of input is compressed to an infinitesimally small duration of time. Note that the delta function has a value only at $t = 0$. Then, the following is the Laplace transform of the delta function:

$$L[\delta(t)] = \int_0^{\infty} e^{st}\delta(t)\,dt = 1$$

The inverse of $F(s)$ is the rainfall input and was defined as

$$f(t) = k_0 + k_1\cos\theta t + k_2\sin\theta t$$

Then

$$g_b(t) = \int_0^t p(t-\tau)f(\tau)d\tau$$

$$= \int_0^t \left[c_1\delta(t-\tau) - (c_1b-c_2)e^{-b(t-\tau)}\right]f(\tau)d\tau$$

$$= c_1\int_0^t \delta(t-\tau)f(\tau)d\tau - (c_1b-c_2)\int_0^t e^{-b(t-\tau)}f(\tau)d\tau$$

where τ is the time scale. As indicated before, the inverse transform of the product of two functions is defined by a convolution integral. Since the Laplace transform of $\delta(t) = 1$, then

$$\int_0^t \delta(\tau)f(t-\tau)d\tau = \int_0^t \delta(t-\tau)f(\tau)d\tau = L^{-1}\left[(1)(F(s))\right] = f(t)$$

Expression $\delta(t - \tau)$ is called a *delayed impulse*.

Note that $f(t) = k_0 + k_1\cos\theta t + k_2\sin\theta t$ .Thus

$$g_b(t) = c_1 f(t) - (c_1b - c_2)\int_0^t e^{-b(t-\tau)}f(\tau)d\tau$$

$$= c_1(k_0 + k_1\cos\theta\tau + k_2\sin\theta\tau) - (c_1b-c_2)\int_0^t e^{-b(t-\tau)}(k_0 + k_1\cos\theta\tau + \sin\theta\tau)d\tau$$

$$= c_1(k_0 + k_1\cos\theta\tau + k_2\sin\theta\tau) -$$

$$(c_1b-c_2)e^{-bt}\left(k_0\int_0^t e^{b\tau}d\tau + k_1\int_0^t e^{b\tau}\cos\theta\tau\, d\tau + k_2\int_0^t e^{b\tau}\sin\theta\tau\, d\tau\right)$$

After computing the integrals and factorizing, the above expression becomes

$$g_b(t) = \frac{c_2k_0}{b} - \left(\frac{c_2k_0}{b} - c_1k_0 - \frac{(c_1b-c_2)(k_1b-k_2\theta)}{b^2+\theta^2}\right)e^{-bt}$$

$$+ \left(c_1k_1 - \frac{(c_1b-c_2)(k_1b-k_2\theta)}{b^2+\theta^2}\right)\cos\theta t + \left(c_1k_2 - \frac{(c_1b-c_2)(k_1\theta+k_2b)}{b^2+\theta^2}\right)\sin\theta t$$

The second term of the transform of the forced response has the following solution:

$$L^{-1}\left[\frac{c_1 f(0)}{s+b}\right] = c_1(k_0 + k_1)e^{-bt}$$

Then, the final expression for the forced response is

$$g_B(t) = \frac{c_2 k_0}{b} - \left(\frac{c_2 k_0}{b} + c_1 k_1 - \frac{(c_1 b - c_2)(k_1 b - k_2 \theta)}{b^2 + \theta^2}\right)e^{-bt}$$
$$+ \left(c_1 k_1 - \frac{(c_1 b - c_2)(k_1 b - k_2 \theta)}{b^2 + \theta^2}\right)\cos\theta t + \left(c_1 k_2 - \frac{(c_1 b - c_2)(k_1 \theta + k_2 b)}{b^2 + \theta^2}\right)\sin\theta t$$

Total Response. As disclosed previously, the total response is the sum of the free response and the forced response. Thus

$$g(t) = g(0)e^{-bt} + \frac{c_2 k_0}{b} - \left(\frac{c_2 k_0}{b} + c_1 k_1 - \frac{(c_1 b - c_2)(k_1 b - k_2 \theta)}{b^2 + \theta^2}\right)e^{-bt}$$
$$+ \left(c_1 k_1 - \frac{(c_1 b - c_2)(k_1 b - k_2 \theta)}{b^2 + \theta^2}\right)\cos\theta t + \left(c_1 k_2 - \frac{(c_1 b - c_2)(k_1 \theta + k_2 b)}{b^2 + \theta^2}\right)\sin\theta t$$

It is now a simple task to replace the above expression with the known numerical values for b, c_1, c_2, k_0, k_1, k_2, $\cos\theta$ and $\sin\theta$. Then

$$y = 45.68 - 23.12\cos 0.809\,t - 13.86\sin 0.809\,t - 22.56e^{-0.492t} + g(0)e^{-0.492t}$$

where $g(0)$ is the pasture initial yield. An educated guess for this initial yield may be obtained from the data, but the final value is better obtained by non linear regression. The following are the results after this procedure:

$$y = 45.68 - 23.12\cos 0.809\,t - 13.86\sin 0.809\,t - 22.56e^{-0.492t} + 22.60e^{-0.492t}$$

where $g(0) = y_0 = 22.60$. The following is a summary of the non linear curve fitting statistics:

Table 6.4.1

Source	DF	Sum of Squares	Mean Square
Regression	1	16948.26085	16948.26085
Residual	9	1024.73915	113.85991
Uncorrected Total	10	17973.00000	
(Corrected Total)	9	4208.90000	
R squared = 1 - Residual SS / Corrected SS = 0.75653			

Parameter	Estimate	Asymptotic Std. Error	"t"
y_0	22.597416584	0.100807556	223.70

The accuracy of the state equation may be improved by including more parameters in the non linear curve fitting process. To preserve the identification of the free response, the total response may be written as follows:

$$y = Ae^{-bt} + B - (B + C)e^{-bt} + C\cos\theta t + D\sin\theta t$$

Then, the following equation was obtained:

$$y = 14.26e^{-0.492t} + 45.82 - 31.29e^{-0.492t} - 14.53\cos 0.809t - 19.09\sin 0.809t$$

where the first term of the above equation corresponds to the free response. Parameter θ was not included in the curve fitting process. The graph of the response functions is shown in Fig. 6.4.2.

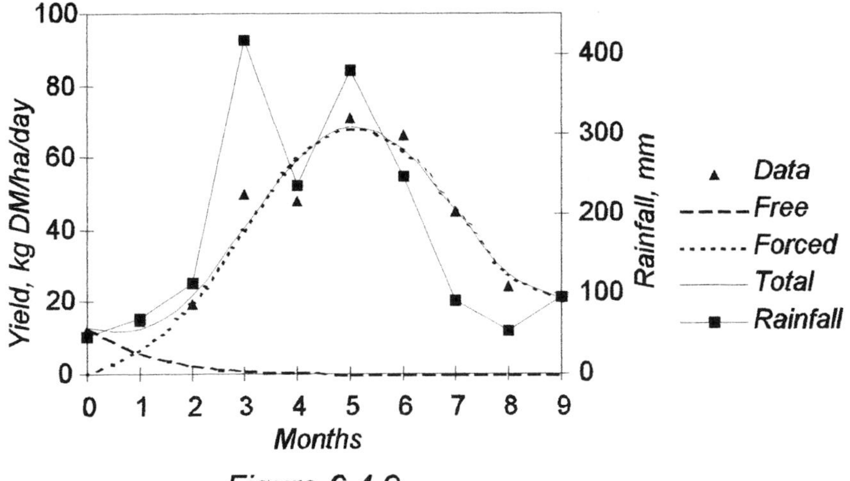

Figure 6.4.2

The following is the summary of the regression statistics:

Table 6.4.2

Source	DF	Sum of Squares	Mean Square
Regression	4	17683.78934	4420.94734
Residual	6	289.21066	48.20178
Uncorrected Total	10	17973.00000	
(Corrected Total)	9	4208.90000	

R squared = 1 - Residual SS / Corrected SS = 0.93129

Parameter	Estimate	Asymptotic Std. Error	95 % Confidence Interval Lower	Upper
A	14.262346109	6.139706127	-0.760973576	29.285665794
B	45.817639700	2.862976589	38.812188354	52.823091046
C	-14.53461940	3.234187731	-22.44839168	-6.620847108
D	-19.08527739	3.394823204	-27.39211052	-10.77844426

In conclusion, since transfer functions relates the system response with a particular input function, the researcher can simulate countless system responses by changing the input trajectory.

Summary

Transfer functions relate the system response with an input trajectory. When all initial values are zero, the transfer function relates the response function of the system and the input by the expression $G(s) = H(s)F(s)$, where $G(s)$, $H(s)$ and $F(s)$ are the Laplace or Z transforms of the system response, the transfer function and the input trajectory. Thus, a transfer function is the relation between the transform of the response function and the transform of the input function, when all the initial values are zero. This definition applies either for continuous or for discrete systems.

6.5 STRUCTURAL PROPERTIES OF SYSTEMS

The notion of structure is related to how the parts of something are put together and organized to form a more complicated arrangement. Then, the structure of systems is related to how *component systems* are coupled to form a more complicated system.

The following structural classification of agricultural systems has been adopted for this book:

- Interactive coupling
- Conjunctive coupling

Interacting component systems may be coupled by means of interconnected differential of difference equations, determining an *interactive coupling*. Difference and differential equations denote the existence of interfaces between the coupled components. In a simpler type of coupling, a set of components may be coupled as one system having no interface relationships between such components. This type of coupling is called *conjunctive coupling*.

Interactive Coupled Systems

Interactive coupled agricultural systems may be arranged into two groups:

• Compartmental systems
• Non compartmental systems

Compartmental Systems. The components of compartmental systems are called *compartments*, a label that is widely used in tracer kinetics. Compartments work as communicating chambers among which a substance is considered to move. A compartment is defined, in a morphological sense, as a chamber with a given substance that occupies the chamber. Compartmental systems are called *closed systems*, if it is assumed that no material enters or leaves the system. If communication with the external environment is permitted, then the system is called an *open system*. Modeling of compartmental systems is called *compartmental analysis*. An abstract representation of a compartmental system is shown in Fig. 6.5.1.

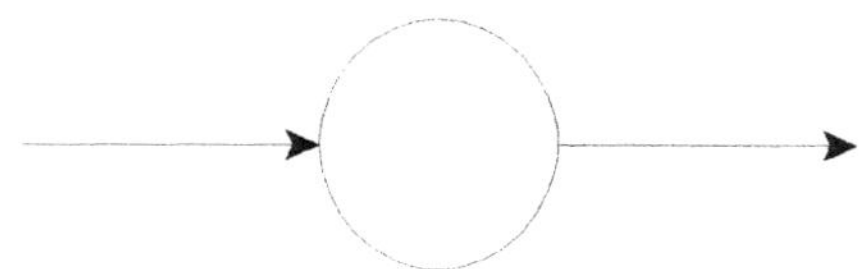

Figure 6.5.1

An open system with two compartments is illustrated in the following example:

Example 6.5.1 As defined in Example 6.2.1, the movement of DDT from plant to soil was 25% per month, from soil to plant 2% and carried out with ground water 5%. This system is pictured in Fig. 6.5.2:

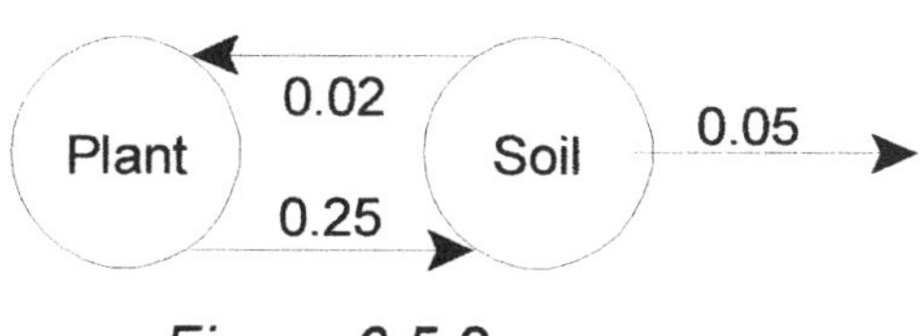

Figure 6.5.2

The following set of differential equations was defined for this system:

$$\frac{dY}{dt} = (A + B)Y = \begin{bmatrix} -0.25 & 0.02 \\ 0.25 & -0.02 \end{bmatrix} Y + \begin{bmatrix} 0 \\ -0.05 \end{bmatrix} Y$$

for $Y = (y_p, y_s)$, where y_p is a state of the plant compartment, y_s is a state of the soil compartment and t is months. The state changes are determined only by the output $(A+B)Y$, because there are no external inputs to the system. The matrix of constant coefficients determining exchange rates between compartments is A. Coefficients with positive signs are compartment inputs and coefficients with negative signs are the compartment outputs. Note that the system is represented by two differential equations, because it has two compartments. Note also that the sum of the coefficients of each column of matrix A should always add up to zero. Matrix B is determined by the system output to the outside environment.

Compartmental analysis may yield information on state changes and exchange rates between compartments. It may provide also information on the distribution volumes and the mass of the system compartments. The next example illustrates this possibility.

Example 6.5.2 It was shown in Example 6.3.1 that a patient with an immunodeficiency problem was dosed with 9.9 grams of gamma globulin intravenously. The blood concentration of the patient gamma globulin was described by the following equation[10],

$$y = 218 + 245 e^{-0.0386t}$$

where y is IgG gamma globulin concentration in mg/dl and t is time in days. Determine the gamma globulin distribution volume for a one-compartment model of the system.

Solution: The distribution volume is given by the relationship $V = D/y_0$, where V is the distribution volume in deciliters, D is the gamma globulin dose in milligrams and y_0 is the blood gamma globulin concentration in milligrams per deciliter at time zero. Then

$$V = \frac{9900}{463} = 21.4 \text{ dl}$$

By knowing the distribution volume of the marker, it is possible to convert the

---

[10]Vohnout, K., Unpublished

state equation from concentration of the marker to amount of the marker, such that $y_w = y\,V$. Then

$$y_w = 4661 + 5239e^{-0.0386t}$$

where y_w is milligrams of gamma globulin. he corresponding differential equation is here

$$\frac{dy_w}{dt} = 180 - 0.0386y_w$$

where 180 is an input and $0.0386y_w$ is the output. The system is represented in Fig. 6.5.3.

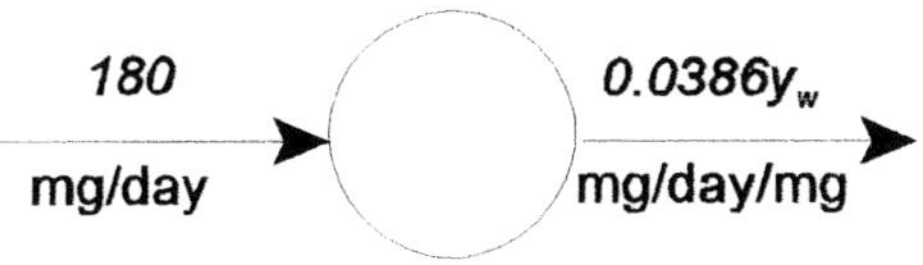

Figure 6.5.3

As shown in the above examples, the following is a fundamental feature of compartmental systems:

- The sum of the coefficients in each column of the matrix representing exchanges among compartments always add up to zero

The reader is advised not to confuse compartments with the states of a finite discrete system. To emphasize the difference, states are represented as the dashed circles shown in Fig. 6.5.4.

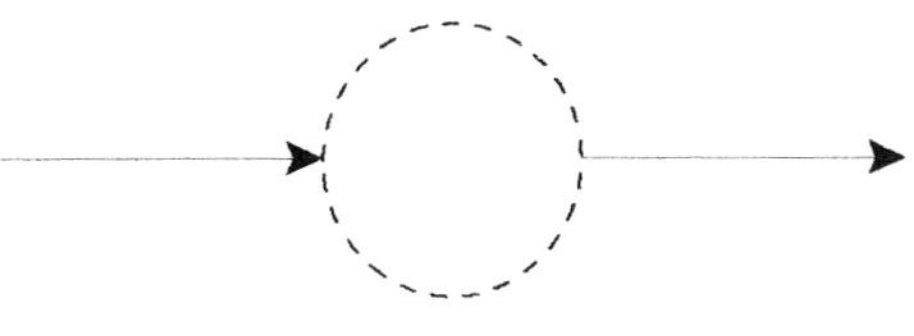

Figure 6.5.4

The state at which the system is operating is called the *mode of operation* of the system.

Example 6.5.3 It was found that, when the trees in a citrus plantation are healthy, 20% may get a disease within a year. When diseased, 30% of the trees may recover and 10% may die. Define the mathematical model representing the system.

Solution: There is a temptation of defining this system as continuous and open, with a compartment of healthy trees and a compartment of diseased trees. As shown in Fig. 6.5.5, this is a finite discrete system with three states, the healthy state, the diseased state and dead state. This is called the *next state diagram of a finite discrete system* or just, a *state transition diagram.*

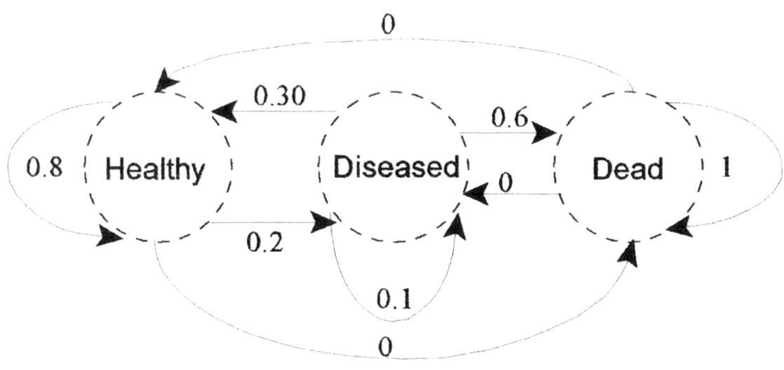

Figure 6.5.5

The corresponding state transition matrix is shown in the following table.

Table 6.5.1

Present State	Next State		
	Healthy	**Diseased**	**Dead**
Healthy	0.8	0.2	0
Diseased	0.3	0.1	0.6
Dead	0	0	1

The first row in the table shows that when the trees are healthy, the probability of remaining healthy in the next state is 0.8 and the probability of becoming diseased is 0.2. The second row shows that when the trees are diseased, the probability of getting healthy in the next state is 0.3, the probability of remaining diseased is 0.1 and the probability of dying is 0.6. The third row shows that dead trees would remain dead. Then, the mathematical model of the system is given by the following next state equation:

$$Y_{n+1} = \begin{bmatrix} 0.8 & 0.2 & 0 \\ 0.3 & 0.1 & 0.6 \\ 0 & 0 & 1 \end{bmatrix} Y_n$$

where y_1, y_2, y_3 are the healthy, the diseased and the dead states and n is years.

The system is represented by three difference equations, because it has three states. Note that no substance is moving between the states, but information that 20% of healthy trees may turn diseased and 80% may remain healthy. Thirty percent of diseased trees may become healthy, 60% may remain diseased and 10% may die. All the dead trees remain dead.

Non Compartmental Systems. Components of a non compartmental system may work as transducers with no chambers among which matter may move, but *black box*es linking the components by inputs and outputs of information. Mathematical models of non compartmental agricultural systems are usually of empirical nature. An abstract representation of a non compartmental system is the black box shown in Fig. 6.5.6.

Figure 6.5.6

Example 6.5.5 Without predators, every year the population of a type of bird doubles. When predators are introduced, the bird population is reduced in proportion to ten times the number of predators. The number of predators increases in proportion to the number of birds by a 0.01 factor. Define the mathematical model of the system.

Solution: Since birds and predators are discrete variables, it is reasonable to define the system as discrete. The following is the set of difference equations representing the system:

$$Y_{n+1} = A Y_n = \begin{bmatrix} 2 & -10 \\ 0.01 & 0 \end{bmatrix} Y_n$$

for $Y_n = (y_b, y_p)$, where y_b is the birds component, y_p is the predators component and n is years. As shown, the birds are affected by birds by a 2 factor and by predators by a factor -10. Predators are affected by birds by a factor 0.01. This system is pictured in Fig. 6.5.7. The system is represented by two difference equations because it has two components. Note that the sum of the columns of matrix A does not have to add to zero, because there is no matter moving between the components, but information on state changes. Note also that no inputs from outside have been defined for this system. Therefore, the state changes are determined only by the output AY_n.

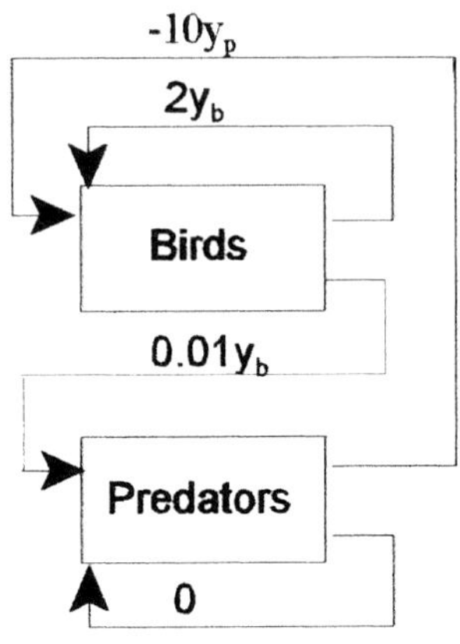

Figure 6.5.7

Example 6.5.6 The leaf growth of Kikuyu pastures and milk production of dairy cows was measured for periods of four weeks, during three consecutive years[11]. The following matrix equation defines the relationship between the state variables:

$$\frac{dY}{dt} = AY + BT + C = \begin{bmatrix} -0.3661 & 0 \\ 0.6311 & -0.7892 \end{bmatrix} Y + \begin{bmatrix} 0 \\ -0.6498 \end{bmatrix} t + \begin{bmatrix} 14.13 \\ 26.37 \end{bmatrix}$$

The state variables are here $Y = (y_p, y_c)$, where y_p is leaf growth in kilograms of dried green leaf per hectare per day and y_c is milk production in kilograms of 4% fat corrected milk per hectare per day. Define the input and output of the system.

Solution: The system input is $BT+C$ and the output is $Z=AY$. The corresponding picture of the model is shown in Fig. 6.5.8. Note that the differential equations of the system

---

[11]Computed from Murtagh, G.J. et.al.

correspond to an empirical model fitted to the data. As opposed to compartmental models, non compartmental models are often empirical in nature. Therefore, the numerical coefficients in the differential or difference equations may not have a physical interpretation other than establishing relationships among variables. For instance, the inputs in Example 6.5.6 may include a group of factors not accounted for explicitly in the mathematical model of the system. In non compartmental systems there is no material flow but a flow of information.

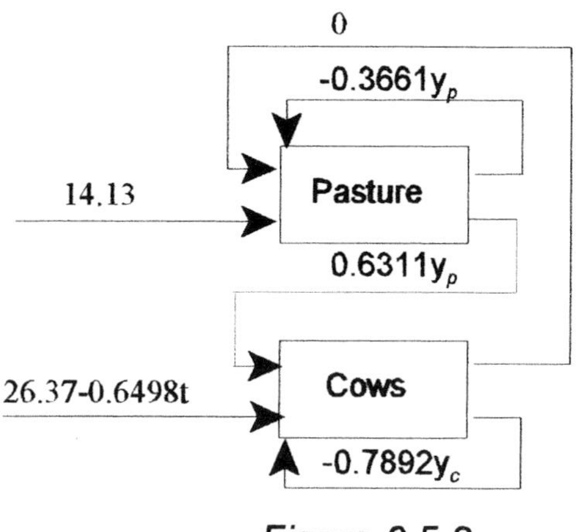

Figure 6.5.8

As shown in the above examples, the following are fundamental features of non compartmental systems:

- The sum of the coefficients in each column of the matrix representing relationships among components may not have to add up to zero
- The coefficients in the differential or difference equations represent information flows and may not have a physical interpretation

Conjunctive Coupled Systems

The notion of conjunctive coupling is that of systems in which each component has its own inputs and operates independently. The essence of this concept is the grouping of the experimental material such that each group is a component system and constitutes a single trial or replication. Grouping determines the sources of variation in the typical analysis of variance.

The "Source of Variation" in the analysis of variance may include input variables and non-input variables. An input is a variable definable as a function of time. Non-input variables are usually qualitative variables, such as blocks, breeds, species or any particular trait. Non- input variables are component systems and are not definable as functions of time.

Example 6.5.7 An experiment was carried out to test the effect of potash on the yield of cotton. The experiment was arranged in 3 randomized blocks and the treatments were five levels of K_2O per acre. Define the conjunctive coupling of the experiment as a system.

Solution: The following is the analysis of variance proposed for the above experiment:

Table 6.5.2

Source of Variation	Degrees of Freedom
Blocks	2
Treatments	4
Error	8

Here the blocks can be portrayed as system components. Each block is a replication of the experiment and operates independently. The treatments are inputs of the system, because applications of potash are scheduled over the time variable.

This system is shown in Fig. 6.5.9.

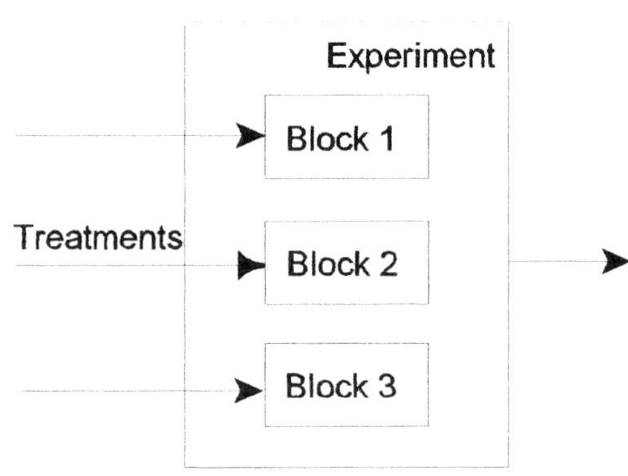

Figure 6.5.9

Note that there are no interfaces between the experimental blocks. Each block is an independent system by itself. However, the three blocks are components of a system called experiment.

The notion of conjunctive coupling is particularly helpful in factorial arrangements of treatments and in split-plot experimental designs.

Example 6.5.8 An experiment was designed to study how the starch content of the diet of steers affects the digestibility of roughage. The experimental roughage was stems of the banana plant, sugarcane leafs and Star grass hay. *In vivo* digestibility procedures were carried out by placing the bags containing the roughage in the rumen of six fistulated steers. The starch was provided by six different amounts of green bananas. Define the experiment as a conjunctive coupled system.

Solution: Each steer is here an independent component of the experiment as a system. There are six steers in the experiment, meaning that the experiment has six components and five degrees of freedom for steers. Each steer received the three types of roughage in bags, placed in the rumen, for *in vivo* digestion. Thus, each roughage is an independent component of a steer as a system and a sub-component of the experiment. There are three roughages per steer, meaning that the experiment has two degrees of freedom for roughage and 10 degrees of freedom for the interaction steers×roughage. There are six levels of green bananas, meaning that the experiment has five degrees of freedom for bananas, plus all the corresponding interactions. Note that each roughage is a treatment, but is not an input because the roughage bags are placed in the rumen of the steers for digestion and are not scheduled over the time scale. Green bananas are also treatments but are not components. Bananas are inputs, because consumption of green bananas is scheduled over the time scale[12].

A formal plan for the analysis of variance in the experiment is shown in the following table:

Table 6.5.3

Source of Variation	Degrees of Freedom
Steers (S)	5
Roughage (R)	2
RS (Error I)	10
Green Bananas (B)	5
BR	10
BS	25
BRS (Error II)	50
Total	107

In the traditional analysis of variance, the steers would be called block, roughage would

---

[12]Computed from San Martin F.A.

be classes and green bananas subclasses.

The following model was proposed for the digestion of crude protein of the experimental roughage, as affected by the input x:

$$\frac{\partial y}{\partial t} + by = f(x)$$

where y is digestibility of crude protein as percentage, t is time in hours and x is percent of dried bananas in the diet. The data should be fitted to this model for each experimental roughage. Then, the constant coefficients of each equation can be compared by a "t" test between roughages.

The experiment as a system, with the steers as components in conjunctive coupling, is shown in Fig. 6.5.10.

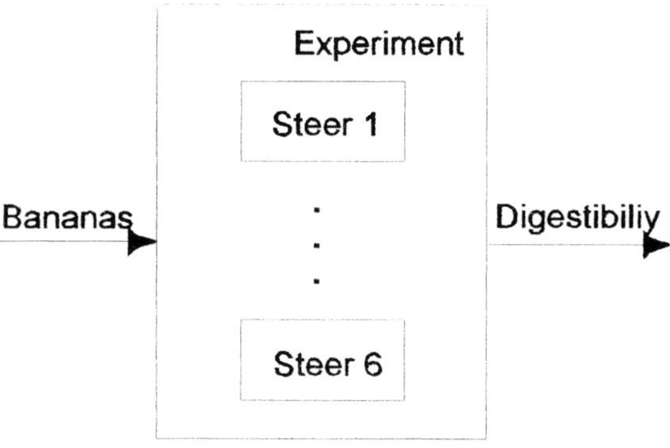

Figure 6.5.10

A steer component with roughage, also in conjunctive coupling, is illustrated in Fig. 6.5.11.

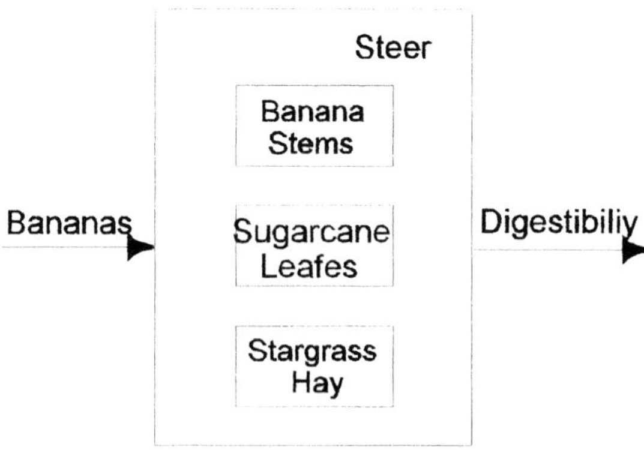

Figure 6.5.11

In conclusion, the following statements apply to conjunctive coupling of agricultural systems, as they affect the design of experiments:

• The notion of conjunctive coupling applies mainly to grouping of experimental material in the design of agricultural experiments
• Input variables should always be subclasses of component variables
• Component variables should never be subclasses of inputs

Summary

Interactive coupled systems are systems interfacing by means of interconnected difference or differential equations. The number of interconnected equations corresponds to the number of components of the system. When components work as communicating chambers among which a material is considered to flow, the system is called a compartmental system. Components of non-compartmental systems may work as transducers with no chambers among which matter may move, but black boxes linked by inputs and outputs of information. In conjunctive coupling, each component works as an independent system. This concept pertains mainly to grouping of experimental material in the design of agricultural experiments.

7

STOCHASTIC MODELS OF SYSTEMS

As disclosed before, the operation of all agricultural systems must be considered subject to some kind of uncertainties. Depending on whether uncertainties are being considered or are ignored, the models of systems are either stochastic or deterministic. Many other sources of uncertainties affect a system, such as

- Uncertainties as to the actual inputs
- Randomness in the arrival of inputs
- Uncertainties in the response of the system
- Uncertainties introduced by the mathematical model of the system

Just a few of the sources of uncertainty affecting a system may be controllable by the researcher by means of experimental designs.

This chapter is related to stochastic models of systems, with an emphasis in *Markov processes* or *Markov chains* that is, processes where the next state of the system is completely determined by the present state.

7.1 MODELING STOCHASTIC AGRICULTURAL SYSTEMS

The basic feature of stochastic models of systems is that state variables are defined as probability distributions. In contrast, state variables in deterministic models are defined as expected values.

For the scope of this book, the following criteria for modeling stochastic processes have been adopted:

- Modeling of Markov chains
- Modeling on non-Markov processes

Markov Chains

Many applications of classical probability theory to the study of systems are based on the assumption that the outcomes of successive trials of an experiment are independent from each other. In contrast, *Markov processes* or *Markov chains* are stochastic processes in which the probability of the next state of the system is completely determined by the probability of the present state. Markov processes can be used to model many agricultural applications. This section is related to the theoretical concept and to the characterization of Markov processes, as they apply to agricultural research. A finite Markov chain is defined as follows:

Definition 7.1.1 A sequence of trials of an experiment is a Markov chain if the outcome of trial $n+1$ depends only on the outcome of trial n and not on the outcomes of earlier trials

Representing state transitions in a Markov process requires:

- Defining *probability vectors* $P = (p_1, p_2, ..., p_m)$ such that $p_1 + p_2 + ... + p_m = 1$, where $p_i \geq 0$ is interpreted as the probability of a state $y_i \in Y$ and Y is the set of states of the system.
- Defining *transition matrix* $Q_{m \times m}$, where each of its rows is a probability vector

Then, if P is the present state vector and Q is the transition matrix, PQ is the next state vector of the system. Matrix Q is also called a *probability matrix*, because the elements of each row ad to one. The following is the formal definition of a probability matrix:

Definition 7.1.2 A matrix $Q_{m \times m}$ is a probability matrix if each element q_{ij} is the probability that the state y_i at time n would change to the state y_j at time $n+1$, for $i = 1,2,...,m$ and $j = 1,2,...,m$ and each row is a m-state probability vector, such that

$$\sum_1 q_{ij} = 1 \; ; \; q_{ij} \geq 0$$

These concepts are illustrated in the following example.

Example 7.1.1 The trees of a citrus farm are surveyed and classified as healthy or diseased. It was found that, when the trees are healthy, 20% get a disease within a year and when the trees are diseased, 30% of them recover. Define the state transitions of the system.

Solution: A tree diagram for state changes of the system during the first two years is shown in Fig. 7.1.1. Note that at zero time, all the trees are supposed to be healthy. After one year, there are 80% healthy and 20% diseased trees. After two years, only 64% of the healthy trees remain healthy and 30% of the diseased trees recover, making a total of 70% healthy trees and 30% diseased trees.

Representing state transitions as tree diagrams is awkward. An easier procedure requires defining the state changes as a Markov chain. The probability vector is here $P = (p_1, p_2)$, where p_1 is the probability of the healthy state and p_2 is the probability of the diseased state.

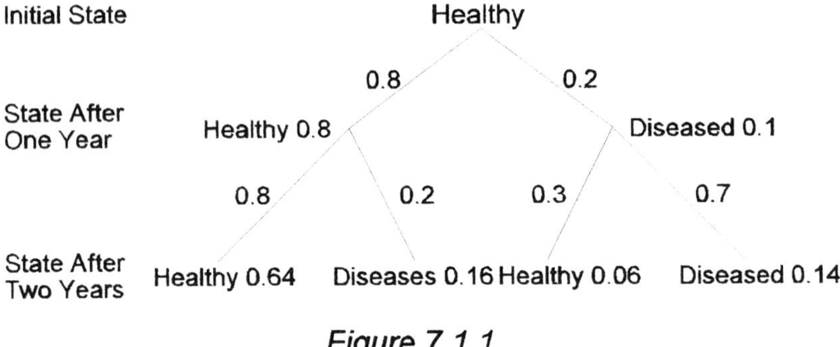

Figure 7.1.1

The transition matrix Q is shown in the following table:

Table 7.1.1

Present State	Next State	
	Healthy	**Diseased**
Healthy	0.80	0.20
Diseased	0.30	0.70

The first row shows that the probability of healthy trees of remaining healthy in the next state is 80% and that the probability of becoming diseased is 20%. The elements of the second row show that the probability of diseased trees of becoming healthy in the next state is 30% and that the probability of remaining diseased is 70%.

The graphic representation of matrix Q, depicting the state changes of the system, is shown in Fig. 7.1.2.

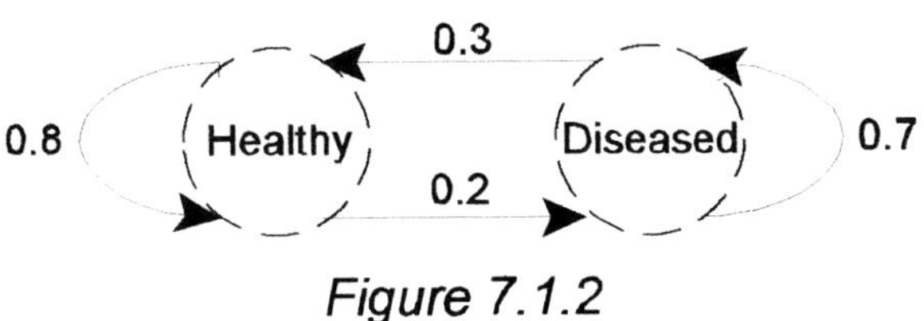

Figure 7.1.2

The diagram in Fig. 7.1.2 is called the *next state diagram* of the system.

The state changes of the system, as defined by the product PQ, are shown in Table 7.1.2. The probability vector for the initial condition of the system is $P = (1,0)$, meaning that all trees are healthy. By knowing the present state P and the probability matrix Q, the next state PQ of the system was predicted. Note that at time 2, the values for the present state are consistent with the values obtained in the tree diagram of Fig. 7.1.1.

Table 7.1.2

Time n	Present State P	Transition Matrix Q	Next State PQ
0	(1, 0)	$\begin{bmatrix} 0.8 & 0.2 \\ 0.3 & 0.7 \end{bmatrix}$	(0.80, 0.20)
1	(0.80, 0.20)	$\begin{bmatrix} 0.8 & 0.2 \\ 0.3 & 0.7 \end{bmatrix}$	(0.70, 0.30)
2	(0.70, 0.30)	$\begin{bmatrix} 0.8 & 0.2 \\ 0.3 & 0.7 \end{bmatrix}$	(0.65, 0.35)
3	(0.65, 0.35)	$\begin{bmatrix} 0.8 & 0.2 \\ 0.3 & 0.7 \end{bmatrix}$	(0.625, 0.375)
$\vdots$	$\vdots$	$\vdots$	$\vdots$
n	(0.60, 0.40)	$\begin{bmatrix} 0.8 & 0.2 \\ 0.3 & 0.7 \end{bmatrix}$	(0.60, 0.40)

If this Markov chain would be extended indefinitely, it would be found that the system may reach a steady state condition. A steady state is reached when the difference between the present state and the next state approaches zero as a limit. Then, the steady state of the system is given by the expression $PQ = P$, that is

$$(p_1, p_2) \begin{bmatrix} 0.8 & 0.2 \\ 0.3 & 0.7 \end{bmatrix} = (p_1, p_2)$$

By solving the above equation, it is found that the steady state probability vector is $P = (60, 40)$.

Note that the next state of the system is completely determined by the present state, defined by vector P and the probability matrix Q. Thus, the system is represented by the following set of next state equations:

$$P_{n+1} = P_n Q$$

where P_n is the set of states at time n and P_{n+1} is the set of states at time $n+1$. This is a difference equation representing a free response of the system.

Non-Markov Processes

Not all stochastic systems can be represented as Markov processes. This is especially true when only partial information on the outcomes of an event is available. Then, this partial information must be taken into account as conditional probability. This approach may often result in countless mathematical difficulties. Modeling of such processes is usually accomplished by reducing the detail being considered in the model. This suggestion is illustrated in the next example.

Example 7.1.2 A type of bird is surveyed for its ability to consume and control caterpillars in a cotton field. Determine a stochastic model for the activity of a bird over the population of caterpillars.

Solution: Many sources of uncertainties affect this system, such as the availability of caterpillars, the length of time needed by the bird to find or select the caterpillars, the presence of other birds, the time required to make the catch, the time of the day, the wind and so on. Having the actual data, it might be possible to describe all these factors by probabilistic equations. However, the resulting mathematical model would be extremely complicated. The approach taken in this example is reducing the details in the mathematical model.

The outcomes of the system are the success and the failure of a catch. Then

$$P(X=x, t \mid X=x-1, t-\Delta t) \qquad success$$
$$P(X=x, t \mid X=x, t-\Delta t) \qquad failure$$

are the probability of a success at a time t, if at a time $t-\Delta t$ the bird had $X=x-1$ successes and the probability of a failure at a time t, if at a time $t-\Delta t$ the bird had $X=x$ successes.

Note that the total number of successes X does not change when the outcome is a failure and that there may not be failures in a strict sense. Note also that during the time interval Δt there is always the probability of more than one catch. However, if the interval chosen is small enough, such probability may be negligible.

It is assumed here that the probability of success is proportional to the time interval Δt. Then, the probabilities of success and failure must be

$$P(X=x, t \mid X=x-1, t-\Delta t) = k\Delta t \quad success$$
$$P(X=x, t \mid X=x, t-\Delta t) = 1 - k\Delta t \quad failure$$

where k is a proportionality constant. These outcomes can be put together by using the multiplication theorem of conditional probability[1]:

$$P(A) = P(B_1)P(A \mid B_1) + P(B_2)P(A \mid B_2)$$

If $A \equiv (X=x, t)$, $B_1 = (X=x-1, t-\Delta t)$ and $B_2 = (X=x, t-\Delta t)$, then

$$P(X=x, t) = P(X=x-1, t-\Delta t) \, P(X=x, t \mid X=x-1, t-\Delta t)$$
$$+ P(X=x, t-\Delta t) \, P(X=x, t \mid X=x, t-\Delta T)$$

After replacing the success and failure expressions in the above equation, the following new expression is obtained:

$$P(X=x, t) = P(X=x-1, t-\Delta t) k\Delta t + P(X=x, t-\Delta t)(1 - k\Delta t)$$
$$= \left[P(X=x-1, t-\Delta t) - P(X=x, t-\Delta t)\right] k\Delta t + P(X=x, t-\Delta t)$$

Finally, after rearranging the above terms, this equation becomes

$$\frac{P(X=x, t) - P(X=x, t-\Delta t)}{\Delta t} = k\left[P(X=x-1, t-\Delta t) - P(X=x, t-\Delta t)\right]$$

By making the time interval Δt approaching zero as a limit, the left-hand side of the equation becomes a derivative and the Δt terms vanish from the right-hand side. Thus

---

[1]See Appendix D

$$\frac{d}{dt}P(X=x,t) = k\left[P(X=x-1,t) - P(X=x,t)\right]$$

Since $P(X = x,t) = f(x,t)$, this differential equation is equivalent to

$$\frac{d}{dt}f(x,t) = k\left[f(x-1,t) - f(x,t)\right]$$

The following procedure was developed for finding the solution of this equation. As a first step, it is assumed that the initial number of successes at time zero is a. Then

$$\frac{d}{dt}f(a,t) = k\left[f(a-1,t) - f(a,t)\right]$$

where $f(a-1, t)$ is zero, because the initial value is a and any value before time zero is zero. Therefore

$$\frac{d}{dt}f(a,t) = -kf(a,t) \quad ; \quad f(a,t) = e^{-kt} \quad ; \quad \ln(f(a,t)) = -kt$$

where $P(X = a, t = 0) = 1$, because it is known that the event took place.

If $f(a)$ is known, it is possible to find $f(a+1)$:

$$\frac{d}{dt}f(a+1,t) = k\left[f(a,t) - f(a+1,t)\right] = k\left[e^{-kt} - f(a+1,t)\right]$$

The above expression is equivalent to

$$\frac{d}{dt}(f(a+1,t) + kf(a+1,t) = ke^{-kt}$$

The following is the Laplace transform of this differential equation:

$$sF(s) - f(a+1,0) + kF(s) = \frac{k}{s+k}$$

Note here that $f(a+1, 0) = 0$ because this event does not exist. Therefore

$$F(s) = \frac{k}{(s+k)^2} \quad ; \quad f(a+1) = \frac{k}{1!} t e^{-kt}$$

By the same approach, knowing $f(a+1, t)$ it is possible to find $f(a+2, t)$:

$$\frac{d}{dt} f(a+2,t) + k f(a+2,t) = k^2 t e^{-kt}$$

The solution of this equation is

$$f(s) = \frac{k^2}{(s+k)^3} \quad ; \quad f(a+2) = \frac{(kt)^2}{2!} e^{-kt}$$

The full pattern for defining the system differential equation and its solution has now emerged. The system differential equation can now be expressed as the nonhomogeneous expression

$$\frac{d}{dt} f(x,t) + f(x,t) = k^x t e^{-kt}$$

This expression has the following solution:

$$F(s) = \frac{k^x}{(s+k)^{x+1}} \quad ; \quad f(x) = \frac{(kt)^x}{x!} e^{-kt}$$

This solution corresponds to a Poisson distribution, for t and X starting at zero.

Fig. 7.1.3 shows the probability distribution curves for the success of catching caterpillars at various time values when $k=3$ and time is in hours.

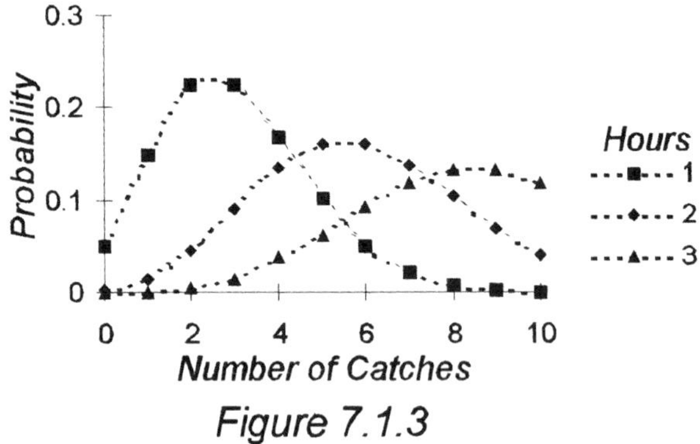

Figure 7.1.3

Summary

State variables in stochastic systems are defined as probability distributions. Markov processes are stochastic processes in which the probability of the next state of the system is completely determined by the probability of the present state, such that $P_{n+1} = P_n Q$, where P is a probability vector and Q is a transition matrix. Modeling of complex non-Markov processes is usually accomplished by reducing the details being considered in the model.

7.2 THE POWERS OF A PROBABILITY MATRIX

Frequently, manipulation of Markov processes requires defining the powers of the probability matrix Q. As disclosed in Chapter 2, when a matrix Q is of order m, the corresponding characteristic equation is a polynomial of degree m in the characteristic root λ and has also m solutions. Corresponding to these solutions, it is expected to find m characteristic vectors v. If this holds true, then it is possible to write

$$Qv_i = \lambda_i v_i$$

The above equation is equivalent to

$$Q(v_1,v_2,...,v_m) = (v_1,v_2,...,v_m) \begin{bmatrix} \lambda_1 & 0 & ... & 0 \\ 0 & \lambda_2 & ... & 0 \\ \vdots & \vdots & \vdots & \vdots \\ 0 & 0 & ... & \lambda_m \end{bmatrix}$$

where $(v_1,v_2,...,v_m) = V$ is a characteristic matrix. Then, defining the following equation is possible:

$$AV = VD$$

where D is a diagonal matrix of order n of the roots $\lambda_1,\lambda_2,...,\lambda_m$. Hence

$$Q = VDV^{-1}$$
$$Q^2 = QQ = (VDV^{-1})(VDV^{-1}) = VD^2V^{-1}$$
$$\vdots \qquad \vdots$$
$$Q^n = VD^nV^{-1}$$

Finding Q^n is now the process of determining the characteristic matrix V and its inverse V^{-1} and determining D^n. Finding D^n is simple:

$$D^n = \begin{bmatrix} \lambda_1 & 0 & ... & 0 \\ 0 & \lambda_2 & ... & 0 \\ \vdots & \vdots & \vdots & \vdots \\ 0 & 0 & ... & \lambda_m \end{bmatrix}^n = \begin{bmatrix} \lambda_1^n & 0 & ... & 0 \\ 0 & \lambda_2^n & ... & 0 \\ \vdots & \vdots & \vdots & \vdots \\ 0 & 0 & ... & \lambda_m^n \end{bmatrix}$$

Matrix D is known as the *canonical form of matrix A under similarity*. The above operation is illustrated in the following example.

Example 7.2.1 It was observed that, within an hour and when resting, some cattle in a herd would remain in this position 90% of the time and would stand up to walk 10% of the time. When walking, they would keep walking 80% of the time and would lie down 20% of the time. Determine the probability that the animals are resting or walking after six hours, when the initial state is resting and when the initial state is walking.

Solution: This analysis can be understood better if pictured as a tree diagram. The tree diagram for the initial state at rest and for the first two hours, is shown in Fig. 7.2.1

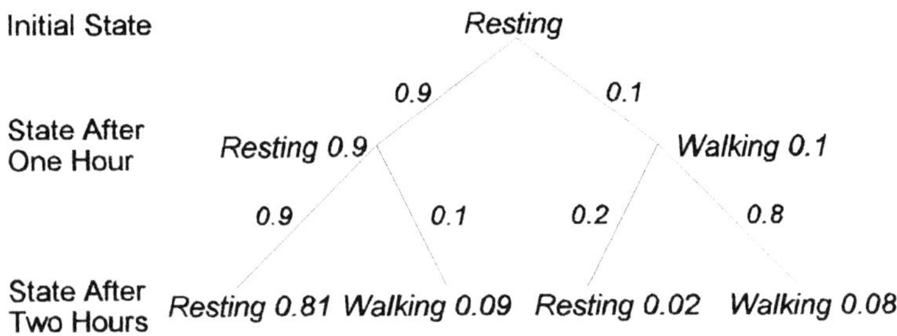

Figure 7.2.1

Note that the initial state of the system was $(1, 0)$ for resting and walking. After one hour, the state of the system was $(0.9, 0.1)$ and after two hours, the state was $(0.81+0.02, 0.09+0.08)$, that is $(0.83, 0.17)$.

The tree diagram for the initial state at walking and also for the first two hours is shown in Fig. 7.2.2. The initial state of the system was here $(0, 1)$ for resting and walking. After one hour, the state of the system was $(0.2, 0.8)$ and after two hours, the state was $(0.34, 0.66)$. The complete picture of the system states for the first two hours is obtained by putting together the data of the two tree diagrams.

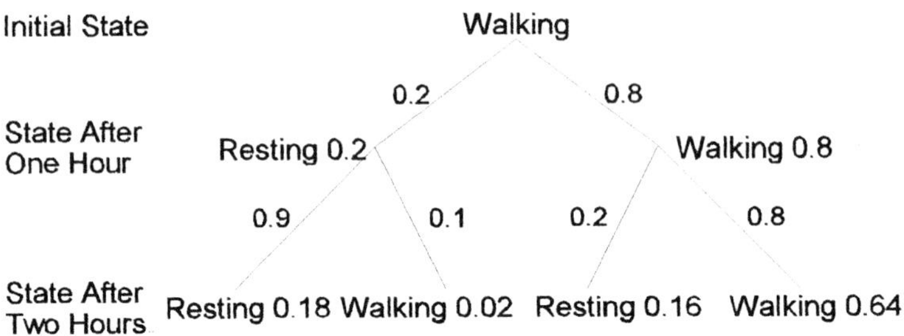

Figure 7.2.2

As shown above, calculations of the system states using tree diagrams are cumbersome, but are useful for understanding and developing the system model. A simple procedure requires determining the powers of the probability matrix Q of the system. The elements of the first row in matrix Q are the probability 0.9 of resting cattle to remain resting and the probability 0.1 that they would stand up. The elements of the second row are the probability 0.2 of walking animals to rest down and the probability 0.8 of remaining in the walking state. Thus

$$Q = \begin{bmatrix} 0.9 & 0.1 \\ 0.2 & 0.8 \end{bmatrix}$$

The following sequence of states is obtained by joining the data in Fig. 7.2.1 and Fig. 7.2.2:

Initial state: $\quad \begin{bmatrix} 1 & 0 \\ 0 & 1 \end{bmatrix} = Q^0$

State after one hour: $\quad \begin{bmatrix} 0.9 & 0.1 \\ 0.2 & 0.8 \end{bmatrix} = Q^1$

State after two hours: $\quad \begin{bmatrix} 0.83 & 0.17 \\ 0.34 & 0.66 \end{bmatrix} = Q^2$

A pattern has evolved here, suggesting that the state of the system at a discrete time n is given by the expression $Q^n = VD^nV^{-1}$, where V is the characteristic matrix of Q and D is the diagonal matrix of the roots of Q. The characteristic equation is here

$$|Q - \lambda I| = \begin{vmatrix} 0.9-\lambda & 0.1 \\ 0.2 & 0.8-\lambda \end{vmatrix} = \lambda^2 - 1.7\lambda + 0.7 = 0$$

and the characteristic roots are $\lambda_1 = 1.0$ and $\lambda_2 = 0.7$. For the first root, the corresponding vector is

$$\begin{bmatrix} 0.9-1.0 & 0.1 \\ 0.2 & 0.8-1.0 \end{bmatrix}\begin{bmatrix} y_1 \\ y_2 \end{bmatrix} = \begin{bmatrix} 0 \\ 0 \end{bmatrix} \ ; \quad \begin{matrix} -0.1y_1 + 0.1y_2 = 0 \\ 0.2y_1 - 0.2y_2 = 0 \end{matrix} \ ; \quad y_1 = y_2 \ ; \quad v_1 = k_1\begin{bmatrix} 1 \\ 1 \end{bmatrix}$$

For the second root, the corresponding vector is

$$\begin{bmatrix} 0.9-0.7 & 0.1 \\ 0.2 & 0.8-0.7 \end{bmatrix}\begin{bmatrix} y_1 \\ y2 \end{bmatrix} = \begin{bmatrix} 0 \\ 0 \end{bmatrix} \ ; \quad \begin{matrix} 0.2y_1 + 0.1y_2 = 0 \\ 0.2y_1 + 0.1y_2 = 0 \end{matrix} \ ; \quad y_1 = -y_2 \ ; \quad v_2 = k_2\begin{bmatrix} -1 \\ 2 \end{bmatrix}$$

Thus, the following are the characteristic matrix, its inverse and the diagonal D^{n} :

$$V = \begin{bmatrix} 1 & -1 \\ 1 & 2 \end{bmatrix} \quad ; \quad V^{-1} = \frac{1}{3}\begin{bmatrix} 2 & 1 \\ -1 & 1 \end{bmatrix} \quad ; \quad D^{n} = \begin{bmatrix} 1^{n} & 0 \\ 0 & 0.7^{n} \end{bmatrix}$$

Now, the following matrix to the n power is obtained for expression $P^{n} = VD^{n}V^{-1}$, where $n = 1,2,\ldots,m$ is time in hours:

$$P^{n} = \frac{1}{3}\begin{bmatrix} 2 + 0.7^{n} & 1 - 0.7^{n} \\ 2 - 2(0.7)^{n} & 1 + 2(0.7)^{n} \end{bmatrix}$$

If time is six hours, then $n=6$. Thus

$$P^{6} = \begin{bmatrix} 0.71 & 0.29 \\ 0.59 & 0.41 \end{bmatrix}$$

According to this result, if the initial state was resting, there is 71% probability that the animals will be resting after six hours and 29% probability that they will be walking. If the initial state was walking, there is 59% probability that they will be resting and 41% probability that they will be walking. This model does not consider the cyclical behavior of cattle and is used here only to illustrate the procedure for determining the powers of a matrix. The interpretation of this result is made easier in the following table:

Table 7.2.1

Initial State	State After Six Hours	
	Resting	**Walking**
Resting	0.71	0.29
Walking	0.59	0.41

Example 7.2.2 The weather in a particular region was classified as sunny, cloudy and rainy. It was found that the probability of being sunny, cloudy and rainy is 0.6, 0.2 and 0.2, when the previous day was sunny. The probabilities are 0.25, 0.50 and 0.25, when the previous day was cloudy. The probabilities are 0.25, 0.25 and 0.50, when the previous day was rainy. Give a five-day forecast for the weather.

Solution: The following is the probability matrix of the system:

$$Q = \begin{pmatrix} 0.60 & 0.20 & 0.20 \\ 0.25 & 0.50 & 0.25 \\ 0.25 & 0.25 & 0.50 \end{pmatrix}$$

Then, the characteristic equation is

$$\begin{vmatrix} 0.60-\lambda & 0.2 & 0.2 \\ 0.25 & 0.50-\lambda & 0.25 \\ 0.25 & 0.25 & 0.50-\lambda \end{vmatrix} \begin{aligned} &= \lambda^3 - 1.6\lambda^2 + 0.6875\lambda - 0.0875 \\ &= (\lambda-0.25)(\lambda-0.35)(\lambda-1) \end{aligned}$$

The following is the characteristic vector for root $\lambda = 0.25$:

$$\begin{pmatrix} 0.35 & 0.20 & 0.20 \\ 0.25 & 0.25 & 0.25 \\ 0.25 & 0.25 & 0.25 \end{pmatrix} \begin{pmatrix} y_1 \\ y_2 \\ y_3 \end{pmatrix} = \begin{pmatrix} 0 \\ 0 \\ 0 \end{pmatrix} \quad ; \quad y_2 = -y_3 \quad ; \quad v_1 = \begin{pmatrix} 0 \\ 1 \\ -1 \end{pmatrix}$$

For the second root $\lambda = 0.35$, the characteristic vector was obtained as follows:

$$\begin{pmatrix} 0.25 & 0.20 & 0.20 \\ 0.25 & 0.15 & 0.25 \\ 0.25 & 0.25 & 0.15 \end{pmatrix} \begin{pmatrix} y_1 \\ y_2 \\ y_3 \end{pmatrix} \sim \ldots \sim \begin{pmatrix} 5 & 0 & 8 \\ 0 & -1 & 1 \\ 0 & 0 & 0 \end{pmatrix} \begin{pmatrix} y_1 \\ y_2 \\ y_3 \end{pmatrix} = \begin{pmatrix} 0 \\ 0 \\ 0 \end{pmatrix} \quad ; \quad v_2 = \begin{pmatrix} -1.6 \\ 1 \\ 1 \end{pmatrix}$$

The characteristic vector for root $\lambda = 1$ was obtained as follows:

$$\begin{pmatrix} -0.40 & 0.20 & 0.20 \\ 0.25 & -0.50 & 0.25 \\ 0.25 & 0.25 & -0.50 \end{pmatrix} \begin{pmatrix} y_1 \\ y_2 \\ y_3 \end{pmatrix} \sim ... \sim \begin{pmatrix} 0 & -0.50 & 0.50 \\ 1 & -1 & 0 \\ 0 & 0.75 & -0.75 \end{pmatrix} \begin{pmatrix} y_1 \\ y_2 \\ y_3 \end{pmatrix} = \begin{pmatrix} 0 \\ 0 \\ 0 \end{pmatrix} \; ; \; v_3 = \begin{pmatrix} 1 \\ 1 \\ 1 \end{pmatrix}$$

Thus

$$V = \begin{pmatrix} 0 & -1.6 & 1 \\ 1 & 1 & 1 \\ -1 & 1 & 1 \end{pmatrix} \; ; \; V^{-1} = \begin{pmatrix} 0 & 0.5 & -0.5 \\ -0.3846 & 0.1923 & 0.1923 \\ 0.3846 & 0.3077 & 0.3077 \end{pmatrix} \; ; \; D^n = \begin{pmatrix} 0.25^n & 0 & 0 \\ 0 & 0.35^n & 0 \\ 0 & 0 & 1^n \end{pmatrix}$$

Expression $Q^n = VD^n V^{-1}$ is, then

$$\begin{pmatrix} 0.3846+0.6154(0.35)^n & 0.3077-0.3077(0.35)^n & 0.3077-0.3077(0.35)^n \\ 0.3846-0.3846(0.35)^n & 0.3077+0.5(0.25)^n+0.1923(0.35)^n & 0.3077-0.5(0.25)^n+0.1923(0.35)^n \\ 0.3846-0.3846(0.35)^n & 0.3077-0.5(0.25)^n+0.1923(0.35)^n & 0.3077+0.5(0.25)^n+0.1923(0.35)^n \end{pmatrix}$$

where n is time in days. Determining the five-day forecast is now a simple task. The first row of the above matrix corresponds for a sunny initial state, the second row for a cloudy initial state and the third row for a rainy initial state.

Summary

Manipulation of Markov processes often requires defining the powers of the probability matrix Q. The powers of matrix Q are given by the expression $Q^n = VD^n V^{-1}$, where V is the characteristic matrix of Q and D is a diagonal matrix of the characteristic roots of Q.

7.3 MARKOV PROCESSES IN AGRICULTURAL RESEARCH

Markov chains are related to many agricultural applications and have been especially useful in the analysis of the genetic makeup as it changes from one generation to another.

Characterization of Markov Chains

Some definitions are needed for a better understanding of the Markov theory and for developing a categorization of the different types of Markov processes.

The first concept is related to the possibility of a given state to be reached from another state. A state is said to be reachable from another state if there is a direct path between the two states. This concept is formally defined as follows:

Definition 7.3.1 Given a transition matrix Q and a set of states Y, a state $y_j \in Y$ is said to be *accessible* from a state $y_i \in Y$, if there is a sequence $i_0, i_1, ..., i_n$ in Y for some integer $n \geq 1$, such that $i_0 = i$, $i_n = j$ and $q_{ij} > 0$.

The sequence $i_0, i_1, ..., i_n$ is called a *direct path from i to j* and the path is said to consist of n steps or periods of time. None of the probabilities in the direct path are zero. The subset of all j states accessible from y_i is denoted $Y_j \subset Y$.

These concepts are illustrated in the following example.

Example 7.3.1 Given the probability matrix

$$Q = \begin{bmatrix} 1/2 & 1/2 & 0 \\ 1/4 & 1/2 & 1/4 \\ 0 & 1/2 & 1/2 \end{bmatrix}$$

determine the accessibility of each state from the others.

Solution: State 3 can not be the next after state 1, because $q_{13} = 0$. In the same way, state 1 can not be the next state after state 3, because $q_{31} = 0$.

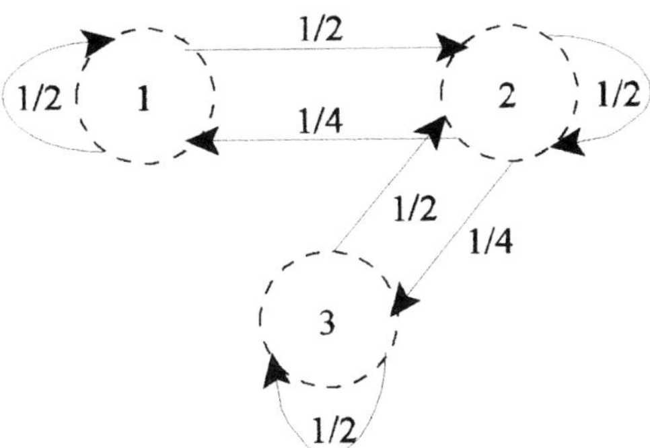

Figure 7.3.1

These relationships are easily seen in the next state graph of Fig. 7.3.1. Note that there is no pathway between states 1 and 3. Note also that there is only one step between states 1 and 2. In addition, there are two steps between states 1 and 3, always going through state 2. The subsets of states reachable from state 1 are states 2 and 3. A system may eventually evolve toward a state or a subset of states from which no escape is possible. This concept is defined as follows:

Definition 7.3.2 Given a matrix Q and a set of states Y, a state $y_i \in Y$ is called an *absorbing state* if, once the system reaches this state on some trial, the system will remain in such state on all future trials. Then, $q_{ii} = 1$ and no escape from the absorbing state is possible.

Definition 7.3.3 Given a transition matrix Q and a set of states Y, a non-empty subset of states $Y_A \subset Y$ is called an *absorbing sub-chain*, when no state $y_i \notin Y_A$ is accessible from a state $y_j \in Y_A$. Then $\sum_{y_j \in Y_A} q_{ij} = 1$ and no escape from the absorbing sub-chain is possible

An absorbing sub-chain is also called a *closed sub-chain*.
   The following example illustrates this definition.

Example 7.3.2 Determine the absorbing states in the following system:

$$Q = \begin{bmatrix} 0 & 1/2 & 0 & 1/2 & 0 \\ 1/2 & 0 & 1/2 & 0 & 0 \\ 0 & 0 & 1 & 0 & 0 \\ 0 & 0 & 0 & 0 & 1 \\ 0 & 0 & 0 & 1/2 & 1/2 \end{bmatrix}$$

Solution: By inspection, state 3 is an absorbing state because the probability of the next state from state 3 to state 3 is $q_{33} = 1$ .This means that no escape is possible from state 3. Thus, a probability of 1 in the diagonal of matrix Q means that the state in that row is an absorbing state. The system features are revealed in the next state diagram of Fig 7.3.2.
   Matrix Q can be partitioned into the following sub-matrices:

$$Q = \begin{bmatrix} A & B \\ 0 & C \end{bmatrix}$$

such that

$$C = \begin{bmatrix} 0 & 1 \\ 1/2 & 1/2 \end{bmatrix}$$

Note that $C \subset Q$ is a probability sub-matrix located in the diagonal of matrix Q. Each row vector $y_4 \subset C$ and $y_5 \subset C$ adds up two one. Therefore, once the system reaches state 4, it will remain cycling between states 4 and 5 forever. Then, states 4 and 5 are a subset of absorbing states.

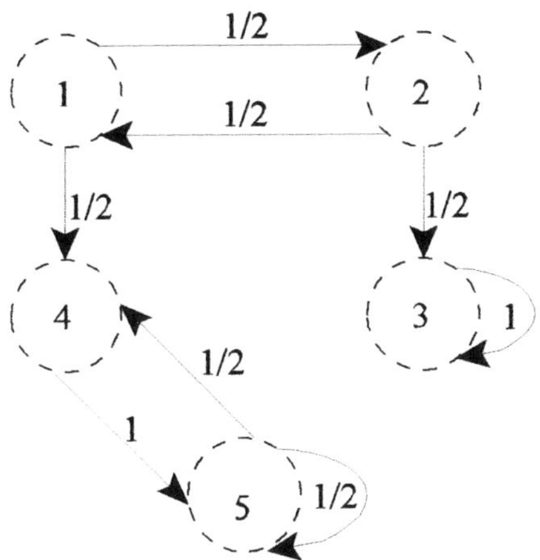

Figure 7.3.2

If no subset of states is absorbing, the system may eventually evolve toward a steady state or toward a condition of periodicity, in which the state of the system alternates between sub-chains. These conditions are defined as follows:

Definition 7.3.4 Given a transition matrix Q and a set of states Y, a subset of states $Y_I \subset Y$ is said to be an *irreducible sub-chain*, if no subset of Y_I is an absorbing subset of states.

If $Y_I = Y$, then matrix Q is a transition matrix of an irreducible Markov chain. As defined below, irreducible Markov chains are of two types.

Definition 7.3.5 Given a transition matrix Q of an irreducible Markov chain, the chain is called *regular* if for some $n \geq 1$, where n is a step or time period and all elements q_{ij} of Q^n are positive. Otherwise, the chain is called *periodic*.

These concepts are portrayed in the following examples.

Example 7.3.3 Given the following matrix, determine if the process is regular or periodic.

$$Q = \begin{bmatrix} 0 & 1/2 & 1/2 \\ 1/2 & 0 & 1/2 \\ 1/2 & 1/2 & 0 \end{bmatrix}$$

Solution: As shown bellow, the square of this matrix has all its elements positive. Therefore, this process is regular.

$$Q^2 = \begin{bmatrix} 1/2 & 1/4 & 1/4 \\ 1/4 & 1/2 & 1/4 \\ 1/4 & 1/4 & 1/2 \end{bmatrix}$$

Example 7.3.4 Given the following matrix, determine if the process is regular or periodic.

$$Q = \begin{bmatrix} 0 & 0 & 1 & 0 \\ 0 & 0 & 0 & 1 \\ 0 & 1 & 0 & 0 \\ 1/3 & 2/3 & 0 & 0 \end{bmatrix}$$

Solution: The following are some powers of matrix Q:

$$Q^2 = \begin{bmatrix} 0 & 1 & 0 & 0 \\ 1/3 & 2/3 & 0 & 0 \\ 0 & 0 & 0 & 1 \\ 0 & 0 & 1/3 & 2/3 \end{bmatrix} \quad Q^3 = \begin{bmatrix} 0 & 0 & 0 & 1 \\ 0 & 0 & 1/3 & 2/3 \\ 1/3 & 2/3 & 0 & 0 \\ 2/9 & 7/9 & 0 & 0 \end{bmatrix} \quad Q^4 = \begin{bmatrix} 1/3 & 2/3 & 0 & 0 \\ 2/9 & 7/9 & 0 & 0 \\ 0 & 0 & 1/3 & 2/3 \\ 0 & 0 & 2/9 & 7/9 \end{bmatrix}$$

At each period, the system alternates from states $\{1, 2\}$ to next states $\{3, 4\}$ and vice versa, such that

$$Q^n = \begin{cases} \begin{bmatrix} 0 & A_n \\ B_n & 0 \end{bmatrix} & ; \quad n = 1,3,5,... \\[2em] \begin{bmatrix} C_n & 0 \\ 0 & D_n \end{bmatrix} & ; \quad n = 2,4,6,... \end{cases}$$

Therefore, the process is periodic. The graph of some powers of matrix Q is shown in Fig. 7.3.3.

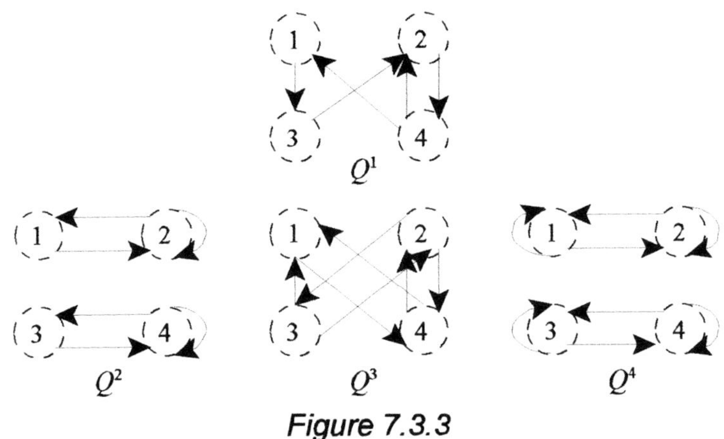

Figure 7.3.3

Regular Processes

A Markov process is called regular if its transition matrix Q is regular. As specified in Definition 7.3.5, a transition matrix is called regular, when all the elements q_{ij} of Q^n are positive for some $n \geq 1$, where $i=1,2,...,n$ and $j=1,2,...,n..$ Then, if all elements of Q^n are positive, the same is true for Q^{n+1}. Thus, a regular process is the progressive change of states of the system, resulting in a final steady state, meaning that a steady state is achieved only if the probability matrix Q is regular.

Example 7.3.5 Find which of the following matrices is regular:

$$Q_1 = \begin{bmatrix} 1/2 & 1/2 \\ 0 & 1 \end{bmatrix} \qquad Q_2 = \begin{bmatrix} 0 & 1/2 & 1/2 \\ 1/2 & 0 & 1/2 \\ 1/2 & 1/2 & 0 \end{bmatrix}$$

$$Q_3 = \begin{bmatrix} 3/5 & 1/5 & 1/5 \\ 1/4 & 1/2 & 1/4 \\ 1/4 & 1/4 & 1/2 \end{bmatrix} \qquad Q_4 = \begin{bmatrix} 1/2 & 1/4 & 1/4 \\ 0 & 0 & 1 \\ 0 & 3/4 & 1/4 \end{bmatrix}$$

Solution: Every power of matrix Q_1 has a zero element. Therefore, this matrix is not regular:

$$Q_1^2 = \begin{bmatrix} 1/4 & 3/4 \\ 0 & 1 \end{bmatrix}, \dots, Q_1^n = \begin{bmatrix} 1/2^n & 1-1/2^n \\ 0 & 1 \end{bmatrix}$$

The square of matrix Q_2 has all its elements positive, therefore is regular:

$$Q_2^2 = \begin{bmatrix} 1/2 & 1/4 & 1/4 \\ 1/4 & 1/2 & 1/4 \\ 1/4 & 1/4 & 1/2 \end{bmatrix}$$

The elements of matrix Q_3 are all positive, therefore is regular. Matrix Q_4 contains the following absorbing sub-chain:

$$Q_{4,A} = \begin{bmatrix} 0 & 1 \\ 3/4 & 1/4 \end{bmatrix}$$

and is therefore, not reducible. Thus, matrix Q_4 is not regular.

The following additional definitions are related to the steady state of a probability matrix:

Definition 7.3.6 If a probability matrix Q is regular, then $\lim\limits_{n \to \infty} Q^n$, if it exists, is called the *steady state matrix*.

The steady state matrix represents the steady state of the system. As indicated before, the steady state of the system is given by the expression $PQ = P$, where P is a steady state vector. The above definition is illustrated in the following examples.

Example 7.3.6 Determine the steady state matrix for the citrus farm of Example 7.1.1.

Solution: The following was the probability matrix defined in Example 7.1.1:

$$Q = \begin{bmatrix} 0.8 & 0.2 \\ 0.3 & 0.7 \end{bmatrix}$$

The first row shows that the probability of healthy trees of remaining healthy is 80% and that the probability of becoming diseased is 20%. The second row shows that the probability of diseased trees of remaining diseased is 70% and that the probability of becoming healthy is 30%. The state changes of the system are shown in the following table, where matrix Q^n represents the state changes over time.

Table 7.3.1

Time	States	Q^n
0	$\begin{bmatrix} 1 & 0 \\ 0 & 1 \end{bmatrix}$	Q^0
1	$\begin{bmatrix} 0.8 & 0.2 \\ 0.3 & 0.7 \end{bmatrix}$	Q^1
2	$\begin{bmatrix} 0.7 & 0.3 \\ 0.45 & 0.55 \end{bmatrix}$	Q^2
3	$\begin{bmatrix} 0.65 & 0.35 \\ 0.475 & 0.525 \end{bmatrix}$	Q^3
$\vdots$	$\vdots$	$\vdots$
n	$\begin{bmatrix} 0.6 & 0.4 \\ 0.6 & 0.4 \end{bmatrix}$	$\lim_{n \to \infty} Q^n$

The steady state of the system is given by the expression $PQ=P$. Then

$$(p_1,p_2)\begin{bmatrix} 0.8 & 0.2 \\ 0.3 & 0.7 \end{bmatrix} = (p_1,p_2)$$

The solution of the above equation is the steady state vector $P = (60, 40)$. Note that the steady state matrix is here

$$Q^n = \begin{bmatrix} 0.6 & 0.4 \\ 0.6 & 0.4 \end{bmatrix}$$

The first row represents the steady state vector for an initial state $P = (1,0)$ that is, all the trees were healthy. The second row represents the steady state vector for the initial state $P = (0,1)$ that is, all the trees were diseased. Therefore, the following statement applies here:

• In regular processes, the same steady state is attained independently of the initial state

Example 7.3.7 A male mouse of genotype Aa is crossed with a female of an unknown genotype. This process is continued for a succession of matings of males Aa with females of unknown genotypes. Determine the long run expected genetic composition of the mice.

Solution: If the genotype of males is always Aa and the genotype of females is any of AA, Aa or aa, the genetic composition of the offspring is as shown in the following table:

Table 7.3.2

Male	Female		Female		Female	
	A	*A*	*A*	*a*	*a*	*a*
A	*AA*	*AA*	*AA*	*Aa*	*Aa*	*Aa*
a	*Aa*	*Aa*	*Aa*	*aa*	*aa*	*aa*

Then, by looking at the female columns of the table, it is possible to define the following transition matrix:

Table 7.3.3

Females	Next Generation		
	AA	*Aa*	*aa*
AA	1/2	1/2	0
Aa	1/4	1/2	1/4
aa	0	1/2	1/2

This matrix is regular because all the elements of the squared transition matrix are positive, that is

$$Q^2 = \begin{bmatrix} 3/8 & 1/2 & 1/8 \\ 1/4 & 1/2 & 1/4 \\ 1/8 & 1/2 & 3/8 \end{bmatrix}$$

If the probability matrix is regular, then the steady state is given by the expression $PQ=P$, that is

$$(p_1,p_2,p_3)\begin{bmatrix} 1/2 & 1/2 & 0 \\ 1/4 & 1/2 & 1/4 \\ 0 & 1/2 & 1/2 \end{bmatrix} = (p_1,p_2,p_3)$$

By solving this equation, it is found that the steady state probability vector is $P = (1/4, 1/2, 1/4)$, meaning that $p(AA) = 1/4$, $p(Aa) = 1/2$ and $p(aa) = 1/4$.

Absorbing Processes

As denoted before, a system may eventually evolve toward a specific state or set of states and remain there forever. Then, such state or set of states are *absorbing states*. The non absorbing states are called *transient states*.

Example 7.3.8 Determine the absorbing and the transient states in the following probability matrix:

$$Q = \begin{bmatrix} 1/2 & 1/4 & 1/4 \\ 0 & 3/4 & 1/4 \\ 0 & 0 & 1 \end{bmatrix}$$

Solution: This system has three states. At first sight, this is not a regular matrix because the probability of the next state from state 3 to state 3 is 1. The probability of 1 in the diagonal always determines the existence of an absorbing state in that row. This is easily seen in the next state diagram of Fig. 7.3.4. Once the system reaches state 3, it will remain there. Thus, state 3 is an absorbing state and states 1 and 2 are transient states.

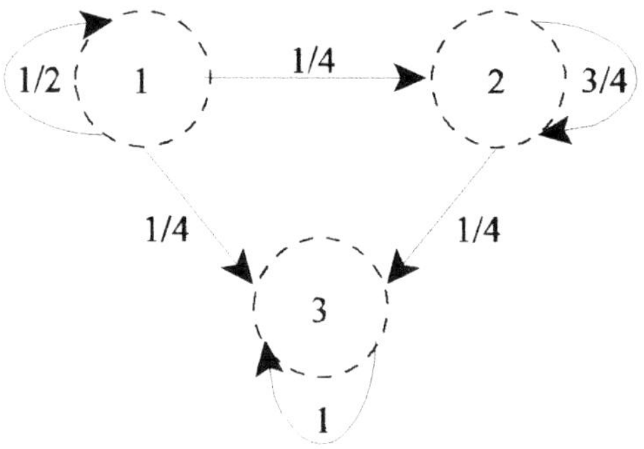

Figure 7.3.4

Example 7.3.9 It was found that, when the trees in a citrus farm are healthy, 20% may get a disease within a year. When diseased, 30% of the trees may recover and 10% may die. Determine the transient and the absorbing states.

Solution: As shown in Fig. 7.3.5, this is a finite discrete system with three states, the healthy state, the diseased state and dead state. The corresponding probability matrix is shown in Table 7.3.4. The first row in the table shows that when the trees are healthy, the probability of remaining healthy in the next state is 0.8 and the probability of becoming diseased is 0.2. The second row shows that when the trees are diseased, the probability of getting healthy in the next state is 0.3, the probability of remaining diseased is 0.1 and the probability of dying is 0.6. The third row shows that dead trees would remain dead. Clearly, the healthy and the diseased are transient states and the dead state is absorbing.

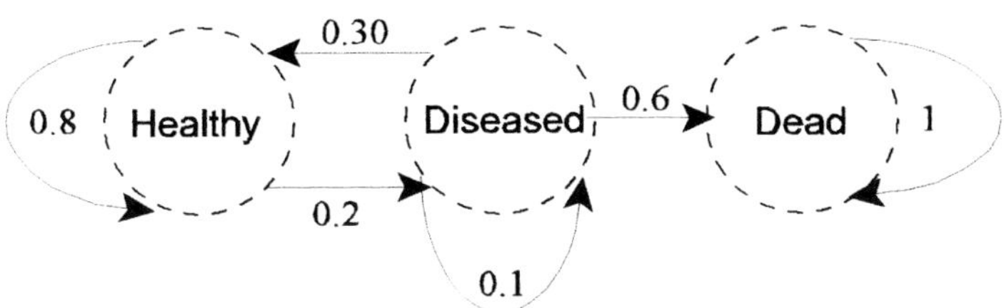

Figure 7.3.5

Table 7.3.4

Present State	Next State		
	Healthy	**Diseased**	**Dead**
Healthy	0.8	0.2	0
Diseased	0.3	0.1	0.6
Dead	0	0	1

The mathematical model of the system is given by the following next state equation:

$$Y_{n+1} = \begin{bmatrix} 0.8 & 0.2 & 0 \\ 0.3 & 0.1 & 0.6 \\ 0 & 0 & 1 \end{bmatrix} Y_n$$

where y_1, y_2, y_3 are the healthy, the diseased and the dead states and n is years.

Usually the problem is determining how long the system is expected to survive in the transient states, before reaching the absorbing state or states. This type of problem is shown in the next examples.

Example 7.3.10 A dog has the choice of selecting three different meals. Once he has tried meal III, he refuses to eat the other two meals. Determine the expected number of steps that would take for the dog before reaching the third meal.

Solution: The following table represents the transition matrix of the system:

Table 7.3.5

Present State	Next State		
	I	**II**	**III**
I	0	1/2	1/2
II	1/2	0	1/2
III	0	0	1

This is not a regular matrix because the probability of transition from meal III to meal III is 1, meaning that the dog would eat only meal III in the next state. Once the system reaches state III, it will remain there. State III is an absorbing state and states I and II are transient states.

The system can reach the absorbing state through four different pathways:

1 - The initial state is I and the final state before reaching the absorbing state is I:

Table 7.3.6

Steps	Probability
I to III	1
I to II to I to III	1/2(1/2)(1) = 1/4
I to II to I to II to I to III	1/2(1/2)(1/2)(1/2)(1) = 1/16
$\vdots$	$\vdots$
I to II to I to II to I to ... to III	1/2(1/2)(1/2)(1/2)...(1)

The following is the expected number of steps n_{11} for the pathway I to I that is, the number of times the dog tried meal I before reaching meal III:

$$n_{11} = 1 + \frac{1}{4} + \frac{1}{16} + ... = 1 + \frac{1}{4} + \frac{1}{4^2} + ... = \frac{1}{1 - 1/4} = \frac{4}{3}$$

2 - The initial state is I and the final state before reaching the absorbing state is II:

Table 7.3.7

Steps	Probability
I to II to III	$1/2(1) = 1/2$
I to II to I to II to III	$1/2(1/2)(1/2)(1) = 1/8$
I to II to I to II to I to II to III	$1/2(1/2)(1/2)(1/2)(1/2)(1) = 1/32$
$\vdots$	$\vdots$
I to II to I to II to I to II to ... to III	$1/2(1/2)(1/2)(1/2)(1/2)...(1)$

The following is the expected number of steps n_{12} for the pathway I to II:

$$n_{12} = \frac{1}{2} + \frac{1}{8} + \frac{1}{32} + ... = \frac{1}{2}\left[1 + \frac{1}{4} + \frac{1}{4^2} + ...\right] = \frac{1}{2}\left[\frac{1}{1 - 1/4}\right] = \frac{2}{3}$$

3 - The initial state is II and the final state before reaching the absorbing state is I. The following is the expected number of steps n_{21} for pathway II to I:

$$n_{21} = \frac{1}{2} + \frac{1}{8} + \frac{1}{32} + ... = \frac{1}{2}\left[1 + \frac{1}{4} + \frac{1}{4^2} + ...\right] = \frac{1}{2}\left[\frac{1}{1 - 1/4}\right] = \frac{2}{3}$$

4 - The initial state is II and the final state before reaching the absorbing state is II. The following is the expected number of steps n_{22} for pathway II to II:

$$n_{22} = 1 + \frac{1}{4} + \frac{1}{16} + ... = 1 + \frac{1}{4} + \frac{1}{4^2} + ... = \frac{1}{1 - 1/4} = \frac{4}{3}$$

The steps related to the four pathways are represented by a matrix N, that is

$$N = \begin{bmatrix} n_{11} & n_{12} \\ n_{21} & n_{22} \end{bmatrix} = \begin{bmatrix} 4/3 & 2/3 \\ 2/3 & 4/3 \end{bmatrix}$$

where each element of matrix N is the expected number of steps needed to reach the absorbing state through each pathway. The total expected number of trials, before reaching the absorbing state, is the sum of the expected number of trials the system was in each transient state. Thus, when the initial state was I, the expected number of times the dog was in state I and in state II, before reaching the absorbing state III, was $4/3 + 2/3 = 2$. Similarly, when the initial state was II, the expected number of steps needed to reach the absorbing state was $2/3 + 4/3 = 2$.

Note that matrix N is the power series of a matrix Q_τ, representing the transient states. Thus

$$N = I + Q_\tau + Q_\tau^2 + Q_\tau^3 + \ldots +$$

where

$$Q_\tau = \begin{bmatrix} 1/2 & 0 \\ 0 & 1/2 \end{bmatrix}$$

is the matrix of transient states and $I = Q_\tau^0$ is an identity matrix. Then

$$N - I = Q_\tau + Q_\tau^2 + Q_\tau^3 + \ldots + = Q_\tau\left[I + Q_\tau + Q_\tau^2 + \ldots + \right] = Q_\tau N$$

Upon rearranging, the above expression becomes

$$N = \left(I - Q_\tau \right)^{-1}$$

The following result is obtained by using this equation in Example 7.3.9:

$$I - Q_\tau = \begin{bmatrix} 1 & 0 \\ 0 & 1 \end{bmatrix} - \begin{bmatrix} 0 & 1/2 \\ 1/2 & 0 \end{bmatrix} = \begin{bmatrix} 1 & -1/2 \\ -1/2 & 1 \end{bmatrix}$$

$$\left(I - Q_\tau \right)^{-1} = \begin{bmatrix} 4/3 & 2/3 \\ 2/3 & 4/3 \end{bmatrix} = N$$

This result is consistent with the results previously obtained.

Example 7.3.11 Suppose that the 70% of the diseased citrus trees in Example 7.1.1 never recover but die. Determine the expected number of years needed for all the trees to die.

Solution: The following is the probability matrix Q of the system:

$$Q = \begin{bmatrix} 0.8 & 0.2 & 0 \\ 0.3 & 0 & 0.7 \\ 0 & 0 & 1 \end{bmatrix}$$

Matrix Q shows that 30% of the trees are expected to recover while the remaining 70% is expected to die. Matrix Q_τ of the transient states is here

$$Q_\tau = \begin{bmatrix} 0.8 & 0.2 \\ 0.3 & 0 \end{bmatrix}$$

Then

$$I - Q_\tau = \begin{bmatrix} 1 & 0 \\ 0 & 1 \end{bmatrix} - \begin{bmatrix} 0.8 & 0.2 \\ 0.3 & 0 \end{bmatrix} = \begin{bmatrix} 0.2 & -0.2 \\ -0.3 & 1 \end{bmatrix}$$

$$(I - Q_\tau)^{-1} = \frac{1}{0.14}\begin{bmatrix} 1 & 0.2 \\ 0.3 & 0.2 \end{bmatrix} = N$$

If the initial state was $(1, 0, 0)$, that is all the trees were healthy, the expected time for the trees to die is $(1 + 0.2)/0.14 = 8.6$ years. Conversely, if the initial state was $(0, 1, 0)$, that is all the trees were diseased, the time required for all the trees to die is $(0.3 + 0.2)/0.14 = 3.6$ years.

Summary

A state is said to be accessible from another state if there is a direct path between the two states. A sub-chain is called absorbing if the system may eventually evolve toward that sub-chain and remain there forever. If no sub-chains are absorbing, the Markov process is called irreducible. Irreducible chains are either regular of periodic. Regular processes eventually evolve toward a steady state. In periodic processes the state of the system alternates between sub-chains. For a process to be regular, there exists a probability matrix Q^n whose elements are all positive for some $n \geq 1$. The steady state is represented by a steady state matrix, such that any row of the steady state matrix is a steady state vector. For a process to be absorbing, the probability that the next state after an absorbing state is 1. The non absorbing states are called transition states. The survival time of a system, before reaching the absorbing condition, is given by the expression $N = (I - Q_\tau)^{-1}$, where matrix Q_τ is a sub-matrix of matrix Q, representing the transition states and N is a power series of Q_τ. The sum of the elements of a row of N represents the survival period for the corresponding initial state of the row.

7.4 RELATIONSHIP BETWEEN STOCHASTIC AND DETERMINISTIC MODELS

As disclosed before, in stochastic models the states of the system are defined as probability distributions. In deterministic models, the states are defined as expected values of the outcomes. The notion of an expected value is related to the idea of an average, in the sense that a given number would summarize and represent certain data. Thus, deterministic models would represent the average of the probability distribution values, as determined for the stochastic model. This conception is presented in the following examples.

Example 7.4.1 Define a deterministic and a stochastic model of the diseased trees in Example 7.1.1.

Solution: As was shown previously, the system is represented by a set of next state equations of the form $P_{n+1} = P_n Q$, where P_n is the set of states at time n, P_{n+1} is the set of states at time $n+1$ and Q is the probability matrix

$$Q = \begin{bmatrix} 0.8 & 0.2 \\ 0.3 & 0.7 \end{bmatrix}$$

The solution of the next state equations of this system is given by the Z transform

$$F(z) = F(0) \frac{z}{z - Q}$$

where $F(0)$ is the initial state. The initial state matrix of the system is here

$$F(0) = \begin{bmatrix} 1 & 0 \\ 0 & 1 \end{bmatrix}$$

Then, the following is the solution of the next state equations:

$$P_n = \begin{bmatrix} 1 & 0 \\ 0 & 1 \end{bmatrix}\begin{bmatrix} 0.8 & 0.2 \\ 0.3 & 0.7 \end{bmatrix}^n$$

This expression is the deterministic model of the system and represents the expected value of probabilities. Note that the first row in the initial state matrix shows that all the trees were healthy. The second row shows that all the trees in the initial state were diseased.

Manipulation of matrix Q^n is made simpler by the procedure outlined in Section 2 of this chapter. As previously defined, a matrix of order n can take the form $Q^n = VD^nV^{-1}$, where V is the characteristic matrix of Q and D is the diagonal matrix of the characteristic roots of Q. The following results were obtained after solving matrix Q for its characteristic roots and vectors:

$$\begin{bmatrix} 0.8 & 0.2 \\ 0.3 & 0.7 \end{bmatrix}^n = \frac{1}{5}\begin{bmatrix} 1 & -2 \\ 1 & 3 \end{bmatrix}\begin{bmatrix} 1^n & 0 \\ 0 & 0.5^n \end{bmatrix}\begin{bmatrix} 3 & 2 \\ -1 & 1 \end{bmatrix} = \frac{1}{5}\begin{bmatrix} 3+2(0.5)^n & 2-2(0.5)^n \\ 3-3(0.5)^n & 2+3(0.5)^n \end{bmatrix}$$

After solving the product $F(0)Q^n$ and assuming that all the trees in the initial state were healthy, the solution P_n is

$$P_n = \frac{1}{5}\left[3+2(0.5)^n,\ 2-2(0.5)^n\right]$$

for $P_n = (p_{1n}, p_{2n})$, where p_{1n} is the proportion of healthy trees, p_{2n} is the proportion of diseased trees and n is time in years. If the total number of trees is m, then the above equation becomes

$$E(X_1, X_2) = \frac{m}{5}\left[3+2(0.5)^n,\ 2-2(0.5)^n\right]$$

where X_1 is the number of healthy trees and X_2 is the number of diseased trees. The graphic representation of expected values is shown in Fig. 7.4.1. The total number of trees is assumed to be 10.

The above expression corresponds to the deterministic model of the system. The state variables may have a binomial distribution, that is

$$f(x_1,x_2) = \frac{m!}{x_1!\,x_2!}\, p_{1n}^{x_1}\, p_{2n}^{x_2}$$

where x_1 and x_2 are the number of healthy and diseased tress out of a total of m trees.

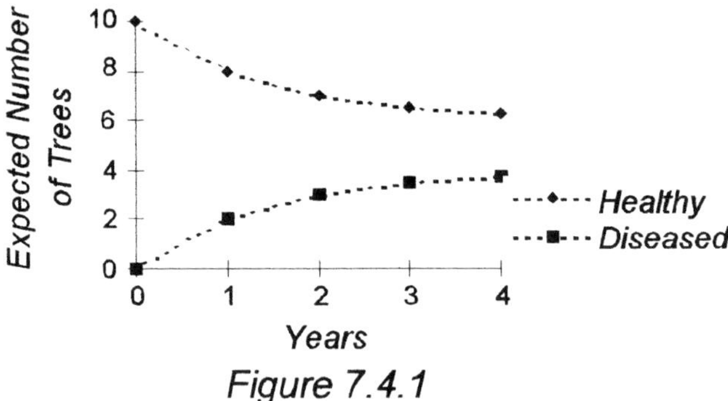

Figure 7.4.1

By replacing the P_n values in the binomial equation for an initial state $P=(1, 0)$, it is possible now to define the following state probability model of the system:

$$f(x_1,x_2) = \frac{m!}{x_1!x_2!}\left(\frac{3+2(0.5)^n}{5}\right)^{x_1}\left(\frac{2-2(0.5)^n}{5}\right)^{x_2}$$

The probability distribution curves of diseased trees are shown in Fig. 7.4.2. The total number of trees was assumed to be 10.

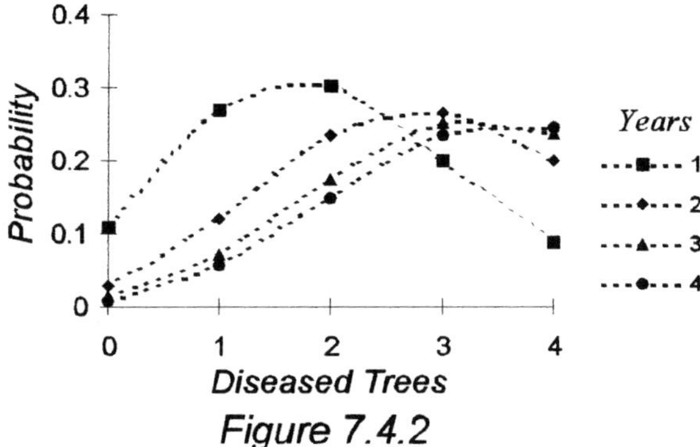

Figure 7.4.2

Example 7.4.2 Define the deterministic and the stochastic models of the system of Example 7.3.7, if the initial female is *AA*.

Solution: As denoted before, a Markov process is represented by a set of next state equations of the form $P_{n+1} = P_n Q$, where P_n is the set of states at period n and P_{n+1} is the set of states at period $n+1$. In this system n is the number of generations. The Z transform for this model is

$$F(z) = F(0)\,\frac{z}{z-Q} \quad ; \quad F(0) = \begin{bmatrix} 1 & 0 & 0 \\ 0 & 1 & 0 \\ 0 & 0 & 1 \end{bmatrix}$$

where $F(0)$ is the initial state for female mice. The first row shows that the initial female is of the *AA* genotype. The second row indicates that the female is an *Aa* and the third row shows that the female is *aa*. Then, the solution of the system is

$$P_n = \begin{bmatrix} 1 & 0 & 0 \\ 0 & 1 & 0 \\ 0 & 0 & 1 \end{bmatrix} \begin{bmatrix} 1/2 & 1/2 & 0 \\ 1/4 & 1/2 & 1/4 \\ 0 & 1/2 & 1/2 \end{bmatrix}^n$$

The above power matrix can take the form $Q^n = VD^nV^{-1}$, where V is the characteristic matrix of Q and D is the diagonal matrix of the characteristic roots of Q. Thus

$$\begin{bmatrix} 1/2 & 1/2 & 0 \\ 1/4 & 1/2 & 1/4 \\ 0 & 1/2 & 1/2 \end{bmatrix}^n = \frac{1}{4} \begin{bmatrix} -1 & 1 & 1 \\ 1 & 1 & 0 \\ -1 & 1 & -1 \end{bmatrix} \begin{bmatrix} 0 & 0 & 0 \\ 0 & 1^n & 0 \\ 0 & 0 & 0.5^n \end{bmatrix} \begin{bmatrix} -1 & 2 & -1 \\ 1 & 2 & 1 \\ 2 & 0 & -2 \end{bmatrix}$$

$$= \frac{1}{4} \begin{bmatrix} 1+2(0.5)^n & 2 & 1-2(0.5)^n \\ 1 & 2 & 1 \\ 1-2(0.5)^n & 2 & 1+2(0.5)^n \end{bmatrix}$$

After solving the product $F(0)Q^n$ and assuming that the initial female is *AA*, the genetic composition of the offspring is

$$P_n = \frac{1}{4}\left[1+2(0.5)^n, \ 2, \ 1-2(0.5)^n\right]$$

for $P_n = (p_{1n}, p_{2n}, p_{3n})$, where p_{1n} is the proportion of AA mice, p_{2n} is the proportion of *Aa* mice, p_{3n} is the proportion of *aa* mice and n is the number of generations. This expression corresponds to the deterministic model of the system that is, the expected genetic composition of the offspring.

The expected genetic composition may be better described by the expression

$$E(X_1, X_2, X_3) = \frac{m}{4}\left[1+2(0.5)^n, \ 2, \ 1-2(0.5)^n\right]$$

where X_1, X_2 and X_3 are the number of *AA*, *Aa* and *aa* offspring mice out of *m* total mice.

The state variables may have a multinomial distribution. Therefore, it is possible to define the stochastic model

$$f(x_1, x_2, x_3) = \frac{m!}{x_1! x_2! x_3!} \left(\frac{1+2(0.5)^n}{4}\right)^{x_1} \left(\frac{1}{2}\right)^{x_2} \left(\frac{1-2(0.5)^n}{4}\right)^{x_3}$$

where x_1, x_2 and x_3 are the number of *AA*, *Aa* and *aa* offspring mice out of a total of *m* mice.

Example 7.4.3 Define a deterministic model for the activity of a bird over the population of caterpillars in Example 7.1.2.

Solution: As denoted in Example 7.1.2, it was assumed that the expected success of the bird in catching caterpillars is proportional to the time interval Δt, that is

$$E(\Delta x, \Delta t) = k\Delta t$$

where X is the total number of successes and Δx is the difference between the number of successes at the end of the interval and the number of successes at the beginning. Note that the expectation of a difference is the difference of the expectations. Then, if $x = f(t)$, it follows that

$$E(\Delta x, \Delta t) = E[f(t+\Delta t)] - E[f(t)] = k\Delta t$$

The following differential equation is obtained by dividing both sides by Δt and taking the limit as Δt approaches zero:

$$\frac{d}{dt} E(X) = k$$

where $E(X)$ is the expected total number of successes by the bird. The deterministic model of the system is the solution of the above differential equation, that is

$$E(X) = kt$$

The stochastic model was defined as a Poisson distribution, that is

$$f(x) = \frac{\mu}{x!} e^{-\mu} \quad ; \quad \mu = kt$$

Thus, as shown in the above expression, the expected value $E(X)$ is the mean μ of the stochastic model of the system.

Note that the deterministic model may be used to generate the k coefficient experimentally. The k coefficient may be easy to generate by measuring the number of catches as a function of time. Then, the time interval Δt may be selected by dividing the time variable into intervals that are small enough, such that the rate of change in the number of catches remains essentially constant. If a sufficient number of events cannot be fitted in such interval, as to make the coefficient of variation acceptably small, a more appropriate function for the expected value may be chosen, that is $\mu \neq kt$. Failures, which are difficult to define or detect, do not need to be recorded.

Note also that the stochastic model could have been determined in a very simple way, just by defining the deterministic model of the system and assuming a Poisson distribution at the beginning of the modeling process. Such procedure is possible only if the assumptions for selecting the deterministic model and the distribution function are acceptable.

In general terms, defining mathematical models for stochastic processes may be accomplished by the following procedure:

- Define an appropriate probability distribution for the problem
- Define the deterministic model
- Determine experimentally the numerical coefficients for a deterministic model of the system
- Fit the probability distribution to the deterministic model

Summary

Mathematical models of stochastic systems may be determined by first defining the deterministic model and then fitting the corresponding probability distribution to the deterministic model. Deterministic models may also be useful in generating numerical coefficients for the stochastic model of the system.

DETERMINISTIC MODELS OF DISCRETE SYSTEMS

As indicated before, deterministic models represent the expected behavior of the system. It was also disclosed that discrete systems are related to qualitative state traits or to state variables representing numbers of individuals.

Discrete state variables cannot be fractionalized, meaning that the system is not differentiable and cannot be represented by differential equations. Thus, discrete systems are represented by difference equations and their solutions. Difference equations define the state changes of the system and their solutions define the state trajectories. The time scale of these systems is the set of non negative integers.

This chapter is related to the process of linking difference equations and their solutions to the system behavior and data. For such, systems are here defined according to their dimension that is, according to the number of the system components and inputs. The number of system components determines the number of first order difference equations in the mathematical model or the order of a single equation representing the system. This second and more empirical approach may be used when separation or identification of the system components is difficult or not possible.

8.1 RELATIONSHIP BETWEEN ORDER AND DIMENSION

The most elementary models are first order expressions of the form

$$y_{n+1} - y_n = x - by_n$$

where by_n is the system output and x is the single input. The above equation is reducible to the following representative form:

$$y_{n+1} + ay_n = x$$

where $a=b-1$. As illustrated in Fig. 8.1, this first order difference equation represents a one component model of a system.

Figure 8.1.1

Example 8.1.1 A rancher sells each month 3.6% of his feedlot steers and buys 90 new animals. Determine the system difference equation.

Solution: The following is the difference representing the system:

$$y_{n+1} - y_n = 90 - 0.036 y_n$$

where 90 is the input and $0.036 y_n$ is the output. This difference may be simplified to

$$y_{n+1} + 0.964 y_n = 90$$

Next in complexity are second order models reducible to the form

$$y_{n+2} + b_1 y_{n+1} + b_2 y_n = x \tag{1}$$

where b_1 and b_2 are constants and x is the system input. As will be shown, this second order difference equation is equivalent to the following set of two first order interconnected equations:

$$\begin{aligned} y_{1(n+1)} &= -c_1 y_{1(n)} + u \\ y_{2(n+1)} &= c_1 y_{1(n)} - c_2 y_{2(n)} \end{aligned} \tag{2}$$

where u is the single input and $c_1 y_{1(n)}$ and $c_2 y_{2(n)}$ are outputs. The corresponding two-component system is shown in Figure 8.1.2.

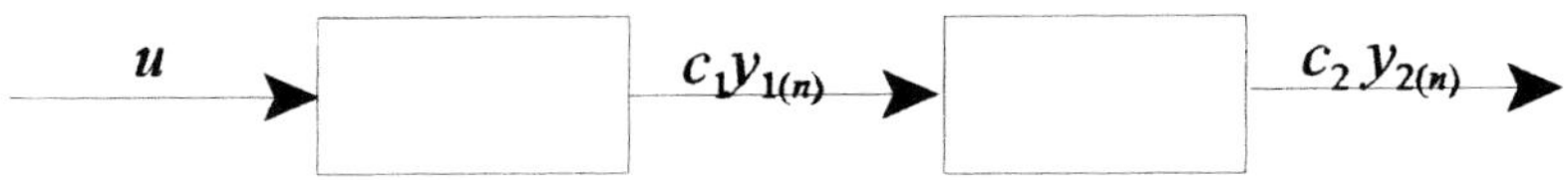

Figure 8.1.2

The equivalent second order form of the first order set (2) is determined by the procedure that follows. First, a second order expression is obtained by taking the second difference of the second equation in (2), that is

$$y_{2(n+2)} = c_1 y_{1(n+1)} - c_2 y_{2(n+1)} \tag{3}$$

Next, the term $y_{1(n+1)}$ in (3) is replaced with the first equation in the set (2):

$$y_{2(n+2)} = c_1(u - c_1 y_{1(n)}) - c_2 y_{2(n+1)}$$

Then, according to the second equation in (2), the state variable $y_{1(n)}$ is

$$y_{1(n)} = \frac{y_{2(n+1)} + c_2 y_{2(n)}}{c_1}$$

Thus

$$y_{2(n+2)} = c_1 [u - (y_{2(n+1)} + c_2 y_{2(n)}] - c_2 y_{2(n+1)} \tag{4}$$

Finally, after factorizing and rearranging terms in (4), the following second order equation is obtained:

$$y_{2(n+2)} + (c_1 + c_2) y_{2(n+1)} + c_1 c_2 y_{2(n)} = c_1 u \tag{5}$$

where $c_1 + c_2 = b_1$, $c_1 c_2 = b_2$ and $c_1 u = x$, in equation (1). Equation (5) is equivalent to equation (1) when operations are carried out on the second component of the system. The reader may want to check that the following second order equation is equivalent to equation (1), when operations are carried out on the first component:

$$y_{1(n+2)} + (c_1 + c_2)y_{1(n+1)} + c_1 c_2 y_{1(n)} = c_2 u$$

Example 8.1.2 The number of individuals of generation n is the sum of the two previous generations. Determine the difference equation for the system and the corresponding set of first order equations.

Solution: The following is the difference equation of the system:

$$y_{n+2} - y_{n+1} - y_n = 0$$

where $c_1 + c_2 = -1$ and $c_1 c_2 = -1$. Then, the following is the set of first order equations:

$$y_{1(n+1)} = -\left(\frac{-1 + \sqrt{5}}{2}\right) y_{1(n)}$$

$$y_{2(n+1)} = \left(\frac{-1 + \sqrt{5}}{2}\right) y_{1(n)} - \left(\frac{-1 - \sqrt{5}}{2}\right) y_{2(n)}$$

Generalizing the model of the system of n first order equations, equivalent to the single input n order difference equation, is now possible:

$$Y_{n+1} = \begin{bmatrix} -c_1 & 0 & \cdots & 0 & 0 \\ c_1 & -c_2 & \cdots & 0 & 0 \\ \vdots & & & & \vdots \\ 0 & 0 & \cdots & c_{n-1} & -c_n \end{bmatrix} Y_n + \begin{bmatrix} 1 \\ 0 \\ \vdots \\ 0 \end{bmatrix} u$$

where the c_i coefficients are the characteristic roots of the system and u is a single input. The following is the corresponding n order difference equation:

$$y_{n+m} + (c_1 + c_2 + ... + c_n)y_{n+(m-1)} + ... + c_1 c_2 ... c_{n y_n} = c_1 c_2 ... c_{n-1} u$$

Summary

An m order difference equation of the form $y_{n+m} + b_1 y_{n+(m-1)} + ... + b_m y_n = x$, where b_i are constants, y is the state variable and the variable x represents the single input of the system, is equivalent to m first order equations.

8.2 SINGLE INPUT LINEAR MODELS

Single input linear models are represented by difference equations reducible to the form

$$g_0(n)y_{n+m} + g_1(n)y_{n+(m-1)} + ... + g_m(n)y_n = f(n)$$

where y is the state variable, $g_i(n)$ and $f(n)$ are functions defined over the discrete time scale or are constants, for $i = 1,2,...,m$. The $g_i(n)$ terms are usually constant coefficients. In such case, the following expression would replace the above equation:

$$y_{n+m} + b_1 y_{n+(m-1)} + ... + b_m y_n = x$$

where a_i are constants and $x = f(n)$ is the single input of the system.

The notion of response functions was presented in Chapter 6. The system reaction to initial conditions, independently of the inputs, was defined as the free response. The reaction of the system to inputs, independently of the initial conditions was defined as the forced response. For conceptual purposes, only simple first order time invariant models of systems were considered. In this section, progressively more complex models will be examined.

The general expression for a first order constant coefficients model was defined as

$$y_{n+1} + b y_n = x$$

The following is the Z transform of the above equation:

$$z\,G(z) - g(0)z - b\,G(z) = F(z)$$

Then

$$G(z) = \frac{z}{z-b}g(0) + \frac{1}{z-b}F(z)$$

The first fraction in the above transform corresponds to the free response of the system, the second fraction is the forced response, $g(0)$ is the initial condition and $z-b$ is the characteristic equation. The inverse of these transform fractions are the state trajectories for the free and the forced responses.

Example 8.2.1 Determine the free and forced responses of the feedlot system of Example 8.1.1, if the rancher had 460 initial animals.

Solution: The following is the difference equation defined for the system:

$$y_{n+1} + 0.964 y_n = 90$$

This is a non homogeneous time invariant difference equation. The corresponding Z transform is

$$G(z) = \frac{460\,z}{z-0.964} + \frac{90\,z}{(z-1)(z-0.964)}$$

The first fraction of the above transform is the free response determined by the initial 460 steers. The following is its inverse:

$$y_{A(n)} = 460\,(0.964)^n$$

The second fraction is the forced response, determined by the purchase of new steers:

$$y_{B(n)} = \frac{90}{1-0.964}[1 - (0.964)^n] = 2500[1 - (0.964)^n]$$

The total response is the sum of the free and forced responses:

$$y_n = 2500 - 2040(0.964)^n$$

The graph of the response functions is shown in Fig. 8.2.1.

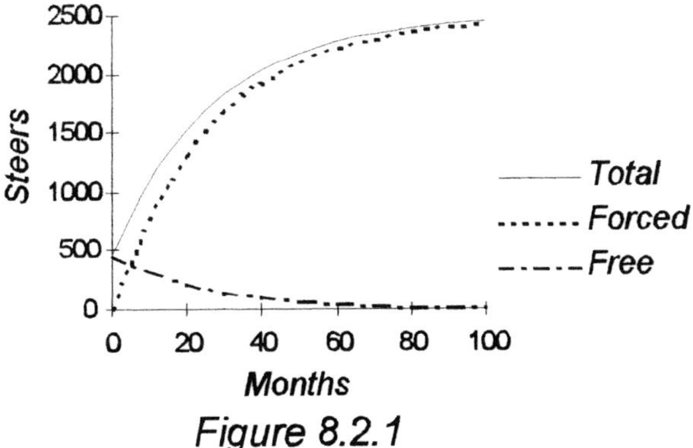

Figure 8.2.1

The following is the general expression for a second order, constant coefficients difference equation:

$$y_{n+2} + by_{n+1} + cy_n = x$$

The corresponding Z transform of this equation is

$$z^2G(z) - z^2g(0) - zg(1) + b[zG(z) - zg(0)] + cG(z) = F(z)$$

where $g(0)$ is a value at a period $n=0$ and $g(1)$ is a value at a period $n=1$. After factorizing and rearranging terms, the Z transform can be written as

$$G(z) = \frac{g(0)z(z + b) + g(1)}{(z - \lambda_1)(z - \lambda_2)} + \frac{F(z)}{(z - \lambda_1)(z - \lambda_2)}$$

where λ_1 and λ_2 are the roots of the characteristic equation $z^2 + bz + c$. The roots of a second degree polynomial are given by the following expression:

$$\lambda = \frac{-b \pm \sqrt{b^2 - 4c}}{2}$$

Then, if $(b^2 - 4c) > 0$ the polynomial has two different real roots, if $(b^2 - 4c) = 0$ the polynomial has two equal real roots and if $(b^2 - 4c) < 0$ the polynomial has two imaginary roots. This third case determines a periodic response of the system.

Note that the coefficients of the characteristic equation are the same coefficients of the difference equation. The first fraction of the transform corresponds to the free response of the system and the second fraction to the forced response.

The $(b^2 - 4c) > 0$ case is illustrated in the following example.

Example 8.2.2 An insect control program was tested for one year in a pasture field. The following difference equation was fitted to the data:

$$\Delta^2 y_n + 0.7955\Delta y_n + 0.1538 y_n = 0$$

This equation has the following expression in the subscript notation:

$$y_{n+2} - 1.2045 y_{n+1} + 0.3583 y_n = 0$$

where y is number of bugs per square meter and n is months. Find the response due to the initial count of insects and the forced response due to the pest control program. The initial count of bugs was 250. The count after one month was 425.

Solution: The above is a homogeneous second order time invariant equation. Therefore, at a first glance, only a free response is possible. However, as will be shown, a second order homogeneous time invariant equation is equivalent to a first order non homogeneous time variant equation. To find the first order expression, the solution of the second order equation is first needed.

The following is the Z transform of the difference equation:

$$z^2 G(z) - 250z^2 - 425z - 1.2045[zG(z) - 250z] + 0.3583G(z) = 0$$

Then

$$G(z) = \frac{250(z^2 - 1.2045z) + 425z}{z^2 - 1.2045z + 0.3583}$$

This transform can be expressed as

$$\frac{G(z)}{z} = 250\left[\frac{A}{z - 0.6686} + \frac{B}{z - 0.5359}\right] + 425\left[\frac{C}{z - 0.6686} + \frac{D}{z - 0.5359}\right]$$

where 0.6686 and 0.5359 are the roots of the characteristic equation. After solving the partial fractions, it was found that $A = -4.0384$, $B = 5.0384$, $C = 7.5358$ and $D = -7.5358$. Thus, the Z transform becomes

$$G(z) = \frac{2193z}{z - 0.6686} - \frac{1943z}{z - 0.5359}$$

Then, the solution of the second order difference equation is

$$y_n = 2193(0.6686)^n - 1943(0.5359)^n$$

The following is the first order difference equation of the system:

$$
\begin{aligned}
y_{n+1} &= 2193(0.6686)^{n+1} - 1943(0.5359^{n+1}) \\
&= 1466(0.6686)^n - 1041(0.5359)^n \\
&= 1466(0.6686)^n - 1041\left[\frac{2193(0.6686)^n - y_n}{1943}\right] \\
&= 290.77(0.6686)^n + 0.5359 y_n
\end{aligned}
$$

Then, after rearranging terms, the following is the first order time variant non homogeneous difference equation of the system:

$$
y_{n+1} - 0.5359 y_n = 290.77(0.6686)^n
$$

where $290.77(0.6686)^n$ is an input and $0.5359 y_n$ is an output. Defining the free and forced responses is now possible.

The new Z transform is

$$
G(z) = \frac{290.77z}{(z-0.6686)(z-0.5359)} + \frac{250z}{z-0.5359}
$$

The first fraction of the above transform is related to the pest control and the second fraction is related to the initial count of insects. Then, the following are the free response $y_{A(n)}$ and the forced response $y_{B(n)}$:

$$
\begin{aligned}
y_{A(n)} &= 250(0.5359)^n \\
y_{B(n)} &= 2193[(0.6686)^n - (0.5359)^n]
\end{aligned}
$$

As expected, the sum of these two equations is the total response determined previously. The graph of the response functions of this system is shown in Fig. 8.2.2.

The procedure for solving the $(b^2 - 4c) > 0$ case is straightforward. That may not be so with the $(b^2 - 4c) < 0$ condition, where the strategy for canceling imaginary terms may require more time and expertise. The following examples illustrate the $(b^2 - 4c) < 0$ case.

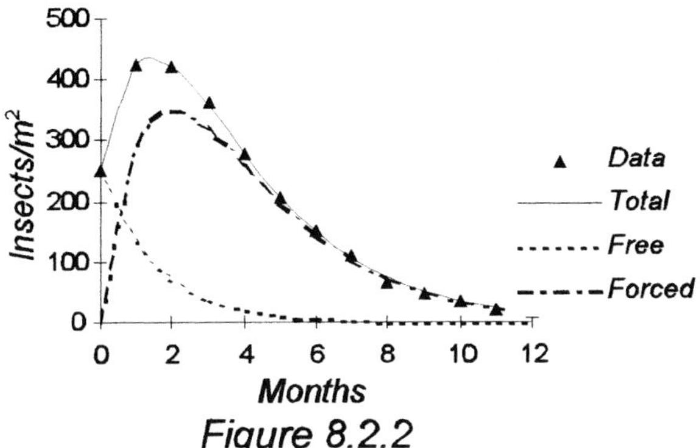

Figure 8.2.2

Example 8.2.3 It was found that the population of an animal species decreases, each generation, by approximately one half the number of animals of the previous generation. It is assumed that each generation is affected by at least the effects of two previous generations. To prevent extinction, 540 new animals are introduced with each new generation. Determine the response corresponding to the initial 1000 animals and the response due to the periodic introduction of new animals.

Solution: The system is represented by the following difference equation:

$$y_{n+2} + 0.5y_{n+1} + 0.25y_n = 540$$

The characteristic equation of the system is here

$$z^2 + 0.5z + 0.25 = \left[z + \left(\frac{1}{4} - i\frac{\sqrt{3}}{4}\right)\right]\left[z + \left(\frac{1}{4} + i\frac{\sqrt{3}}{4}\right)\right]$$

$$= (z - \lambda_1)(z - \lambda_2)$$

Thus

$$\lambda_1 = -\frac{1}{4} + \frac{i\sqrt{3}}{4} = \alpha + i\beta \quad \text{and} \quad \lambda_1 = -\frac{1}{4} - \frac{i\sqrt{3}}{4} = \alpha - i\beta$$

where $\alpha = \dfrac{-0.5}{2} = -\dfrac{1}{4}$ and $\beta = \dfrac{\sqrt{1 - 0.25}}{2} = \dfrac{\sqrt{3}}{4}$.

Then, the Z transform of the difference equation becomes

$$G(z) = \frac{1000(z^2 + 0.5z) + 375z}{(z - \lambda_1)(z - \lambda_2)} + \frac{540z}{(z - 1)(z - \lambda_1)(z - \lambda_2)}$$

where 375 is the number of original animals found after the first period. The first fraction of the transform is related to the response due to the 1000 initial animals. The second fraction is related to the response due to the introduced animals.

The first fraction of the transformed equation may be expressed as follows:

$$\frac{G(z)}{z} = \frac{1000z + 875}{(z - \lambda_1)(z - \lambda_2)} = \frac{A}{z - \lambda_1} + \frac{B}{z - \lambda_2}$$

After solving for A and B, the Z transform of the free response becomes

$$G(z) = \left[\frac{1000\lambda_1 + 875}{\lambda_1 - \lambda_2}\right]\frac{z}{z - \lambda_1} - \left[\frac{1000\lambda_2 + 875}{\lambda_1 - \lambda_2}\right]\frac{z}{z - \lambda_2}$$

Thus, the following is the free response of the system:

$$y_{A(n)} = \left[\frac{1000\lambda_1 + 875}{\lambda_1 - \lambda_2}\right]\lambda_1^n - \left[\frac{1000\lambda_2 + 875}{\lambda_1 - \lambda_2}\right]\lambda_2^n$$

As disclosed in Chapter 4.2, by the Euler's formula:

$$\lambda_1 = \alpha + i\beta = r(\cos\theta + i\sin\theta) = re^{i\theta},$$
$$\lambda_2 = \alpha - i\beta = r(\cos\theta - i\sin\theta) = re^{-i\theta}$$

where $r = \sqrt{\alpha^2 + \beta^2} = 1/2$. Then, $\lambda_1 - \lambda_2 = 2i\beta$, $\lambda_1 = e^{i\theta}/2$ and $\lambda_2 = e^{-i\theta}/2$.

After replacing values, the free response becomes

$$y_{A(n)} = \left[\frac{1000(\alpha + i\beta) + 875}{2i\beta}\right](re^{i\theta})^n - \left[\frac{1000(\alpha - i\beta) + 875}{2i\beta}\right](re^{-i\theta n})$$

where

$$e^{in\theta} = \cos n\theta + i\sin n\theta$$

$$e^{-in\theta} = \cos n\theta - i\sin n\theta$$

$$\cos\theta = \frac{\alpha}{r} = -\frac{1}{2}$$

$$\sin\theta = \frac{\beta}{r} = \frac{\sqrt{3}}{2}$$

$$\theta = \cos^{-1}\left(\frac{-1}{2}\right) = \sin^{-1}\left(\frac{\sqrt{3}}{2}\right) = \frac{2\pi}{3}$$

Finally, after replacing values and factorizing, the free response is

$$y_{A(n)} = \left(\frac{1}{2}\right)^{n}\left[1000\cos n\frac{2\pi}{3} + \frac{2500}{\sqrt{3}}\sin n\frac{2\pi}{3}\right]$$

The following is the fraction of the Z transform representing the forced response:

$$\frac{G(z)}{z} = \frac{540}{(z-1)(z-\lambda_1)(z-\lambda_2)} = 540\left[\frac{A}{z-1} + \frac{B}{z-\lambda_1} + \frac{C}{z-\lambda_2}\right]$$

After solving for A, B and C, the Z transform of the forced response becomes

$$G(z) = 540\left[\frac{z}{(1-\lambda_1)(1-\lambda_2)(z-1)} - \frac{z}{(\lambda_1-\lambda_2)(1-\lambda_1)(z-\lambda_1)} + \frac{z}{(\lambda_1-\lambda_2)(1-\lambda_2)(z-\lambda_2)}\right]$$

Thus, the following is the forced response of the system:

$$y_{B(n)} = 540\left[\frac{1}{(1-\lambda_1)(1-\lambda_2)} - \frac{\lambda_1^{n}}{(\lambda_1-\lambda_2)(1-\lambda_1)} + \frac{\lambda_2^{n}}{(\lambda_1-\lambda_2)(1-\lambda_2)}\right]$$

$$= \frac{540}{(1-\lambda_1)(1-\lambda_2)}\left[1 - \frac{(1-\lambda_2)\lambda_1^{n}}{\lambda_1-\lambda_2} + \frac{(1-\lambda_1)\lambda_2^{n}}{\lambda_1-\lambda_2}\right]$$

Note that the characteristic equation of the system is

$$z^2 - (\lambda_1 + \lambda_2)z + \lambda_1\lambda_2 = z^2 + 0.50z + 0.25$$

Then, $(1 - \lambda_1)(1 - \lambda_2) = 1 - (\lambda_1 + \lambda_2) + \lambda_1\lambda_2 = 1.75$. After replacing the λ values with Euler's equations, the forced response becomes

$$y_{B(n)} = \frac{540}{1.75}\left[1 + \left(\frac{1}{2}\right)^n\left[\frac{(\alpha - i\beta - 1)(\cos n\theta + i\sin n\theta)}{2i\beta} - \frac{(\alpha + i\beta - 1)(\cos n\theta - i\sin n\theta)}{2i\beta}\right]\right]$$

Finally, after replacing the α, β and θ values and factorizing, the resulting forced response of the system is

$$y_{B(n)} = 309\left[1 - \left(\frac{1}{2}\right)^n\left(\cos n\frac{2\pi}{3} + \frac{5}{\sqrt{3}}\sin n\frac{2\pi}{3}\right)\right]$$

The total response is the sum of the free and the forced responses, that is

$$y_n = 309 + \left(\frac{1}{2}\right)^n\left[691\cos n\frac{2\pi}{3} + 662\sin n\frac{2\pi}{3}\right]$$

The graph of the system responses is shown in Fig. 8.2.3.

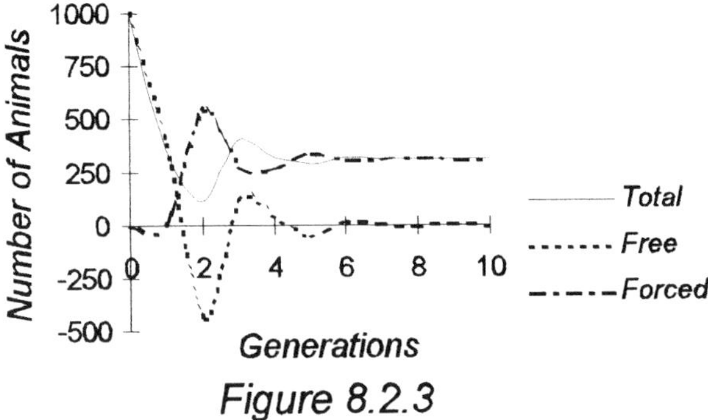

Figure 8.2.3

The above responses may be defined by expressions with a very precise geometrical meaning. Note that, by the addition formula of sinus and cosines

$$\cos(A \pm B) = \cos A \cos B \mp \sin A \sin B$$
$$a\cos[\theta(c \pm n)] = a\cos\theta c \cos\theta n \mp a\sin\theta c \sin\theta n$$
$$= 691\cos n\frac{2\pi}{3} + 662\sin n\frac{2\pi}{3}$$

where $a\cos\theta c = 691 = K_1$, $a\sin\theta c = 662 = K_2$, $\theta = 2\pi/3$ and a and c are unknowns. Then

$$a = \frac{691}{\cos\dfrac{2\pi}{3}c} = \frac{662}{\sin\dfrac{2\pi}{3}c} \quad \text{and} \quad \frac{\sin\dfrac{2\pi}{3}c}{\cos\dfrac{2\pi}{3}c} = \frac{662}{691} = 0.9580$$

such that

$$\tan^{-1}(0.9580) = 0.7640 = \frac{2\pi}{3}c \quad ; \quad c = 0.3648$$

Finding a is now possible:

$$a = \frac{691}{\cos\left(\dfrac{2\pi}{3}0.3348\right)} = \frac{662}{\sin\left(\dfrac{2\pi}{3}0.3648\right)} = 957$$

Note that $a\cos[\theta(c-n)] = a\cos[\theta(n-c)]$. Then, the new expression for the system response has the form

$$y_n = K_0 + ar^n\cos[\theta(n-c)]$$

After replacing values, the total response of the system becomes

$$y_n = 309 + 957\left(\frac{1}{2}\right)^n\cos\left[\frac{2\pi}{3}(n - 0.3648)\right]$$

The following definitions apply for this response model:

- Parameter θ modulates the frequency response of the system
- Parameter r modulates the amplitude response of the system
- Coefficient a modifies parameter r
- Coefficient c is an out of phase parameter
- Coefficient d is the distance between the abscissa and the response axes
- A cycle is equal to $2\pi/\theta$

Note that when $r < 1$, the amplitude decreases over time and if $r > 1$, the amplitude increases. Thus, for $r = 1/2$, the amplitude of the system response decreases over generations. Note also that, if one full cycle is 2π, then $\theta = 2\pi/3$ means that one cycle is here three generations. In this example, the function is $c = 0.3218$ generations out of phase and has an asymptotic value of $d = 309$ animals.

The reader is encouraged to determine why the following relationships hold here:

$$y_n = K_0 + (r)^n[K_1\cos n\theta + K_2\sin n\theta]$$
$$y_{n+2} = K_0(1 - 2r\cos\theta + r^2) + 2r\cos\theta y_{n+1} - r^2 y_n$$
$$\lambda_1 + \lambda_2 = 2r\cos\theta$$
$$\lambda_1\lambda_2 = r^2$$

Summary

Single input linear models with constant coefficients are represented by equations reducible to the form $y_{n+m} + b_1 y_{n+(m-1)} + ... + b_m y_n = x$, where b_i are constants, y is the state variable and the variable x represents the single input of the system.

8.3 MULTIDIMENSIONAL FIRST ORDER LINEAR MODELS

First order multidimensional linear models are represented by difference equations reducible to the form

$$Y_{n+1} = AY_n + X$$

where Y is a set of state variables, A is a matrix of coefficients defining the relations between the state variables and X is the set of input functions of the system.

The Z transform of the above equation is written as follows:

$$zG(z) - zG(0) = AG(z) + F(z)$$
$$(zI - A)G(z) = zG(0) + F(z)$$

where $(zI - A)$ is the characteristic equation of the system, $G(z)$ is the set of Z transforms of the state variables, $G(0)$ is a set of initial conditions and $F(z)$ is the set of transforms of the input functions. Then, the free response $Y_{A(n)}$ and the forced response $Y_{B(n)}$ are

$$Y_{A(n)} = Z^{-1}\left[(zI - A)^{-1} z G(0)\right]$$
$$Y_{B(n)} = Z^{-1}\left[(zI - A)^{-1} F(z)\right]$$

Example 8.3.1 The population of a type of bird doubles every year. The introduction of predators reduces the number of birds by ten times the number of predators. The number of predators also doubles every year. Some 200 new birds move into the ecosystem each year and some 30 predators are hunted down. Determine the response functions of the system, assuming 1000 initial birds and 50 initial predators.

Solution: The difference equation of the system is as follows:

$$Y_{n+1} = \begin{bmatrix} 2 & -10 \\ 0 & 2 \end{bmatrix} Y_n + \begin{bmatrix} 200 \\ -30 \end{bmatrix}$$

where y_A is birds, y_B is predators and n is years. The following is the corresponding Z transform:

$$\begin{bmatrix} z-2 & 10 \\ 0 & z-2 \end{bmatrix} G(z) = \begin{bmatrix} 1000 \\ 50 \end{bmatrix} z + \begin{bmatrix} 200 \\ -30 \end{bmatrix} \frac{z}{z-1}$$

The characteristic equation of the system is here

$$|zI - A| = \begin{vmatrix} z-2 & 10 \\ 0 & z-2 \end{vmatrix} = (z-2)^2$$

Free Response. The following is the Z transform for the free response of the birds:

$$G_{A1}(z) = \frac{1}{(z-2)^2} \begin{vmatrix} 1000z & 10 \\ 50z & z-2 \end{vmatrix} = \frac{z(1000z - 2500)}{(z-2)^2}$$

$$\frac{G_{A1}(z)}{z^2} = \frac{1000z - 2500}{z(z-2)^2} = \frac{A}{z(z-2)} + \frac{B}{(z-2)^2}$$

where $A = 1250$ and $B = -250$. Then

$$G_{A1}(z) = \frac{1250z}{z-2} - \frac{250z^2}{(z-2)^2}$$

The following is the Z transform for the free response of the predators:

$$G_{A2}(z) = \frac{1}{(z-2)^2} \begin{vmatrix} z-2 & 1000z \\ 0 & 50z \end{vmatrix} = \frac{50z}{z-2}$$

Thus, the inverse of the above transforms is the free response of the system:

$$Y_{A(n)} = \begin{bmatrix} 250(4 - n) \\ 50 \end{bmatrix} 2^n$$

Forced Response. The following is the Z transform for the forced response of the birds:

$$G_{B1}(z) = \frac{1}{(z-2)^2} \begin{vmatrix} \dfrac{200z}{z-1} & 10 \\ \dfrac{-30z}{z-1} & z-2 \end{vmatrix} = \frac{200z}{(z-2)(z-1)} + \frac{300z}{(z-2)^2(z-1)}$$

$$= \frac{200z}{(z-2)(z-1)} + 300z\left(\frac{A}{(z-2)^2} + \frac{B}{z-1} + \frac{C}{z-2} \right)$$

where $A = 1$, $B = 1$ and $C = -1$. Then

$$G_{BI}(z) \; = \; \frac{200z}{(z-2)(z-1)} + 150\left(\frac{2z}{(z-2)^2} + \frac{2z}{z-1} - \frac{2z}{z-2}\right)$$

The following is the Z transform for the forced response of the predators:

$$G_{B2}(z) \; = \; \frac{1}{(z-2)^2}\begin{vmatrix} z-2 & \dfrac{200z}{z-1} \\ 0 & \dfrac{-30z}{z-1} \end{vmatrix} = \frac{-30z}{(z-2)(z-1)}$$

The inverse transform of the birds and the predators is the forced response of the system:

$$Y_{B(n)} \; = \; \begin{bmatrix} 100 \\ 30 \end{bmatrix} + \begin{bmatrix} 50(3n-2) \\ -30 \end{bmatrix}2^n$$

The following is the total response of the system:

$$Y_n \; = \; \begin{bmatrix} 100 \\ 30 \end{bmatrix} + \begin{bmatrix} 100(9-n) \\ 20 \end{bmatrix}2^n$$

The graph of the response functions of the birds is shown in Fig. 8.3.1:

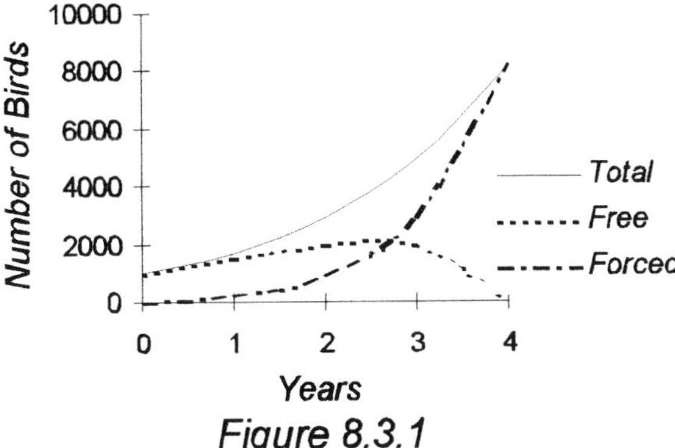

Figure 8.3.1

Testing that the solution is correct is possible by equating the system solution with the difference equations of the system, such that

$$Y_n = \begin{bmatrix} 100 \\ 30 \end{bmatrix} + \begin{bmatrix} 100(9-n) \\ 20 \end{bmatrix} 2^n = \begin{bmatrix} 2 & -10 \\ 0 & 2 \end{bmatrix}^{-1} \left[Y_{n+1} - \begin{bmatrix} 200 \\ -30 \end{bmatrix} \right]$$

The reader may wish to check that the above expression holds true.

The graph of the response functions of the predators is shown in Fig. 8.3.2:

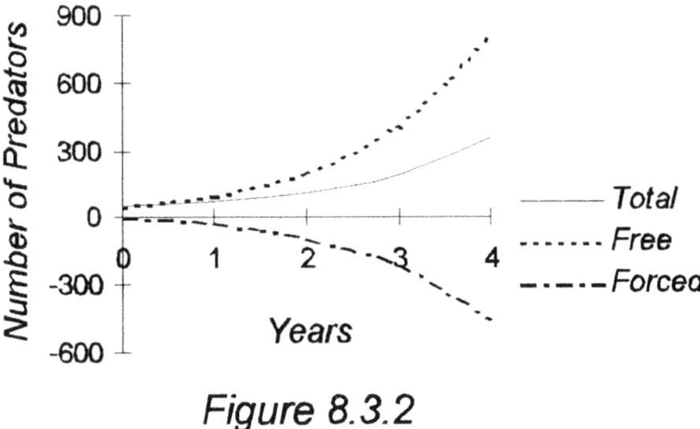

Figure 8.3.2

Example 8.3.2 Two species of birds share the same ecosystem and food sources, affecting each other's reproductive and survival rates. Their relationship is represented by the following set of difference equations:

$$Y_{n+1} = \begin{bmatrix} 0.15934 & 2.20254 \\ -0.45281 & 0.87340 \end{bmatrix} Y_n + \begin{bmatrix} 0 \\ 224.27 \end{bmatrix} + \begin{bmatrix} 0 \\ 5.3174 \end{bmatrix} n$$

where y_1 and y_2 is each of the bird species and n is years. Find the response functions of the system.

Solution: The above set of difference equations has the form

$$y_{n+1} = \begin{bmatrix} b_{11} & b_{12} \\ b_{21} & b_{22} \end{bmatrix} Y_n + \begin{bmatrix} c_1 \\ c_2 \end{bmatrix} + \begin{bmatrix} d_1 \\ d_2 \end{bmatrix} n$$

Then, the following is the Z transform of the system difference equations:

$$\begin{bmatrix} z-b_{11} & -b_{12} \\ -b_{21} & z-b_{22} \end{bmatrix} G(z) = \begin{bmatrix} g_1(0) \\ g_2(0) \end{bmatrix} z + \begin{bmatrix} c_1 \\ c_2 \end{bmatrix} \frac{z}{z-1} + \begin{bmatrix} d_1 \\ d_2 \end{bmatrix} \frac{z}{(z-1)^2}$$

where $g_1(0) = 400$ and $g_2(0) = 200$ are the initial number of birds for each of the two species. The characteristic equation of the system is given by the expansion of the following determinant:

$$|zI - A| = \begin{vmatrix} z-b_{11} & -b_{12} \\ b_{21} & z-b_{22} \end{vmatrix} = \begin{vmatrix} z-0.15934 & -2.20154 \\ 0.45281 & z-0.87340 \end{vmatrix}$$

$$= z^2 - 1.03274z + 1.13605 = (z-\lambda_1)(z-\lambda_2)$$

Then, the following are the characteristic roots:

$$\lambda = \frac{1.03274 \pm i1.86484}{2} = \alpha \pm i\beta$$

where $\alpha = 0.51637$ and $\beta = 0.93242$. Clearly, the system response is represented by periodic functions.

Free Response. The following determinant is the Z transform of the free response for the first species:

$$G_{A1}(z) = \frac{1}{(z-\lambda_1)(z-\lambda_2)} \begin{vmatrix} g_1(0)z & -b_{12} \\ g_2(0)z & z-b_{22} \end{vmatrix}$$

$$= \frac{g_1(0)z^2 - [g_1(0)b_{22} - g_2(0)b_{12}]z}{(z-\lambda_1)(z-\lambda_2)} = A\frac{z}{z-\lambda_1} + B\frac{z}{z-\lambda_2}$$

where $A = \dfrac{g_1(0)\lambda_1 - [g_1(0)b_{22} - g_2(0)b_{12}]}{\lambda_1 - \lambda_2}$ and $B = -\dfrac{g_1(0)\lambda_2 - [g_1(0)b_{22} - g_2(0)b_{12}]}{\lambda_1 - \lambda_2}$.

Then, the inverse of the above transform is the free response of the first species:

$$y_{AI(n)} = \frac{g_1(0)\lambda_1 - [g_1(0)b_{22} - g_2(0)b_{12}]}{\lambda_1 - \lambda_2}\lambda_1^n - \frac{g_1(0)\lambda_2 - [g_1(0)b_{22} - g_2(0)b_{12}]}{\lambda_1 - \lambda_2}\lambda_2^n$$

After replacing the λ terms, the free response becomes

$$y_{AI(n)} = \frac{r^n}{2i\beta}\left[\begin{array}{l}[g_1(0)(\alpha+i\beta) - [g_1(0)b_{22} - g_2(0)b_{12}]](\cos n\theta + i\sin n\theta) - \\ [g_1(0)(\alpha-i\beta) - [g_1(0)b_{22} - g_2(0)b_{12}]](\cos n\theta - i\sin n\theta)\end{array}\right]$$

where

$$(\alpha \pm i\beta)^n = r^n(\cos n\theta \pm i\sin n\theta)$$

$$r = \sqrt{\alpha^2 + \beta^2} = 1.066$$

$$\theta = \cos^{-1}\left(\frac{\alpha}{r}\right) = \sin^{-1}\left(\frac{\beta}{r}\right) = 1.065$$

Finally, after replacing the r, α, β and θ values and factorizing, the free response for the first species is

$$y_{AI(n)} = (1.066)^n[400\cos n(1.065) + 319\sin n(1.065)]$$

The following determinant is the Z transform of the free response of the second species:

$$\begin{aligned}G_{A2}(z) &= \frac{1}{(z-\lambda_1)(z-\lambda_2)}\begin{vmatrix} z-b_{11} & g_1(0)z \\ -b_{21} & g_2(0)z \end{vmatrix} \\ &= \frac{g_2(0)z^2 - [g_2(0)b_{11} - g_1(0)b_{21}]z}{(z-\lambda_1)(z-\lambda_2)} = A\frac{z}{z-\lambda_1} + B\frac{z}{z-\lambda_2}\end{aligned}$$

After solving the partial fractions, and replacing the symbols by the corresponding initial values, the transform becomes

$$\bar{y}_{A2}(z) = \frac{200\lambda_1 - 212.992}{\lambda_1 - \lambda_2}\frac{z}{z-\lambda_1} - \frac{200\lambda_2 - 212.992}{\lambda_1 - \lambda_2}\frac{z}{z-\lambda}$$

Then, the free response of the second species is the inverse transform of the above expression:

$$y_{A2(n)} = \frac{200\lambda_1 - 212.992}{\lambda_1 - \lambda_2}\lambda_1^n - \frac{200\lambda_2 - 212.992}{\lambda_1 - \lambda_2}\lambda_2^n$$

After replacing the λ values, the free response is

$$y_{A1(n)} = \frac{r^n}{2i\beta}\left[\begin{array}{l}[200(\alpha+i\beta)-149.256](\cos n\theta + i\sin n\theta) + \\ [149.256-200(\alpha-i\beta)](\cos n\theta - i\sin n\theta)\end{array}\right]$$

Finally, after replacing the r, α, β and θ values and factorizing, the free response of the second species becomes

$$y_{A2(n)} = (1.066)^n[200\cos n(1.065) - 118\sin n(1.065)]$$

Then, the following is free response of the system:

$$Y_{A(n)} = (1.066)^n\begin{vmatrix}400 & -319 \\ 200 & -119\end{vmatrix}\begin{vmatrix}\cos n(1.065) \\ \sin n(1.065)\end{vmatrix}$$

Forced Response The Z transform of the forced response for the first species is given by the expansion of the following determinants:

$$G_{BI}(z) = \frac{1}{(z-\lambda_1)(z-\lambda_2)}\left[\begin{vmatrix} c_1 & -b_{12} \\ \dfrac{c_2 z}{z-1} & z-b_{21} \end{vmatrix} + \begin{vmatrix} d_1 & -b_{12} \\ \dfrac{d_2 z}{(z-1)^2} & z-b_{21} \end{vmatrix}\right]$$

where $c_1 = 0$ and $d_1 = 0$. Then

$$G_{BI}(z) = \frac{b_{12}z}{(z-\lambda_1)(z-\lambda_2)}\left(\frac{c_2}{z-1} + \frac{d_2}{(z-1)^2}\right)$$

The above transform can be expressed as partial fractions, such that

$$G_{BI}(z) = b_{12}\left[\begin{array}{l} c_2\left[A\dfrac{z}{z-\lambda_1} + B\dfrac{z}{z-\lambda_2} + C\dfrac{z}{z-1}\right] + \\[2em] d_2\left[D\dfrac{z}{(z-\lambda_1)(z-1)} + E\dfrac{z}{(z-\lambda_2)(z-1)} + F\dfrac{z}{(z-1)^2}\right]\end{array}\right]$$

where $A = -\dfrac{1}{(\lambda_1-\lambda_2)(1-\lambda_1)}$, $B = \dfrac{1}{(\lambda_1-\lambda_2)(1-\lambda_2)}$, $C = \dfrac{1}{(1-\lambda_1)(1-\lambda_2)}$, $D=A$, $E=B$ and $F=C$.

The inverse of the above transform is the forced response of the first species:

$$y_{BI(n)} = b_{12}c_2\left[\frac{-\lambda_1^n}{(\lambda_1-\lambda_2)(1-\lambda_1)} + \frac{\lambda_2^n}{(\lambda_1-\lambda_2)(1-\lambda_2)} + \frac{1}{(1-\lambda_1)(1-\lambda_2)}\right] +$$

$$b_{12}d_2\left[-\frac{1-\lambda_1^n}{(\lambda_1-\lambda_2)(1-\lambda_1)^2} + \frac{1-\lambda_2^n}{(\lambda_1-\lambda_2)(1-\lambda_2)^2} + \frac{n}{(1-\lambda_1)(1-\lambda_2)}\right]$$

Manipulation is easier if the above equation is expressed as follows:

$$y_{BI(n)} = \frac{b_{12}c_2}{(1-\lambda_1)(1-\lambda_2)}\left[\frac{(1-\lambda_1)\lambda_2^n - (1-\lambda_2)\lambda_1^n}{(\lambda_1-\lambda_2)} + 1\right] +$$

$$\frac{b_{12}d_2}{(1-\lambda_1)(1-\lambda_2)}\left[\frac{(1-\lambda_2)^2\lambda_1^n - (1-\lambda_1)^2\lambda_2^n}{(\lambda_1-\lambda_2)(1-\lambda_1)(1-\lambda_2)} + \frac{\lambda_1+\lambda_2-2}{(1-\lambda_1)(1-\lambda_2)} + n\right]$$

The characteristic equation of the system is

$$z^2 - 1.03274z + 1.13605 = (z-\lambda_1)(z-\lambda_2)$$

Then

$$(1-\lambda_1)(1-\lambda_2) = 1 - 1.03274 + 1.13605 = 1.10331$$

Thus, after replacing the λ values, the forced response becomes

$$y_{BI(n)} = b_{12}c_2 r^n\left[\frac{(1-\alpha-i\beta)(\cos n\theta - i\sin n\theta) - (1-\alpha+i\beta)(\cos n\theta + i\sin n\theta)}{2(1.1033)i\beta}\right] +$$

$$b_{12}d_2 r^n\left[\frac{[(1-\alpha)^2+\beta^2+2i\beta](\cos n\theta+i\sin n\theta) - [(1-\alpha)^2+\beta^2-2i\beta](\cos n\theta-i\sin n\theta)}{2(1.1033)^2 i\beta}\right] +$$

$$\frac{b_{12}c_2}{1.1033} - \frac{b_{12}d_2}{1.1033}\left[\frac{2(1-\alpha)}{1.1033} - n\right]$$

Finally, after replacing the r, α, β and θ values and factorizing, the following is the forced response for the first species:

$$y_{BI(n)} = 438 + 10.62n - (1.066)^n\left[438\cos n(1.065) + 218\sin n(1.065)\right]$$

The Z transform of the forced response for the second species is the expansion of the following determinants:

$$G_{B2}(z) = \frac{1}{(z-\lambda_1)(z-\lambda_2)}\left[\begin{vmatrix} z-b_{11} & c_1 \\ -b_{21} & \dfrac{c_2 z}{z-1} \end{vmatrix} + \begin{vmatrix} z-b_{11} & d_1 \\ -b_{21} & \dfrac{d_2 z}{(z-1)^2} \end{vmatrix}\right]$$

where $c_1 = 0$ and $d_1 = 0$. Then

$$G_{B2}(z) = \frac{c_2 z(z-b_{11})}{(z-\lambda_1)(z-\lambda_2)(z-1)} + \frac{d_2 z(z-b_{11})}{(z-\lambda_1)(z-\lambda_2)(z-1)^2}$$

By using partial fractions expansion, the above transform becomes

$$G_{B2}(z) = c_2\left[-\frac{\lambda_1-b_{11}}{(\lambda_1-\lambda_2)(1-\lambda_1)}\frac{z}{z-\lambda_1} + \frac{\lambda_2-b_{11}}{(\lambda_1-\lambda_2)(1-\lambda_2)}\frac{z}{z-\lambda_2} + \frac{1-b_{11}}{(1-\lambda_1)(1-\lambda_2)}\frac{z}{z-1}\right] +$$

$$d_2\left[-\frac{\lambda_1-b_{11}}{(\lambda_1-\lambda_2)(1-\lambda_1)}\frac{z}{(z-\lambda_1)(z-1)} + \frac{\lambda_2-b_{11}}{(\lambda_1-\lambda_2)(1-\lambda_2)}\frac{z}{(z-\lambda_2)(z-1)} + \frac{1-b_{11}}{(1-\lambda_1)(1-\lambda_2)}\frac{z}{(z-1)^2}\right]$$

The inverse of this expression is the forced response of the second species. Manipulation is made easier if the above equation is expressed as follows:

$$y_{B2(n)} = \frac{c_2}{(1-\lambda_1)(1-\lambda_2)}\left[\frac{(b_{11}-\lambda_1)(1-\lambda_2)\lambda_1^n - (b_{11}-\lambda_2)(1-\lambda_1)\lambda_2^n}{(\lambda_1-\lambda_2)} + 1-b_{11}\right] +$$

$$\frac{d_2}{(1-\lambda_1)(1-\lambda_2)}\left[\frac{(b_{11}-\lambda_1)(1-\lambda_2)^2(1-\lambda_1^n) - (b_{11}-\lambda_2)(1-\lambda_1)^2(1-\lambda_2^n)}{(\lambda_1-\lambda_2)(1-\lambda_1)(1-\lambda_2)} + (1-b_{11})n\right]$$

The reader is encouraged to determine the missing steps in the above inversion process. By using the Euler's equations, it is possible to write

$$y_{B2(n)} = c_2 r^n \left[\frac{(b_{11}-\lambda_1)(1-\lambda_2)(\cos n\theta + i\sin n\theta) - (b_{11}-\lambda_2)(1-\lambda_1)(\cos n\theta - i\sin n\theta)}{2[(1-\lambda_1)(1-\lambda_2)]i\beta} \right]$$

$$- d_2 r^n \left[\frac{(b_{11}-\lambda_1)(1-\lambda_2)^2(\cos n\theta + i\sin n\theta) - (b_{11}-\lambda_2)(1-\lambda_1)^2(\cos n\theta - i\sin n\theta)}{2[(1-\lambda_1)(1-\lambda_2)]^2 i\beta} \right]$$

$$+ d_2 \left[\frac{(b_{11}-\lambda_1)(1-\lambda_2)^2 - (b_{11}-\lambda_2)(1-\lambda_1)^2}{2[(1-\lambda_1)(1-\lambda_2)]^2 i\beta} \right] + \frac{c_2(1-b_{11}) + d_2(1-b_{11})n}{(1-\lambda_1)(1-\lambda_2)}$$

Replacing the λ values is now easier. After some factorization, the forced response of the second species becomes

$$y_{B2(n)} = 224.27 r^n \left[\frac{2i\beta(0.15934-1)\cos n\theta + 2[0.15934(1-\alpha)-\alpha+\alpha^2+\beta^2]i\sin n\theta}{2i\beta(1.1033)} \right] -$$

$$5.3174 r^n \left[\frac{\dfrac{2i\beta[2(0.15934)(1-\alpha)+\alpha^2+\beta^2-1]\cos n\theta}{2i\beta(1.1033)^2} +}{\dfrac{2[0.15934[(1-\alpha)^2+\beta^2]+2(\alpha^2+\beta^2)(1-\alpha)-\alpha]i\sin n\theta}{2i\beta(1.1033)^2}} \right] + \frac{224.27(1-0.15934)}{1.1033} +$$

$$5.5154 \left[\frac{2i\beta[2(0.15934)(1-\alpha)+\alpha^2+\beta^2-1]}{2i\beta(1.1033)^2} \right] + \frac{5.3174(1-0.15934)n}{1.1033}$$

Finally, after replacing the r, α, β and θ values and factorizing, the forced response of the second species is written as follows:

$$y_{B2(n)} = 172 + 4.052n + (1.066)^n \left[-172\cos n(1.065) + 148\sin n(1.065) \right]$$

The forced response of the system is, then

$$Y_{B(n)} = \begin{bmatrix} 438 \\ 172 \end{bmatrix} + \begin{bmatrix} 10.62 \\ 4.05 \end{bmatrix} n - (1.066)^n \begin{bmatrix} 438 & 218 \\ 172 & -148 \end{bmatrix} \begin{bmatrix} \cos n(1.065) \\ \sin n(1.065) \end{bmatrix}$$

The following is the total response of the system:

$$Y_n = \begin{bmatrix} 438 \\ 172 \end{bmatrix} + \begin{bmatrix} 10.62 \\ 4.05 \end{bmatrix} n + (1.066)^n \begin{bmatrix} -38.3 & 100.8 \\ 27.9 & 30.3 \end{bmatrix} \begin{bmatrix} \cos n(1.065) \\ \sin n(1.065) \end{bmatrix}$$

The graph of the response functions of the first species is shown in Fig. 8.3.3. Note that one full cycle is $2\pi/\theta$ that is, the frequency is here $6.28/1.065 = 5.90$ years. Note also that $r = 1.066 > 1$ that is, the amplitude of the system response increases with time.

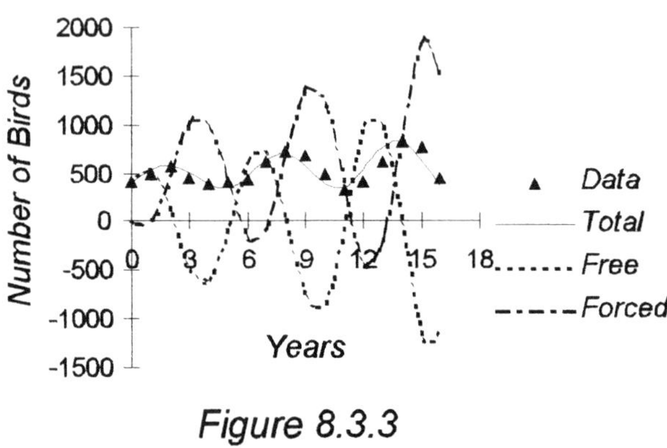

Figure 8.3.3

The graph of the response functions of the second species is shown in Fig. 8.3.4:

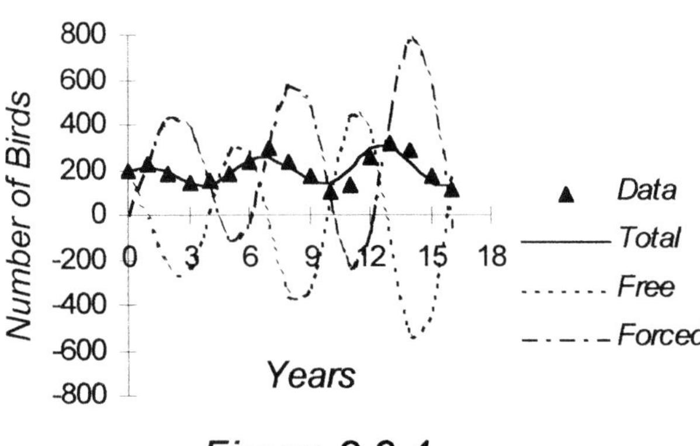

Figure 8.3.4

Summary

First order multidimensional linear models are represented by equations reducible to the form $Y_{n+1} = AY_n + X$, where Y is the set of state variables, A is a matrix of constant coefficients and X is the set of input functions of the system.

8.4 FITTING MODELS TO DATA OF DISCRETE SYSTEMS

As disclosed previously, successive differences can be expressed as a difference table. The entries in each column after the second are placed between two successive entries of the preceding column and are equal to the difference between those entries. As shown in Table 8.4.1, $\Delta y_0 = y_1 - y_0$, $\Delta y_1 = y_2 - y_1$, $\Delta^2 y_0 = \Delta y_1 - \Delta y_0$, $\Delta^3 y_0 = \Delta^2 y_1 - \Delta^2 y_0$, and so on.

Table 8.4.1

A Table of Finite Differences						
n	y	Δy	$\Delta^2 y$	$\Delta^3 y$	$\Delta^4 y$	$\Delta^5 y$
0	y_0					
		Δy_0				
1	y_1		$\Delta^2 y_0$			
		Δy_1		$\Delta^3 y_0$		
2	y_2		$\Delta^2 y_1$		$\Delta^4 y_0$	
		Δy_2		$\Delta^3 y_1$		$\Delta^5 y_0$
3	y_3		$\Delta^2 y_2$		$\Delta^4 y_1$	
		Δy_3		$\Delta^3 y_2$		
4	y_4		$\Delta^2 y_3$			
		Δy_4				
5	y_5					

For processing data, the entries in the difference table may be rearranged as in the following table. Each column in the table is a variable and as such, least squares procedures are feasible for fitting difference equations to the data. Thus, the following linear regression models would fit data with one dependent variable:

$$\Delta y \;=\; a + bn + cy \qquad\qquad\qquad \textit{First order}$$
$$\Delta^2 y \;=\; a + bn + c_1 y + c_2 \Delta y \qquad\qquad \textit{Second order}$$
$$\vdots \qquad\qquad \vdots$$
$$\Delta^n y \;=\; a + bn + c_1 y + c_2 \Delta y + \ldots + c_n \Delta^{n-1} y \quad \textit{n-order}$$

Table 8.4.2

A Modified Table of Finite Differences						
n	y_n	Δy_n	$\Delta^2 y_n$	$\Delta^3 y_n$	$\Delta^4 y_n$	$\Delta^5 y_n$
0	y_0	Δy_0	$\Delta^2 y_0$	$\Delta^3 y_0$	$\Delta^4 y_0$	$\Delta^5 y_0$
1	y_1	Δy_1	$\Delta^2 y_1$	$\Delta^3 y_1$	$\Delta^4 y_1$	
2	y_2	Δy_2	$\Delta^2 y_2$	$\Delta^3 y_2$		
3	y_3	Δy_3	$\Delta^2 y_3$			
4	y_4	Δy_4				
5	y_5					

Example 8.4.1 As indicated in Example 8.2.2, an insect control program was tested during one year in a pasture field. The following is the corresponding data:

n	0	1	2	3	4	5	6	7	8	9	10	11	12
y	250	425	421	362	279	205	151	108	65	49	35	22	18

where y is the number of insects per square meter and n is months. Find first and second order linear models for the data.

Solution: The following first order difference equation was fitted to the data in Table 8.4.3:

$$\Delta y = 323.24 - 33.95 n - 0.7883 y$$

The following is the difference table of the data:

Table 8.4.3

n	y	Δy	$\Delta^2 y$
0	250	175	-179
1	425	- 4	- 55
2	421	-59	- 24
3	362	-83	9
4	279	-74	20
5	205	-54	11
6	151	-43	0
7	108	-43	27
8	65	-16	2
9	49	-14	1
10	35	-13	9
11	22	- 4	
12	18		

The statistics for the regression coefficients of the difference equation was as follows:

Table 8.4.4

Variable	Coefficient	Standard Error	"t"
y	-0.7883	0.2070	-3.812
n	-33.95	8.56	-3.965
Constant	323.24	86.74	3.727

The coefficient of determination was $R^2 = 0.640$ and the standard deviation was $s = 44.36$. The following second order equation was also obtained from the data:

$$\Delta^2 y = -9.4407 + 1.0737n - 0.1323y - 0.7748\Delta y$$

and the following are the corresponding statistics for the regression coefficients:

Table 8.4.5

Variable	Coefficient	Standard Error	"t"
Δy	-0.7748	0.0637	-12.162
y	-0.1323	0.0632	-2.094
n	1.0737	2.9104	0.369 n.s.
Constant	-9.4407	26.7830	-0.352 n.s.

The coefficient of determination and the standard deviation were $R^2 = 0.991$ and $s = 6.926$. The second order is clearly more accurate than the first order model. However, as shown in the above table, the intercept and the coefficient for the time variable n are not significant. Thus, the following is the new regression equation, with those coefficients deleted:

$$\Delta^2 y = -0.7955\Delta y - 0.1538y$$

The corresponding statistics is as follows:

Table 8.4.6

Variable	Coefficient	Standard Error	"t"
Δy	-0.7955	0.0279	-28.546
y	-0.1538	-0.6750	-20.116

As shown in the above table, the significance of the model was improved. The standard deviation was reduced to $s = 6.171$.

The subscript notation form of the second order difference equation is

$$y_{n+2} - 1.2045y_{n+1} + 0.3583y = 0$$

where $\Delta y = y_{n+1} - y_1$ and $\Delta^2 y = y_{n+2} - 2y_{n+1} - y_n$. The following is the Z transform for this equation:

$$G(z) = \frac{250(z^2 - 1.2045z) + 425z}{z^2 - 1.2045z + 0.3583}$$

where 250 and 425 are the first and second differences in the data table. The inverse of this transform is the solution of the difference equation, as defined in Example 8.2.2:

$$y_n = 2193(0.6686)^n - 1943(0.5359)^n$$

The reader should be aware that each of the numerical values of the first and second terms of the y_n sequence, which are 250 and 425, include an error term. Therefore, the accuracy of the above solution is affected by the size of such errors. In the present example, the fit of the difference equation has a coefficient of determination (R^2) of 0.99 and the goodness of fit is high, as shown in Fig. 8.4.1.

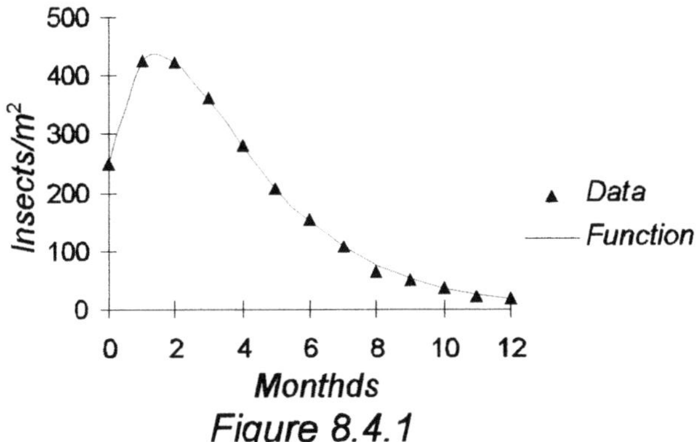

Figure 8.4.1

Example 8.4.2 Two species of birds share the same ecosystem and interact with each other. The following is the data for number of birds:

n	0	1	2	3	4	5	6	7	8	9	10	11	12	13	14	15	16
y_1	400	500	580	450	390	400	420	610	730	680	500	330	400	620	830	770	450
y_2	200	230	180	140	150	180	240	300	240	170	100	130	260	320	290	170	110

where y_1 and y_2 are the number of birds of each of the two species and n is years. Find a first order model for the data.

Solution: The following first order difference expression was obtained from the data in the difference table bellow:

$$\Delta Y = \begin{bmatrix} -0.8982 & 2.1437 \\ -0.4528 & -0.1266 \end{bmatrix} Y + \begin{bmatrix} 28.33 \\ 224.27 \end{bmatrix} + \begin{bmatrix} 2.136 \\ 5.317 \end{bmatrix} n$$

The following is the difference table for the data:

Table 8.4.7

n	y_1	Δy_1	y_2	Δy_2
0	400	100	200	30
1	500	80	230	- 50
2	580	-130	180	- 40
3	450	- 60	140	10
4	390	10	150	30
5	400	20	180	60
6	420	190	240	60
7	610	120	300	- 60
8	730	- 50	240	- 70
9	680	-180	170	- 70
10	500	-170	100	30
11	330	70	130	130
12	400	220	260	60
13	620	210	320	- 30
14	830	-60	290	-120
15	770	-320	170	-60
16	450		110	

The statistical evaluation for the regression coefficients is shown in Table 8.4.8. The coefficients of determination are here $R^2 = 0.966$ and $R^2 = 0.949$ for each of the to state variables and the standard deviations are $s = 31.489$ and $s = 16.818$.

Table 8.4.8

Variable	Coefficient	Standard Error	"t"
y_1	-0.8982	0.0641	-14.023
	-0.4528	0.0342	-13.240
y_2	2.1437	0.1363	15.730
	-0.1266	0.0728	-1.740
n	2.1356	1.9630	1.088 n.s.
	5.3174	1.0481	5.073
Constant	28.3260	33.6027	0.843 n.s.
	224.2747	17.9418	12.500

As shown in the table, there are two non significant coefficients. The following is the new expression with those coefficients deleted:

$$\Delta Y = \begin{bmatrix} -0.8407 & 2.2015 \\ -0.4528 & 0.1266 \end{bmatrix} Y + \begin{bmatrix} 0 \\ 224.27 \end{bmatrix} + \begin{bmatrix} 0 \\ 5.317 \end{bmatrix} n$$

The corresponding equation in subscript notation for the above mathematical model is

$$Y_{n+1} = \begin{bmatrix} 0.1593 & 2.2015 \\ -0.4528 & 0.8734 \end{bmatrix} Y_n + \begin{bmatrix} 0 \\ 224.27 \end{bmatrix} + \begin{bmatrix} 0 \\ 5.317 \end{bmatrix} n$$

As determined in Example 8.3.2, the Z transform of this model is

$$\begin{bmatrix} z-0.1593 & -2.2015 \\ 0.4528 & z-0.8734 \end{bmatrix} F(z) = \begin{bmatrix} 400 \\ 200 \end{bmatrix} z + \begin{bmatrix} 0 \\ 224.27 \end{bmatrix} \frac{z}{z-1} + \begin{bmatrix} 0 \\ 5.317 \end{bmatrix} \frac{z}{(z-1)^2}$$

Note that 400 and 200 are initial values for each of the two species of birds. Note also that these values include the error terms. Therefore, the accuracy of the free response is affected by these error terms.

The total response of the system was defined as

$$Y_n = \begin{bmatrix} 438 \\ 172 \end{bmatrix} + \begin{bmatrix} 10.62 \\ 4.05 \end{bmatrix} n + (1.066)^n \begin{bmatrix} -38.3 & 100.8 \\ 27.9 & 30.3 \end{bmatrix} \begin{bmatrix} \cos n(1.065) \\ \sin n(1.065) \end{bmatrix}$$

Because the free response is affected by how much the first differences deviate from regression, the system solution would need some fine tuning using non linear curve fitting procedures. The "goodness" of fit can be appreciated in Fig. 8.4.2.

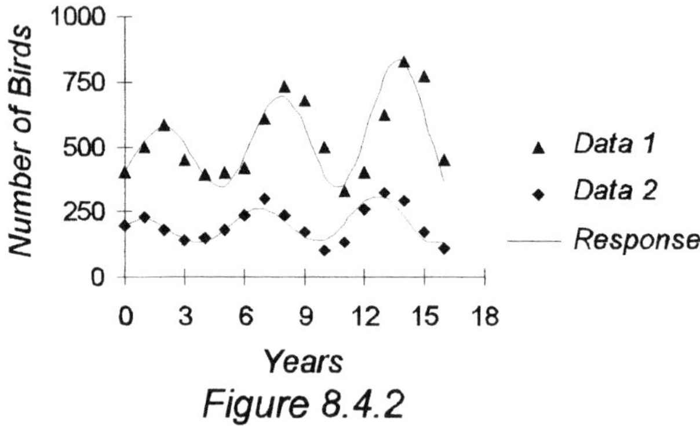

Figure 8.4.2

In conclusion, the following procedure is recommended for fitting linear models to the data of discrete systems:

- Express the data as a difference table
- Use a least squares procedure to determine the most appropriate model
- Define the set of difference equations for the system
- Determine the state equations
- Use non linear regression for fine tuning the state equations

Summary

Fitting linear models to data of discrete systems may be accomplished by linear regression, using data from difference tables. Constant coefficients of resulting state equations may be fine tuned by non linear least squares curve fitting.

9

DETERMINISTIC MODELS OF CONTINUOUS SYSTEMS

Continuous systems are also called *differentiable systems* because they may be represented by differential equations and their solutions. The time scale of these systems is the set of non negative real numbers.

This chapter is related to the process of linking differential equations to the system behavior and data, using constant coefficients linear models. Systems are here classified by their structure and by their dimension. By structure, systems are either compartmental or non compartmental. As with discrete systems, the number of system components also determines the dimension of the system.

9.1 RELATIONSHIP BETWEEN ORDER AND DIMENSION

The following first order model describes one of the most elementary types of systems:

$$\frac{dy}{dt} + by = x$$

where by is the output of the system and x is the single input. As shown in Fig. 9.1.1, a first order differential equation represents a one component system.

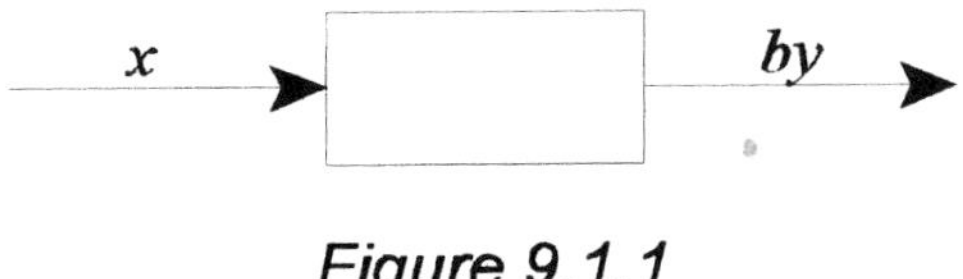

Figure 9.1.1

Example 9.1.1 The concentration of bacteria in the rumen of a group of calves was found to increase with age. The following is the fitted equation describing this process[1]:

$$y = 4.61 - 3.71 e^{-0.142t}$$

where y is concentration of bacteria in millions/G*10^4 and t is age in weeks. Determine the components of the system, as represented by the above equation and define the single input and the output.

Solution: The following is the differential equation of the system:

$$\frac{dy}{dt} + 0.142y = 0.655$$

This is a first order differential equation. Therefore, the system has only one component. The input is 0.655 and the output is 0.142y.

More complex are second order models reducible to the form

$$\frac{d^2y}{dt^2} + b_1 \frac{dy}{dt} + b_2 y = x \tag{1}$$

where b_1 and b_2 are constant coefficients and x is the single input. As will be shown, the above second order differential equation is equivalent to a set of two first order interconnected differential equations of the form

$$\frac{dy_1}{dt} = -a_1 y_1 + u$$
$$\frac{dy_2}{dt} = a_1 y_1 - a_2 y_2 \tag{2}$$

where u is the single input and $a_1 y_1$ and $a_2 y_2$ are outputs. The corresponding two-component system is pictured in Fig. 9.1.2.

---

[1]Computed from Lengeman F.W. and N.N. Allen

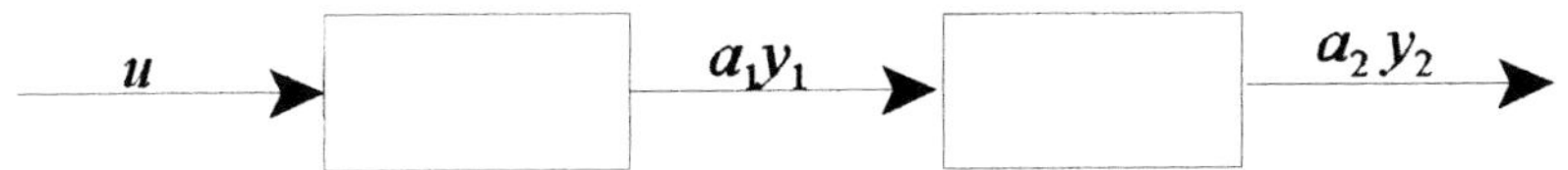

Figure 9.1.2

The procedure that follows determines the equivalent second order form of system (2).
First, a second order equation is obtained by differentiating the second equation in (2):

$$\frac{d^2y_2}{dt^2} = a_1\frac{dy_1}{dt} - a_2\frac{dy_2}{dt} \tag{3}$$

Next, the dy_1/dt term in (3) is replaced with the first equation in (2):

$$\frac{d^2y_2}{dt^2} = a_1(u - a_1y_1) - a_2\frac{dy_2}{dt}$$

Then, according to the second equation in (2), the variable y_1 is

$$y_1 = \frac{1}{a_1}\left[\frac{dy_2}{dt} + a_2y_2\right]$$

Thus

$$\frac{d^2y_2}{dt^2} = a_1\left[u - \left(\frac{dy_2}{dt} + a_2y_2\right)\right] - a_2\frac{dy_2}{dt} \tag{4}$$

Finally, after factorizing and rearranging (4), the following second order equation is obtained:

$$\frac{d^2y_2}{dt^2} + (a_1 + a_2)\frac{dy_2}{dt} + a_1 a_2 y_2 = a_1 u$$

where $a_1 + a_2 = b_1$, $a_1 a_2 = b_2$ and $a_1 u = x$ in equation (1). Equation (5) is equivalent to equation (1), when operations are carried out on the second component of the system. The reader may want to check that the following second order equation is equivalent to equation (1) when operations are carried out on the first component:

$$\frac{d^2y_1}{dt^2} + (a_1 + a_2)\frac{dy_1}{dt} + a_1 a_2 y_1 = a_2 u \qquad (5)$$

As demonstrated, a second order differential equation represents a two-component model of the system.

Example 9.1.2 The following fitted equation represents the growth curve of a group of steers[2]:

$$y = 780 + 265 e^{-1.427t} - 1065 e^{-0.553t}$$

where y is the steers' body weight and t is the steers' age in years. Define the single input, the outputs, the components of the system and the set of equivalent first order differential equations.

Solution: The following are the first and second derivatives of the state equation:

$$\frac{dy}{dt} = -1.427(265)e^{-1.427t} + 0.553(1065)e^{-0.553t}$$

$$\frac{d^2y}{dt^2} = (1.427)^2(265)e^{-1.427t} - (0.553)^2(1065)e^{-0.553t}$$

The second order differential equation is obtained by solving for the exponential terms in the above set of equations and replacing the solutions in the state equation. The reader is invited to prove the following shortcut:

---

[2]Vohnout, K., Unpublished

If $y = b_1 e^{a_1 t} + b_2 e^{a_2 t} + c$, then $\dfrac{d^2 t}{dt^2} + (a_1 + a_2)\dfrac{dy}{dt} + a_1 a_2 y = a_1 a_2 c$

By either procedure, the following is the second order differential equation of the system:

$$\frac{d^2 y}{dt^2} + 1.980\frac{dy}{dt} + 0.789y = 616$$

If $a_1 + a_2 = 1.980$, $a_1 a_2 = 0.789$ and $a_1 u = 616$, then $a_1 = 1.427$, $a_2 = 0.553$, the outputs are $1.427y_1$ and $0.553y_2$ and the input is $u = 432$. Thus, the following is the equivalent first order system:

$$\frac{dy_1}{dt} = -1.427y_1 + 432$$

$$\frac{dy_2}{dt} = 1.427y_1 - 0.553y_2$$

Note that the output coefficients 1.427 and 0.553 are also the characteristic roots of the system. As expected and shown in Fig. 9.1.3, the model of this system has two components.

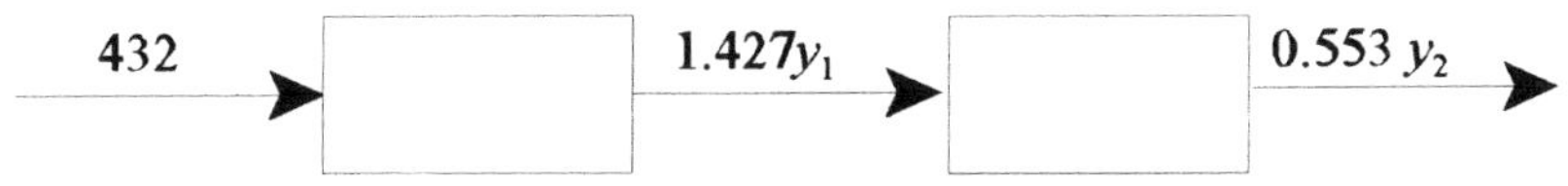

Figure 9.1.3

Generalizing the model is now possible. Thus, the following is the system of n first order equations equivalent to an n order constant coefficients differential equation:

$$\frac{dY}{dt} = \begin{bmatrix} -a_1 & 0 & \cdots & 0 & 0 \\ a_1 & -a_2 & \cdots & 0 & 0 \\ \vdots & & & & \vdots \\ 0 & 0 & \cdots & a_{n-1} & -a_n \end{bmatrix} Y + \begin{bmatrix} 1 \\ 0 \\ \vdots \\ 0 \end{bmatrix} u$$

where the a_i output coefficients are the characteristic roots of the system and u is the single input. The following is the corresponding n order differential equation:

$$\frac{d^n y_n}{dt^n} + (a_1 + a_2 + ... + a_n)\frac{d^{n-1} y_n}{dt^{n-1}} + ... + a_1 a_2 ... a_n y_n = a_1 a_2 ... a_{n-1} u$$

Clearly, the above system has n components.

Summary

An n order differential equation of the form

$$\frac{d^n y}{dt^n} + b_1 \frac{d^{n-1} y}{dt^{n-1}} + ... + b_n y = x$$

where $x = f(t)$ is the single input of the system, y is the system response and b_i are constant coefficients, is equivalent to n first order equations.

9.2 SINGLE INPUT LINEAR MODELS

Single input non compartmental linear models with constant coefficients are represented by differential equations reducible to the form

$$\sum_0^n b_i \frac{d^i y}{dt^i} = \sum_0^m c_i \frac{d^i x}{dt^i}$$

where $x = f(t)$ is an input trajectory, y is the system response and b_i and c_i are constants. The general expression for a first order constant coefficients model is

$$\frac{dy}{dt} + by = c_1 \frac{dx}{dt} + c_2 x$$

and the following is the Laplace transform of the above system:

$$sG(s) - y_0 + bG(s) = c_1[sF(s) - f(0)] + c_2 F(s)$$

Then

$$G(s) = \frac{y_0}{s+b} + \frac{c_1 s + c_2}{s+b} F(s) - \frac{c_1 f(0)}{s+b}$$

where y_0 is the initial condition of the system, $f(0)$ is the input initial value, $(c_1 s + c_2)/(s+b)$ is the transfer function and $s+b$ is the characteristic equation of the system.

Definitions for the response functions of the system were presented in Chapter 6. The reader is reminded that the system reaction to initial conditions, independently of the inputs, is known as the free response. The reaction of the system to inputs, independently of the initial conditions, was defined as the forced response. The solution of the Laplace transform is the free and forced responses. In the above example, the first fraction of the transform corresponds to the free response and the other two fractions to the forced response.

Example 9.2.1 The following equation was fitted to the energy content of milk from a group of cows[3]:

$$y = 2.821 + 0.965 e^{-0.0423t}$$

where y is the energy content in MJoules/ Kg and t is days after calving. Determine the free and forced responses of the system.

Solution: The following is the differential equation representing this system:

$$\frac{dy}{dt} = 0.1193 - 0.0423y$$

where 0.1193 is the input and 0.0423y is the output. An asymptotic value is obtained when the input and the output are equal, that is $y = 0.1193/0.0423 = 2.82$. The above is a non homogeneous time invariant differential equation. The corresponding Laplace transform is

$$G(s) = \frac{y_0}{s+0.0423} + \frac{0.1193}{s(s+0.0423)}$$

---

[3]Computed from B.G. Lowman, R.A. Edwards and S.H. Somerville

The first fraction of the Laplace transform is the free response, determined by the cows condition before calving. The following is its inverse:

$$y_A = 3.786e^{-0.0423t}$$

The second fraction is the forced response, determined by the system inputs after calving and the following is the corresponding trajectory:

$$y_B = 2.821(1 - e^{-0.0423t})$$

The total response of the system is the fitted equation, which is the sum of the free and the forced responses.

The graph of the response functions of the system is shown in Fig. 9.2.1.

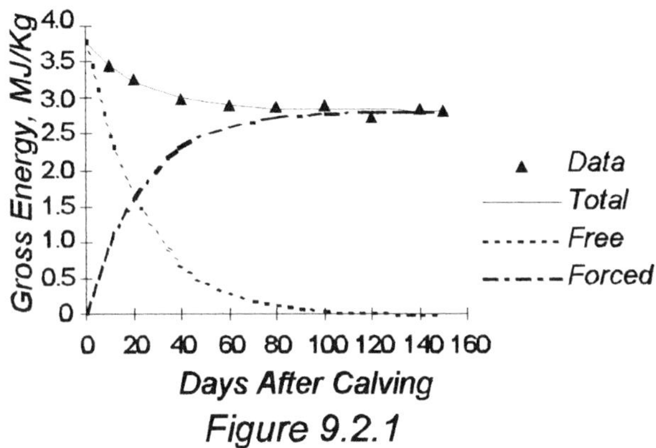

Figure 9.2.1

Example 9.2.2 The following is the equation fitted to the lactation curve of a group of dairy cows[4]:

$$y = e^{-0.484t}(298 + 411t)$$

where y is milk production in kilograms/month and t is months. Define the free and forced responses of the system.

---

[4]Vohnout, K., Unpublished

Solution: The following is the differential equation of the system:

$$\frac{dy}{dt} = 411e^{-0.484t} - 0.484y$$

where $411e^{-0.484t}$ is the input and $0.484y$ is the output. This is a first order time variant non homogeneous equation.

The differential expression of the free response is here

$$\frac{dy}{dt} + 0.484y = 0$$

which has as its solution

$$y_A = 298e^{-0.484t}$$

This free response represents exclusively the milk production expected from the physical condition of the cows before calving.

The fitted equation minus the free response is the forced response:

$$y_B = 411te^{-0.484t}$$

The forced response represents milk production due to the system input after calving. The graph of the system response functions is shown in Fig. 9.2.2.

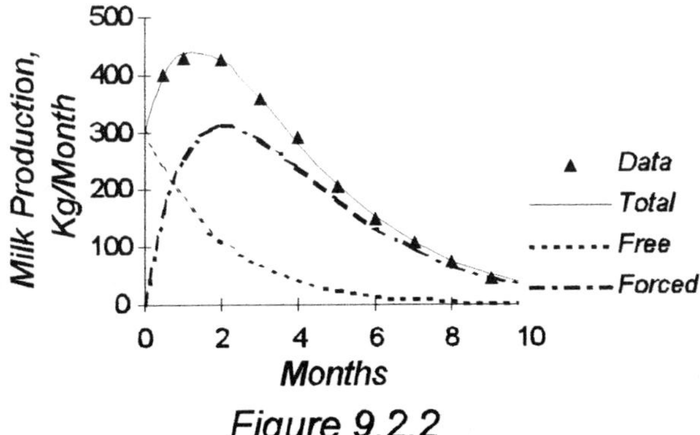

Figure 9.2.2

Example 9.2.3 The following differential equation represents the response of Carpet grass to rainfall[5]:

$$\frac{dy}{dt} + 0.7067y = 0.04818\frac{dx}{dt} + 0.05340x$$

where y is pasture yield, as kg/ha/day of green dry leaves and x is rainfall in mm/month. The following equation was fitted to rainfall data[5]:

$$x = 206 - 152.6\cos 0.809t - 43.5\sin 0.809t$$

Determine the response functions of the system.

Solution: As defined before, the Laplace transform of the system has the form

$$G(s) = \frac{y_0}{s+b} + \frac{c_1 s + c_2}{s+b}F(s) - \frac{c_1 f(0)}{s+b}$$

and the rainfall equation has the symbolic expression

$$x = k_0 + k_1\cos\theta t + k_2\sin\theta t$$

The first term of the Laplace transform corresponds to the free response. Then, the free response is simply

$$y_A = y_0 e^{-0.707t}$$

where y_0 is the initial state of the system. An approximate value, obtained from the data, is 3 and should be further fine tuned by non linear regression.

The remaining terms of the Laplace transform are related to the forced response. This model was used and solved in Example 6.4.3 for a Kikuyu pasture experiment.

---

[5]Computed from Murtagh, G.J. et.al.

Therefore, there is no need for repeating all the steps of the procedure and only the solution is presented here:

$$y_B = \frac{c_2 k_0}{b} - \left(\frac{c_2 k_0}{b} + k_1 c_1 - \frac{(c_1 b - c_2)(k_1 b - k_2 \theta)}{b^2 + \theta^2} \right) e^{-bt}$$
$$+ \left(c_1 k_1 - \frac{(c_1 b - c_2)(k_1 b - k_2 \theta)}{b^2 + \theta^2} \right) \cos \theta t + \left(c_1 k_2 - \frac{(c_1 b - c_2)(k_1 \theta + k_2 b)}{b^2 + \theta^2} \right) \sin \theta t$$

The following is the resulting forced response after replacing the symbols with the corresponding numerical values:

$$y_B = 15.57 - 7.00 e^{-0.707t} - 8.57 \cos 0.809 t - 4.68 \sin 0.809 t$$

For this particular data, the forced response is also the total response, because the initial value y_0 is not statistically significant. The following final equation for the total response was fitted by non linear regression:

$$y = 15.63 - 7.63 e^{-0.707t} - 8.00 \cos 0.809 t - 8.98 \sin 0.809 t$$

The graph of this equation is shown in Fig. 9.2.3.

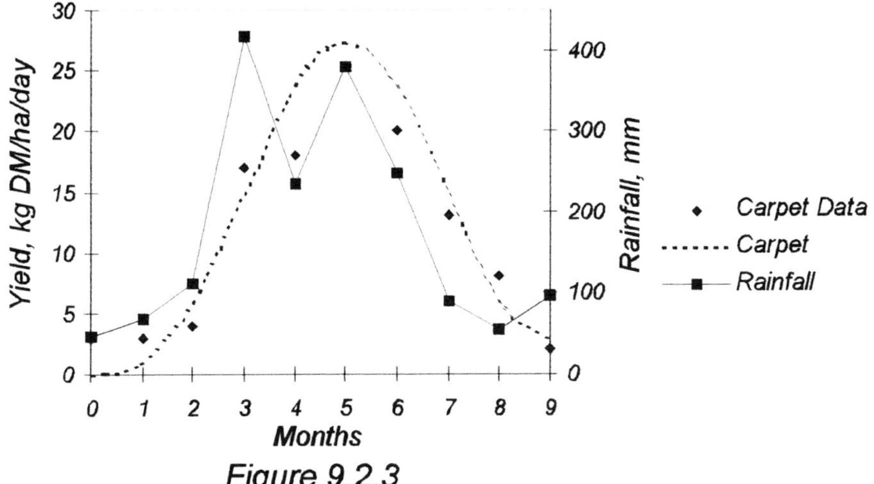

Figure 9.2.3

The following is the general expression for a second order, constant coefficients model:

$$\frac{d^2y}{dt^2} + b_1\frac{dy}{dt} + b_2y = c_1\frac{dx^2}{dt^2} + c_2\frac{dx}{dt} + c_3x$$

The corresponding Laplace transform is here

$$s^2G(s) - sg(0) - g'(0) + b_1[sG(s) - g(0)] + b_2G(s)$$
$$= c_1[s^2F(s) - sf(0) - f'(0)] + c_2[sF(s) - f(0)] + c_3F(s)$$

where $g(0)$, $g'(0)$, $f(0)$ and $f'(0)$ are the initial values of the solution and of the first derivatives of the response and the input. Then, the following is the Laplace transform of the system response after factorization:

$$G(s) = \frac{g(0)(s+b) + g'(0)}{(s+\lambda_1)(s+\lambda_2)} + \frac{c_1s^2 + c_2s + c_3}{(s+\lambda_1)(s+\lambda_2)}F(s) - \frac{f(0)(s+c_2) + f'(0)}{(s+\lambda_1)(s+\lambda_2)}$$

where λ_1 and λ_2 are the roots of the characteristic equation $s^2 + bs + c$ of the system. The transfer function is here

$$H(s) = \frac{c_1s^2 + c_2s + c_3}{(s+\lambda_1)(s+\lambda_2)}$$

The first fraction of the transform represents the free response and the other fractions the forced response. Note that the coefficients in the characteristic equation are the same coefficients of the differential equation. Note also that the degree of the characteristic equation is the same as the order of the differential equation.

The following expression gives the roots of a second degree polynomial:

$$\lambda = \frac{-b \pm \sqrt{b^2 - 4c}}{2}$$

Then, if $(b^2-4c)>0$ the polynomial has two different real roots, if $(b^2-4c)=0$ the polynomial has two equal real roots and if $(b^2-4c)<0$ the polynomial has two imaginary roots. The $(b^2-4c)>0$ case is illustrated in the following example.

Example 9.2.4 The following differential equation corresponds to the fitted equation of the growth curve of a group of steers, as defined in Example 9.1.2:

$$\frac{d^2y}{dt^2} + 1.980\frac{dy}{dt} + 0.789y = 616$$

Determine the response functions of this system.

Solution: The characteristic equation of the system

$$s^2 + 1.980s + 0.789$$

has two real different roots, 1.427 and 0.553. Then, the following is the Laplace transform of the above differential equation:

$$G(s) = \frac{g(0)(s+1.98) + g'(0)}{(s+1.427)(s+0.553)} + \frac{616}{s(s+1.427)(s+0.553)}$$

The first fraction represents the free response and the second fraction represents the forced response of the system. The corresponding inverses are easily obtained from standard tables of Laplace transforms. Then, if $g(0)=30$ and $g'(0)=183$, the free response y_A and the forced response y_B are

$$y_A = 258e^{-0.553t} - 229e^{-1.427t}$$
$$y_B = 780 - 1274e^{-0.553t} + 494e^{-1.427t}$$

Note that the characteristic roots are also the exponents in the exponential terms of the response functions.

The graph of the system response functions is shown in Fig. 9.2.4.

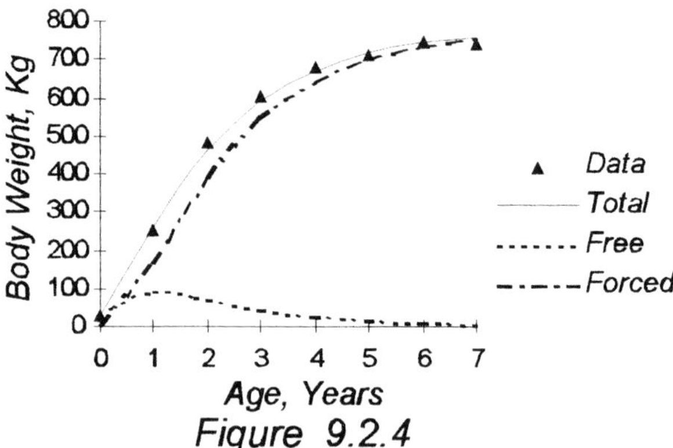

Figure 9.2.4

The reader is encouraged to show that, for a state equation of the form

$$y = k_0 + k_1 e^{-\lambda_1 t} + k_2 e^{-\lambda_2 t}$$

the corresponding second order differential equation has the form

$$\frac{d^2 y}{dt^2} + (\lambda_1 + \lambda_2)\frac{dy}{dt} + \lambda_1 \lambda_2 y = \lambda_1 \lambda_2 k_0$$

Then, by the above expression, a second order differential equation with constant coefficients may be determined, in a single step, from the solution.

The $(b^2 - 4c) = 0$ case is portrayed in the next example.

Example 9.2.5 The lactation curve of a group of cows was defined in Example 9.1.4 as

$$y = e^{-0.484 t}(298 + 411 t)$$

Find the second order differential equation of the system and determine the corresponding characteristic roots.

Solution: The following first order, time variant non homogeneous equation, represented this system:

$$\frac{dy}{dt} = 411e^{-0.484t} - 0.484y$$

Differentiating the above expression determines a second order differential equation such that

$$\frac{d^2y}{dt^2} + 0.968\frac{dy}{dt} + 0.234y = 0$$

Note that the first order non homogeneous time variant model in this example is equivalent to the following second order homogeneous time invariant model:

$$\frac{d^2y}{dt^2} + 2(0.484)\frac{dy}{dt} + (0.484)^2y = 0$$

Clearly, the characteristic equation has two equal roots, that is $\lambda = 0.484$, which is the exponent of the exponential term of the lactation curve.

The following important conclusion is achieved from this example:

A first order non homogeneous time variant equation, is equivalent to a second order homogeneous time invariant equation

The first order and the second order models are equivalent in mathematical terms. However, in some cases, the second order model may not be appropriate, because it determines only a free response of the system. This type of considerations will be discussed in Chapter 10.

The reader is encouraged to prove that, for a response equation of the form

$$y = K_0 + K_1e^{-b_1t} + K_2e^{-b_2t}$$

the corresponding differential equation has the form

$$\frac{d^2y}{dt^2} + (b_1 + b_2)\frac{dy}{dt} + b_1b_2y = K_0b_1b_2$$

Finally, the $b^2 - 4c < 0$ is illustrated in the following example.

Example 9.2.6 The following equation was fitted to the dry matter production of a Kikuyu pasture field[6]:

$$y = 1170t - 839 + 839\cos(0.686t) - 915\sin(0.686t)$$

where y is the accumulated dry matter yield in Kg/Ha and t is months. Determine the response functions of the system.

Solution: The following are the first and second derivatives of the above equation:

$$\frac{dy}{dt} = 1169.8 - 627.8\cos(0.6861t) - 575.5\sin(0.6861t)$$

$$\frac{d^2y}{dt^2} = -394.9\cos(0.6861t) + 430.7\sin(0.6861t)$$

By solving for the unknowns $\cos(0.6861t)$ and $\sin(0.6861t)$, replacing the corresponding values in the state equation and factorizing, the differential equation of the system is the following expression:

$$\frac{d^2y}{dt^2} - 0.00003316\frac{dy}{dt} + 0.4707y = 550.7t - 394.9$$

Note that the coefficient of the first order differential is extremely small and can be dropped. Thus, the following is the new form of the system differential equation:

$$\frac{d^2y}{dt^2} + 0.4707y = 550.7t - 394.9$$

Then, the following is the characteristic equation of the system:

---

[6]Computed from Murtagh, G.J. et.al.

$$s^2 + 0.4707 = (s - 0.6861i)(s + 0.6861i)$$

The corresponding Laplace transform is here

$$G(s) = \frac{g(0)s}{(s+\lambda_1)(s+\lambda_2)} + \frac{g'(0)}{(s+\lambda_1)(s+\lambda_2)} + \frac{550.7}{s^2(s+\lambda_1)(s+\lambda_2)} - \frac{349.9}{s(s+\lambda_1)(s+\lambda_2)}$$

where $g(0) = 0$, $g'(0) = 542$, $\lambda_1 = -0.6861i$ and $\lambda_2 = 0.6861i$.

The first two terms of the above transform corresponds to the free response of the system and the following expression is their inverse:

$$y_A = \frac{542}{\lambda_2 - \lambda_1}\left(e^{-\lambda_1 t} - e^{-\lambda_2 t}\right)$$

where $e^{-\lambda_1 t} = \cos(\beta t) + i\sin(\beta t)$, $e^{-\lambda_2 t} = \cos(\beta t) - i\sin(\beta t)$, $\lambda_2 - \lambda_1 = 2\beta i$ and $\beta = 0.6861$. Then, after replacing values and factorizing, the free response becomes

$$y_A = 790.0\sin(0.6861t)$$

The following expression is the Laplace transform of the forced response of the system:

$$\begin{aligned}
G_B(s) &= \frac{-394.9}{s(s+\lambda_1)(s+\lambda_2)} + \frac{550.7}{s^2(s+\lambda_1)(s+\lambda_2)} \\
&= \frac{-394.9}{s(s+\lambda_1)(s+\lambda_2)} + 550.7\left[\frac{1}{\lambda_1\lambda_2 s^2} + \frac{1}{\lambda_1(\lambda_1-\lambda_2)s(s+\lambda_1)} - \frac{1}{\lambda_2(\lambda_1-\lambda_2)s^2(s+\lambda_2)}\right]
\end{aligned}$$

The inverse of the above transform is the forced response:

$$y_B = \frac{-349.9}{\lambda_1\lambda_2}\left[1 + \frac{\lambda_2 e^{-\lambda_1 t} - \lambda_1 e^{-\lambda_2 t}}{\lambda_1-\lambda_2}\right] + 550.7\left[\frac{t}{\lambda_1\lambda_2} + \frac{1-e^{-\lambda_1 t}}{\lambda_1^2(\lambda_1-\lambda_2)} - \frac{1-e^{-\lambda_2 t}}{\lambda_2^2(\lambda_1-\lambda_2)}\right]$$

$$= \frac{-349.9}{\lambda_1\lambda_2} - \frac{550.7(\lambda_1+\lambda_2)}{\lambda_1^2\lambda_2^2} + \frac{550.7t}{\lambda_1\lambda_2} + e^{-\lambda_2 t}\left[\frac{349.9\lambda_2+550.7}{\lambda_2^2(\lambda_1-\lambda_2)}\right] - e^{-\lambda_1 t}\left[\frac{349.9\lambda_1+550.7}{\lambda_1^2(\lambda_1-\lambda_2)}\right]$$

where $e^{-\lambda_1 t} = \cos(\beta t) + i\sin(\beta t)$, $e^{-\lambda_2 t} = \cos(\beta t) - i\sin(\beta t)$ and $\lambda_1+\lambda_2 = 0$. Then

$$y_B = \frac{-349.9}{\lambda_1\lambda_2} + \frac{550.7t}{\lambda_1\lambda_2} + (\cos\beta t - i\sin\beta t)\left[\frac{394.9\lambda_2+550.7}{\lambda_2^2(\lambda_1-\lambda_2)}\right] - (\cos\beta t + i\sin\beta t)\left[\frac{349.9\lambda_1+550.7}{\lambda_1^2(\lambda_1-\lambda_2)}\right]$$

After factorizing and replacing λ with the corresponding numerical values, the forced response becomes

$$y_B = 1170t - 743 + 743\cos(0.6961t) - 1705\sin(0.6861t)$$

The graph of the system response functions is shown in Fig. 9.2.5.

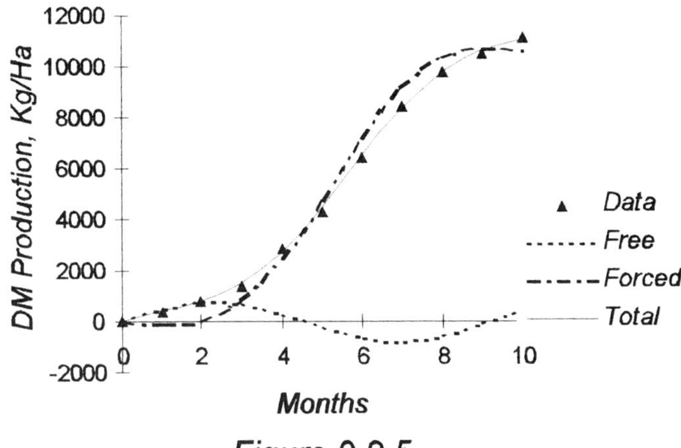

Figure 9.2.5

The reader is encouraged to prove that, for a state equation of the form

$$y = K_0 + K_1 t + e^{-\alpha t}\left(K_2\cos(\beta t) + K_3\sin(\beta t)\right)$$

the corresponding second order differential equation has the form

$$\frac{d^2y}{dt^2} + (\lambda_1 + \lambda_2)\frac{dy}{dt} + \lambda_1\lambda_2 y = \lambda_1\lambda_2 K_0 + \lambda_1\lambda_2 K_1 t$$

where $\lambda = \alpha \mp i\beta$.

Example 9.2.7 Determine the response functions as the monthly dry matter yield for the Kikuyu pasture field of the previous example.

Solution: The response functions, as the monthly yield, are simply the first derivatives of the response functions of the accumulated yield:

$$y_A' = 542\cos(0.6861t)$$
$$y_B' = 1170 - 1170\cos(0.6861t) - 576\sin(0.6861t)$$
$$y' = 1170 - 628\cos(0.6861t) - 576\sin(0.6861t)$$

The corresponding graph is shown in Fig. 9.2.6.

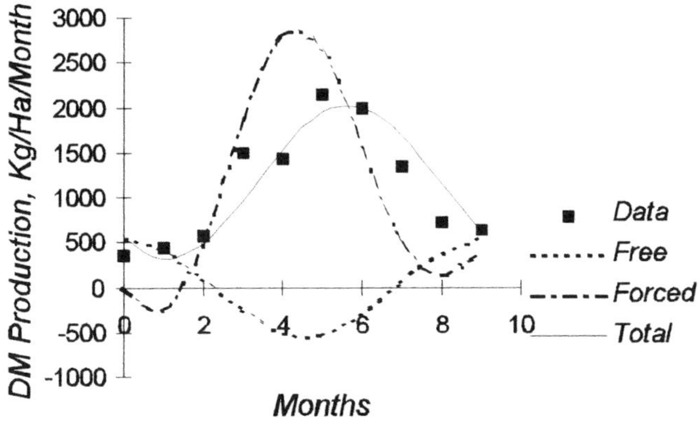

Figure 9.2.6

The total response of the production rate of the Kikuyu field was defined by the following type of mathematical model:

$$y = K_0 + e^{-\alpha t}\left(K_1\cos(\beta t) + K_2\sin(\beta t)\right) \tag{6}$$

Note that, by the addition formula of sines an cosines,

$$\begin{aligned}
\cos(A \pm B) &= \cos A\cos B \mp \sin A\sin B \\
&= a\cos(\beta b)\cos(\beta t) \mp a\sin(\beta b)\sin(\beta t) \\
&= K_1\cos(\beta t) \mp K_2\sin(\beta t)
\end{aligned}$$

Note also that $K_1 = a\cos(\beta b)$ and $K_2 = a\sin(\beta b)$ where a and b are unknowns. Then

$$a = \frac{K_1}{\cos(\beta b)} = \frac{K_2}{\sin(\beta b)}, \quad \frac{K_2}{K_1} = \frac{\sin(\beta b)}{\cos(\beta b)} = \tan(\beta b) \text{ and } b = \frac{\tan^{-1}(K_2/K_1)}{\beta}$$

After replacing values in equation (6), the following new equivalent expression emerges:

$$y = K_0 + ae^{-\alpha t}\cos\left[\beta(t-b)\right] \tag{7}$$

This form of the mathematical model is geometrically very useful. The following definitions apply here:

- Parameter β modulates the frequency response of the system
- The term $ae^{-\beta t}$ modulates the amplitude response of the system
- Coefficient b is the out of phase parameter
- Coefficient K_0 is the distance between the abscissa and the axes of the response curve
- A cycle is equal to $2\pi/\beta$

Note that when $\alpha < 1$ the amplitude decreases over time, when $\alpha > 1$ the amplitude of the curve increases and when $\alpha = 0$ the amplitude is a constant.

Example 9.2.8 Find the equivalent equation, as defined in (7), for the total response of the production rate of the Kikuyu pasture field in the previous example.

Solution: The state equation was defined as

$$y = 1169.8 - 627.8\cos(0.6861t) + 575.5\sin(0.6861t)$$

where $a\cos\beta b = 627.8,\ a\sin\beta b = 575.5$ and $\beta = 0.6861$. Then

$$a = \frac{-627.8}{\cos(0.6861b)} = \frac{575.5}{\sin(0.6861b)} \quad \text{and} \quad \tan(0.6861b) = \frac{575.5}{-627.8} = -0.9167$$

Thus

$$b = \frac{\tan^{-1}(-0.9167)}{0.6861} = -1.0814$$

$$a = \frac{-627.8}{\cos[(0.6861)(-1.0814)]} = \frac{575.5}{\sin[(0.6861)(-1.0814)]} = -851.7$$

The new expression is now

$$y = 1170 - 852\cos\left[0.686(t+1.081)\right]$$

The cycle of the system is here $2\pi/0.6861 = 9.16$ months, with 1.08 months out of phase. The value 1170 is the intersection between the abscissa and the axes of the curve and 852 is the amplitude.

By now, the reader should be aware that increasing the order of the system makes determining the symbolic solution progressively more difficult. Thus, after developing a model for the solution, a numerical procedure using non linear regression may often be more practical.

Summary

Single input constant coefficients linear models are represented by equations reducible to the n order form

$$\sum_0^n b_i \frac{d^i y}{dt^i} = \sum_0^m c_i \frac{d^i x}{dt^i}$$

where $x = f(t)$ is an input trajectory, y is the system response and b_i and c_i are constant coefficients.

9.3 MULTIDIMENSIONAL NON COMPARTMENTAL FIRST ORDER LINEAR MODELS

As indicated before, the components of multidimensional non compartmental systems may work as transducers, linking inputs and outputs of such components. The following multidimensional first order linear model will be addressed in this section:

$$\frac{dY}{dt} = BY + X$$

where Y is a set of state variables, B is a matrix of constant coefficients defining relationships between state variables and X is the set of input functions of the system.

The following is the Laplace transform of the above equation:

$$sG(s) - G(0) = BG(s) + F(s)$$
$$(sI - B)G(s) = G(0) + F(s)$$

where $(sI-B)$ is the characteristic equation of the system, $G(s)$ is the set of Laplace transforms corresponding to the set of state variables, $F(s)$ is the Laplace transform of the input functions, and $G(0)$ is the set of initial conditions of the system Then, the following expressions are the transforms of the free response Y_A and the forced response Y_B of the system:

$$Y_A = L^{-1}\left[(sI - B)^{-1}G(0)\right]$$
$$Y_B = L^{-1}\left[(sI - B)^{-1}F(s)\right]$$

Example 9.3.1 Body weight and efficiency of milk production in a group of Holstein cows were related by the following set of differential equations[6]:

$$\frac{dY}{dt} = \begin{bmatrix} -0.399526 & 0 \\ -0.007892 & 0 \end{bmatrix} Y + \begin{bmatrix} 238.890 \\ 4.569 \end{bmatrix} + \begin{bmatrix} 4.17028 \\ 0.07011 \end{bmatrix} t$$

---

[6]Computed from Miller R.H and N.W. Hooven Jr.

where y_1 is body weight in kilograms, y_2 is kilograms of milk per Mcal of metabolizable energy, t is months after calving and matrix B defines the relations between the state variables. Determine the system responses.

Solution: The following is the Laplace transform of the above equations:

$$\begin{bmatrix} s+0.3995 & 0 \\ 0.007892 & s \end{bmatrix} G(s) = G(0) + \begin{bmatrix} 238.890 \\ 4.569 \end{bmatrix}\frac{1}{s} + \begin{bmatrix} 4.170 \\ 0.07011 \end{bmatrix}\frac{1}{s^2}$$

where the characteristic equation of the system is

$$|sI - B| = \begin{vmatrix} s+0.3995 & 0 \\ 0.007892 & s \end{vmatrix} = s(s+0.3995)$$

Then, the following is the Laplace transform for body weight:

$$G_1(s) = \frac{1}{s(s+0.3995)} \begin{vmatrix} g_1(0)+\dfrac{238.890}{s}+\dfrac{4.170}{s^2} & 0 \\ g_2(0)+\dfrac{4.569}{s}+\dfrac{0.07011}{s^2} & s \end{vmatrix}$$

$$= \frac{g_1(0)}{s+0.3995} + \frac{238.890}{s(s+0.3995)} + \frac{4.170}{s^2(s+0.3995)}$$

where $g_1(0) = 607$ and $g_2(0) = 1.25$ are initial values. The free response y_{A1} for body weight is clearly the inverse transform of the first fraction of the above equation, that is

$$y_{A1} = 607e^{-0.400t}$$

The forced response y_{B1} is the inverse of the two remaining fractions, that is

$$y_{B1} = \frac{238.890}{0.3995}(1 - e^{-0.3995t}) + 4.170\left[\frac{t}{0.3995} - \frac{1}{0.3995^2}(1 - e^{-0.3995t})\right]$$

After rearranging terms, the above equation becomes

$$y_{B1} = 10.4t + 572(1 - e^{-0.400t})$$

The total response y_1 is the sum of the free and forced responses, that is

$$y_1 = 572 + 10.4t + 35.1e^{-0.400t}$$

The graph of the response functions of body weight is shown in Fig. 9.3.1.

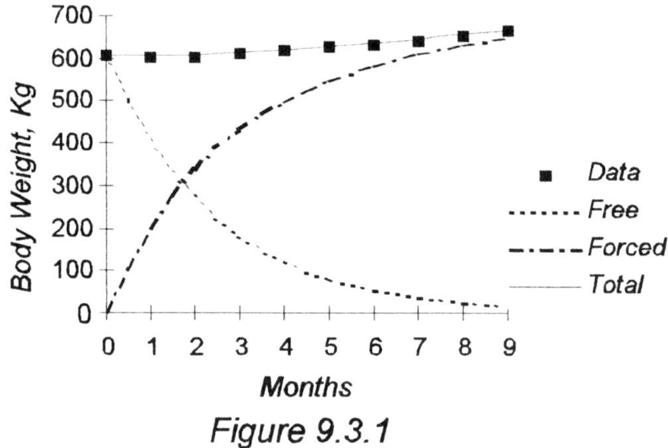

Figure 9.3.1

The following is the Laplace transform of efficiency:

$$G_2(s) = \frac{1}{s(s+0.3995)} \begin{vmatrix} s+0.3995 & g_1(0)+\dfrac{238.890}{s}+\dfrac{4.170}{s^2} \\[2ex] 0.007892 & g_2(0)+\dfrac{4.569}{s}+\dfrac{0.07011}{s^2} \end{vmatrix}$$

$$= \frac{g_2(0)}{s} + \frac{4.569}{s^2} + \frac{0.07011}{s^3} - \frac{0.007892}{s(s+0.3995)}\left[g_1(0) + \frac{238.890}{s} + \frac{4.170}{s^2}\right]$$

Then, the inverse of the terms with the initial values gives the free response of efficiency, that is

$$y_{A2} = 11.98e^{-0.400t} - 10.73$$

The inverse of the remaining terms gives the forced response, that is

$$y_{B2} = 0.05634t - 0.00613t^2 + 11.29(1 - e^{-0.400t})$$

Finally, the total response is the sum

$$y_2 = 0.5577 + 0.05634t + 0.02892t^2 + 0.6923e^{-0.400t}$$

The reader is encouraged to check the above solutions.
The graph of the efficiency response functions is shown in Fig. 9.3.2

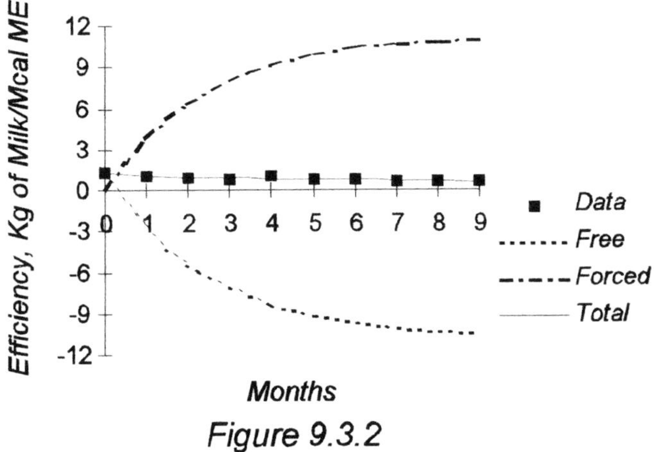

Figure 9.3.2

Example 9.3.2 Protein nitrogen and ammonia nitrogen in the rumen of steers fed a soy meal diet, were found related by the following set of differential equations[7]:

---

[7]Computed from Davis, G.V. and O.T. Stallcup

$$\frac{dY}{dt} = \begin{bmatrix} -1.0249 & 0.1888 \\ -1.1692 & -0.6931 \end{bmatrix} Y + \begin{bmatrix} 0 \\ 132.4006 \end{bmatrix} - \begin{bmatrix} 1.0911 \\ 1.6694 \end{bmatrix} t$$

where y_1 is protein nitrogen and y_2 is ammonia nitrogen in Mg/100 Ml of ruminal fluid, t is hours after feeding and matrix B defines the relations between the state variables. Determine the response functions of the system.

Solution: The following is the Laplace transform of the system:

$$\begin{bmatrix} s+1.0249 & -0.1888 \\ 1.1692 & s+0.6931 \end{bmatrix} G(s) = G(0) + \begin{bmatrix} 0 \\ 132.4006 \end{bmatrix} \frac{1}{s} - \begin{bmatrix} 1.0911 \\ 1.6694 \end{bmatrix} \frac{1}{s^2}$$

where $G(0)$ is initial values, such that $g_1(0) = 15$ and $g_2(0) = 127$. The characteristic equation of the system is

$$\begin{aligned} |sI - B| &= s^2 + 1.7180s + 0.9311 = (s+\lambda_1)(s+\lambda_2) \\ &= [s + (0.8590 - 0.4396i)][s + (0.8590 + 0.4396i)] \end{aligned}$$

Then $\lambda = \alpha \mp \beta i = 0.8590 \mp 0.4390i$.

The following transform expression defines the free response of the system:

$$G_A(s) = \begin{bmatrix} s+1.0249 & -0.1888 \\ 1.1692 & s+0.6931 \end{bmatrix}^{-1} \begin{bmatrix} 15 \\ 127 \end{bmatrix}$$

Then, the following determinant gives the free response for protein nitrogen:

$$G_{A1}(s) = \frac{1}{(s+\lambda_1)(s+\lambda_2)} \begin{vmatrix} 15 & -0.1888 \\ 127 & s+0.6931 \end{vmatrix} = \frac{15s + 34.3741}{(s+\lambda_1)(s+\lambda_2)}$$

The solution of the above transform is

$$y_{A1} = \frac{15}{\lambda_1 - \lambda_2}\left(\lambda_1 e^{-\lambda_1 t} - \lambda_2 e^{-\lambda_2 t}\right) - \frac{34.3741}{\lambda_1 - \lambda_2}\left(e^{-\lambda_1 t} - e^{-\lambda_2 t}\right)$$

$$= \frac{1}{\lambda_1 - \lambda_2}\left[(15\lambda_1 - 34.3741)e^{-\lambda_1 t} - (15\lambda_2 - 34.3741)e^{-\lambda_2 t}\right]$$

where $\lambda = \alpha \mp \beta i$. Then, the solution may be written as

$$y_{A1} = \frac{1}{\lambda_1 - \lambda_2}\left[(15\lambda_1 - 34.3741)e^{-(\alpha - \beta i)t} - (15\lambda_2 - 34.3741)e^{-(\alpha + \beta i)t}\right]$$

$$= \frac{e^{-\alpha t}}{\lambda_1 - \lambda_2}\left[(15\lambda_1 - 34.3741)e^{\beta i t} - (15\lambda_2 - 34.3741)e^{-\beta i t}\right]$$

$$= \frac{e^{-\alpha t}}{\lambda_1 - \lambda_2}\left[(15\lambda_1 - 34.3741)(\cos\beta t - i\sin\beta t) - (15\lambda_2 - 34.3741)(\cos\beta t + i\sin\beta t)\right]$$

After rearranging terms and replacing the α, β and λ values, the protein free response becomes

$$y_{A1} = e^{-0.859t}\left[15\cos(0.440t) + 48.88\sin(0.440t)\right]$$

The following determinant gives the free response of the ammonia nitrogen:

$$G_{A2}(s) = \frac{1}{(s+\lambda_1)(s+\lambda_2)}\begin{vmatrix} s+1.0249 & 15 \\ 1.1692 & 127 \end{vmatrix} = \frac{127s + 112.6243}{(s+\lambda_1)(s+\lambda_2)}$$

The inverse of the above transform is

$$y_{A2} = \frac{127}{\lambda_1 - \lambda_2}\left(\lambda_1 e^{-\lambda_1 t} - \lambda_2 e^{-\lambda_2 t}\right) - \frac{112.6243}{\lambda_1 - \lambda_2}\left(e^{-\lambda_1 t} - e^{-\lambda_2 t}\right)$$

The reader is encouraged to check that the final form of the ammonia free response is

$$y_{A2} = e^{-0.859t}\left[127\cos(0.440t) + 8.03\sin(0.440t)\right]$$

Then, the following expression gives the free response of the system:

$$
Y_A = \begin{bmatrix} 15 & 48.88 \\ 127 & 8.03 \end{bmatrix} \begin{bmatrix} \cos(0.440t) \\ \sin(0.440t) \end{bmatrix} e^{-0.859t}
$$

The free response of the system is shown in Fig. 9.3.3

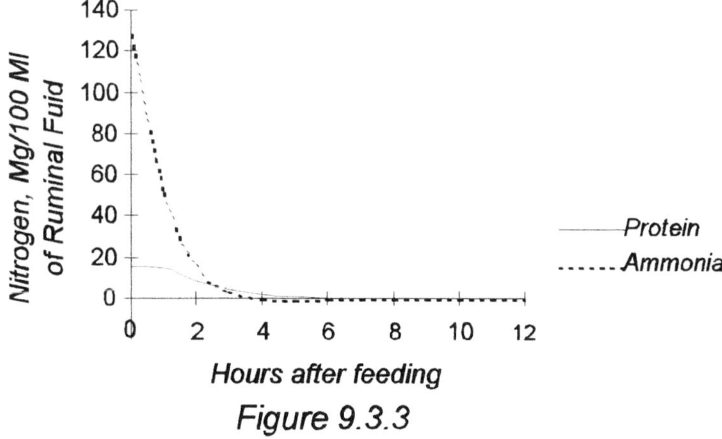

Figure 9.3.3

The following Laplace transform represents the forced response of the system:

$$
G_B(s) = \begin{bmatrix} s+1.0249 & -0.1888 \\ 1.1692 & s+0.6931 \end{bmatrix}^{-1} \left(\begin{bmatrix} 0 \\ 132.4006 \end{bmatrix} \frac{1}{s} - \begin{bmatrix} 1.0911 \\ 1.6694 \end{bmatrix} \frac{1}{s^2} \right)
$$

The following determinant gives the corresponding forced response for protein nitrogen:

$$
G_{BI}(s) = \frac{1}{(s+\lambda_1)(s+\lambda_2)} \left(\begin{vmatrix} 0 & -0.1888 \\ 132.4006/s & s+0.6931 \end{vmatrix} + \begin{vmatrix} -1.0911/s^2 & -0.1888 \\ -1.6694/s^2 & s+0.6931 \end{vmatrix} \right)
$$

$$
= \frac{1}{(s+\lambda_1)(s+\lambda_2)} \left[\frac{23.9061}{s} - \frac{1.0714}{s^2} \right]
$$

After a partial fractions expansion, the above equation becomes

$$G_{BI}(s) = \frac{23.9061}{s(s+\lambda_1)(s+\lambda_2)} - 1.0714\left[\frac{1}{\lambda_1\lambda_2 s} + \frac{1}{\lambda_1(\lambda_1-\lambda_2)[s(s+\lambda_1)]} - \frac{1}{\lambda_2(\lambda_1-\lambda_2)[s(s+\lambda_2)]}\right]$$

The following is the inverse of this equation:

$$y_{BI} = \frac{23.9061}{\lambda_1\lambda_2}\left[1 + \frac{\lambda_2 e^{-\lambda_1 t} - \lambda_1 e^{-\lambda_2 t}}{\lambda_1 - \lambda_2}\right] - 1.0714\left[\frac{t}{\lambda_1\lambda_2} + \frac{1-e^{-\lambda_1 t}}{\lambda_1^2(\lambda_1-\lambda_2)} - \frac{1-e^{-\lambda_2 t}}{\lambda_2^2(\lambda_1-\lambda_2)}\right]$$

$$= 27.7983 - 1.1507t + \frac{\lambda_2^2(23.9061\lambda_1 + 1.0714)e^{-\lambda_1 t} - \lambda_1^2(23.9061\lambda_2 + 1.0714)e^{-\lambda_2 t}}{\lambda_1^2\lambda_2^2(\lambda_1-\lambda_2)}$$

where $\lambda = \alpha \mp \beta i$. Thus

$$y_{BI} = 27.7983 - 1.1507t + e^{-\alpha t}\left[\frac{\lambda_2^2 e^{-\beta i t}(23.9061\lambda_1 + 1.0714) - \lambda_1^2 e^{\beta i t}(23.9061\lambda_2 + 1.0714)}{\lambda_1^2\lambda_2^2(\lambda_1-\lambda_2)}\right]$$

By using the Euler's theorem, the above equation becomes

$$y_{BI} = 27.7983 - 1.1507 + \frac{e^{-\alpha t}}{\lambda_1^2\lambda_2^2(\lambda_1-\lambda_2)}\left[\begin{array}{l}\lambda_2^2(\cos\beta t - i\sin\beta t)(23.9061\lambda_1 + 1.0714) \\ - \lambda_1^2(\cos\beta t + i\sin\beta t)(23.9061\lambda_2 + 1.0714)\end{array}\right]$$

After rearranging terms and replacing the α, β and λ values, the final expression for the forced response of protein nitrogen is

$$y_{BI} = 27.80 - 1.15t - e^{-859 t}[27.80\cos(0.440t) + 51.70\sin(0.440t)]$$

The following Laplace expression represents the forced response for ammonia nitrogen:

$$G_{B2}(s) = \frac{1}{(s+\lambda_1)(s+\lambda_2)}\left\|\begin{matrix} s+1.0249 & 0 \\ 1.1692 & 132.4006/s \end{matrix}\right| + \left|\begin{matrix} s+1.0249 & -1.0911/s^2 \\ 1.1692 & -1.6694/s^2 \end{matrix}\right\|$$

$$= \frac{1}{(s+\lambda_1)(s+\lambda_2)}\left[132.4006 + \frac{134.0280}{s} - \frac{0.4353}{s^2}\right]$$

After a partial fractions expansion, the inverse of the above transform is the state equation of the forced response for ammonia nitrogen:

$$y_{B2} = \frac{132.4006}{\lambda_2-\lambda_1}\left[e^{-\lambda_1 t}-e^{-\lambda_2 t}\right] + \frac{134.0280}{\lambda_1\lambda_2}\left[1 - \frac{\lambda_2 e^{-\lambda_1 t}-\lambda_1 e^{-\lambda_2 t}}{\lambda_2-\lambda_1}\right] -$$

$$0.4353\left[\frac{t}{\lambda_1\lambda_2} + \frac{1-e^{-\lambda_1 t}}{\lambda_1^2(\lambda_1-\lambda_2)} - \frac{1-e^{-\lambda_2 t}}{\lambda_2^2(\lambda_1-\lambda_2)}\right]$$

By rearranging terms, the forced response becomes

$$y_{B2} = 144.8085 - 0.4675t - \frac{e^{-\lambda_1 t}}{\lambda_1-\lambda_2}\left[132.4006 - \frac{134.0280\lambda_2}{\lambda_1\lambda_2} + \frac{0.4353\lambda_2^2}{\lambda_1^2\lambda_2^2}\right] +$$

$$\frac{e^{-\lambda_2 t}}{\lambda_1-\lambda_2}\left[132.4006 - \frac{134.0280\lambda_1}{\lambda_1\lambda_2} + \frac{0.4353\lambda_1^2}{\lambda_1^2\lambda_1^2}\right]$$

where $\lambda = \alpha \mp \beta i$. As discussed in Chapter 4.2, by using the Euler's theorem, the forced response may be written as

$$y_{B2} = 144.8085 - 0.4675t +$$

$$\frac{e^{-\alpha t}}{\lambda_1-\lambda_2}\left[\begin{matrix}(\cos\beta t + i\sin\beta t)\left[132.4006 - \frac{134.0280\lambda_2}{\lambda_1\lambda_2} + \frac{0.4353\lambda_2^2}{\lambda_1^2\lambda_2^2}\right] \\ \\ -(\cos\beta t - i\sin\beta t)\left[132.4006 - \frac{134.0280\lambda_1}{\lambda_1\lambda_2} + \frac{0.4353\lambda_1^2}{\lambda_1^2\lambda_1^2}\right]\end{matrix}\right]$$

After rearranging terms and replacing the α, β and λ values, the following is the final form of the forced response of ammonia nitrogen:

$$y_{B2} = 144.8 - 0.468t - e^{-0.859t}\left[144.8\cos(0.440t) - 19.3\sin(0.440t)\right]$$

Then, the following expression defines the forced response of the system:

$$Y_B = \begin{bmatrix} 27.80 \\ 144.81 \end{bmatrix} - t\begin{bmatrix} 1.151 \\ 0.468 \end{bmatrix} - e^{-0.859t}\begin{bmatrix} 27.80 & 51.70 \\ 144.81 & -19.28 \end{bmatrix}\begin{bmatrix} \cos(0.440t) \\ \sin(0.440t) \end{bmatrix}$$

The forced response of the system is shown in Fig. 9.3.4.

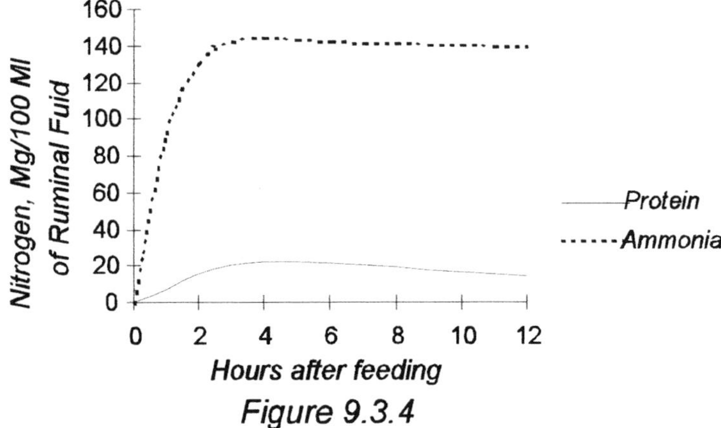

Figure 9.3.4

The following expression is the set of state equations for the total response of the system:

$$Y = \begin{bmatrix} 27.8 \\ 144.8 \end{bmatrix} - \begin{bmatrix} 1.151 \\ 0.468 \end{bmatrix}t - e^{-0.859t}\begin{bmatrix} 12.80 & 2.82 \\ 17.81 & -27.32 \end{bmatrix}\begin{bmatrix} \cos(0.440t) \\ \sin(0.440t) \end{bmatrix}$$

As shown below, testing that the above solution is correct is possible by differentiating the state equations and equating them with the set of differential equations of the system. Thus

$$\frac{dY}{dt} = \begin{bmatrix} -1.0249 & 0.1889 \\ -1.1692 & -0.6931 \end{bmatrix} \hat{Y} + \begin{bmatrix} 0 \\ 132.4006 \end{bmatrix} - \begin{bmatrix} 1.0911 \\ 1.6694 \end{bmatrix} t$$

$$= \begin{bmatrix} -1.1507 \\ -0.4675 \end{bmatrix} + e^{-0.859t} \begin{bmatrix} -12.2326 & 3.2052 \\ -3.2886 & 31.2946 \end{bmatrix} \begin{bmatrix} \cos(0.440t) \\ \sin(0.440t) \end{bmatrix}$$

Then

$$Y = \begin{bmatrix} -1.0249 & 0.188 \\ -1.1692 & -0.6931 \end{bmatrix}^{-1} \left[\begin{bmatrix} -1.1507 \\ -132.8681 \end{bmatrix} + \begin{bmatrix} 1.0911 \\ 1.6694 \end{bmatrix} t + e^{-0.859t} \begin{bmatrix} -12.2326 & 3.2052 \\ -3.2886 & 31.2946 \end{bmatrix} \begin{bmatrix} \cos(0.440t) \\ \sin(0.440t) \end{bmatrix} \right]$$

It can be easily shown that this expression is the set of state equations of the system:

$$Y = \begin{bmatrix} 27.8 \\ 144.8 \end{bmatrix} - \begin{bmatrix} 1.151 \\ 0.468 \end{bmatrix} t - e^{-0.859t} \begin{bmatrix} 12.80 & 2.82 \\ 17.81 & -27.32 \end{bmatrix} \begin{bmatrix} \cos(0.440t) \\ \sin(0.440t) \end{bmatrix}$$

The following equations represents the system, after a fine tunning by non linear regression, when only those coefficients related to initial conditions were allowed to change:

$$Y = \begin{bmatrix} 27.80 \\ 144.81 \end{bmatrix} - \begin{bmatrix} 1.151 \\ 0.468 \end{bmatrix} t + e^{-0.859t} \begin{bmatrix} 14.90 - 27.80 & 68.47 - 51.70 \\ 128.11 - 144.81 & 52.80 + 19.28 \end{bmatrix} \begin{bmatrix} \cos 0.440\,t \\ \sin 0.440\,t \end{bmatrix}$$

Clearly, the new free response is here

$$Y_A = e^{-0.859t} \begin{bmatrix} 14.90 & 68.47 \\ 128.11 & 52.80 \end{bmatrix} \begin{bmatrix} \cos 0.440\,t \\ \sin 0.440\,t \end{bmatrix}$$

The forced response was left the same. The graph of the total response of the system, as defined by the new equations, is shown in Fig. 9.3.5.

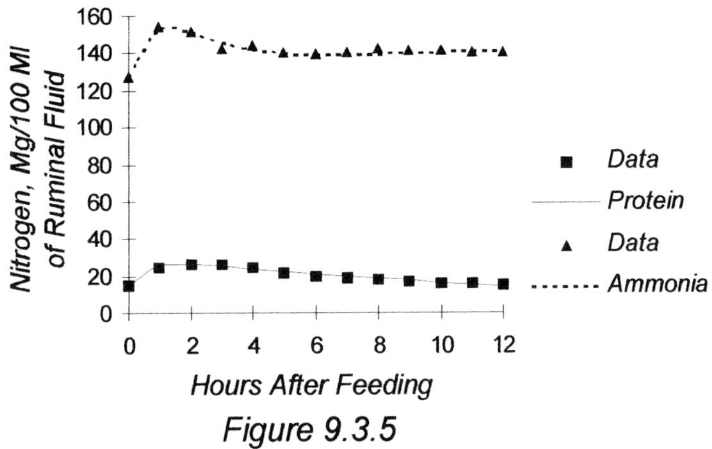

Figure 9.3.5

Summary

First order multidimensional linear models are represented by differential equations reducible to the form $dY/dt = AY + X$, where Y is a set of state variables, A is a matrix of constant coefficients determining the relations between variables and X is the set of input functions of the system.

9.4 COMPARTMENTAL FIRST ORDER LINEAR MODELS

As disclosed before, compartmental systems work as communicating chambers, among which a substance is considered to move. Compartmental first order linear models are represented by differential equations reducible to the form

$$\frac{dY}{dt} = (A + B)Y + X$$

where Y is the set of state variables, A is a matrix of constant coefficients defining the exchange of a substance between compartments, B is a matrix defining the system output to the outside environment and X is the set of input functions of the system. The sum of the coefficients of each column of matrix A should always ad up to zero and B is a diagonal matrix. The model represents a closed system if B is a null matrix, otherwise the system is open.

The following is the Laplace transform of the system:

$$[sI - (A + B)]\, G(s) \;=\; G(0) + F(s)$$

where $[sI - (A+B)]$ is the characteristic equation of the system, $G(s)$ is the set of Laplace transforms corresponding to the set of state variables, $G(0)$ is the set of initial conditions and $F(s)$ is the set of transforms of the input functions.

Example 9.4.1 The movement of DDT from plant to soil is 25% per month, from soil to plant is 2% and carried out with ground water is 5%. Define the set of state equations representing the system.

Solution: The movement of DDT between compartments is shown in Fig.9.4.1

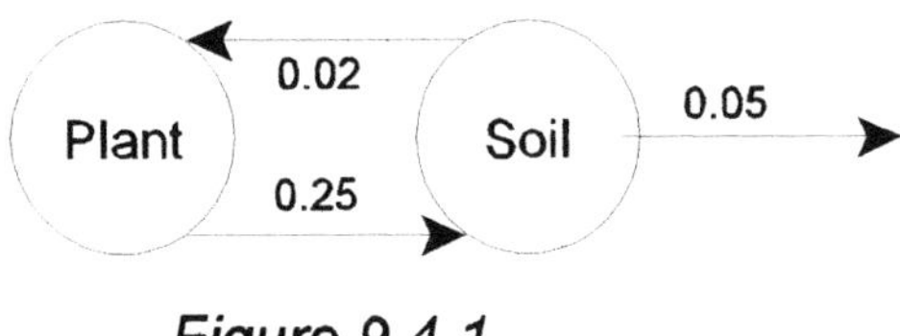

Figure 9.4.1

The following is the corresponding set of differential equations:

$$\frac{dY}{dt} = \begin{bmatrix} -0.25 & 0.02 \\ 0.25 & -0.02 \end{bmatrix} Y + \begin{bmatrix} 0 & 0 \\ 0 & -0.05 \end{bmatrix} Y$$

$$= \begin{bmatrix} -0.25 & 0.02 \\ 0.25 & -0.07 \end{bmatrix} Y$$

for $Y = (y_p, y_s)$, where y_p is the plant compartment and y_s is the soil compartment. The state changes are determined by the exchange rates in matrix A and by the output rates leaving the system in matrix B. There are no external inputs to the system. Coefficients with positive signs are input rates and coefficients with negative signs are output rates. Note that two differential equations represent the system, because it has two compartments. Note also that the sum of the coefficients of each column of matrix A ad up to zero. The system is open because matrix B is not a null matrix.

The following is the Laplace transform of the system differential equations:

$$\begin{bmatrix} s+0.25 & -0.02 \\ -0.25 & s+0.07 \end{bmatrix} G(s) = G(0)$$

where $G(0) = (0.6, 0.4)$ are initial values.

The following is the characteristic equation of the system:

$$\begin{vmatrix} s+0.25 & -0.02 \\ -0.25 & s+0.07 \end{vmatrix} = (s+0.0455)(s+0.2745)$$

Then, the Laplace transforms for the plant and soil compartments are expressed as follows:

$$G_p(s) = \frac{1}{(s+0.0455)(s+0.2745)} \begin{vmatrix} g_p(0) & -0.02 \\ g_s(0) & s+0.07 \end{vmatrix} = \frac{0.6(s+0.0833)}{(s+0.0455)(s+0.2745)}$$

and

$$G_s(s) = \frac{1}{(s+0.0455)(s+0.2745)} \begin{vmatrix} s+0.25 & g_p(0) \\ -0.25 & g_s(0) \end{vmatrix} = \frac{0.4(s+0.6250)}{(s+0.0455)(s+0.2745)}$$

The inverse of the above transforms are the state equations of the system:

$$y_p = \frac{0.6}{0.2745-0.0455}\left[(0.0833-0.0455)e^{-0.0455t} - (0.0833-0.2745)e^{-0.2745t}\right]$$

and

$$y_s = \frac{0.4}{0.2745-0.0455}\left[0.6250-0.0455)e^{-0.0455t} - (0.6250-0.2745)e^{-0.2745t}\right]$$

The system response is shown in Fig. 9.4.2.

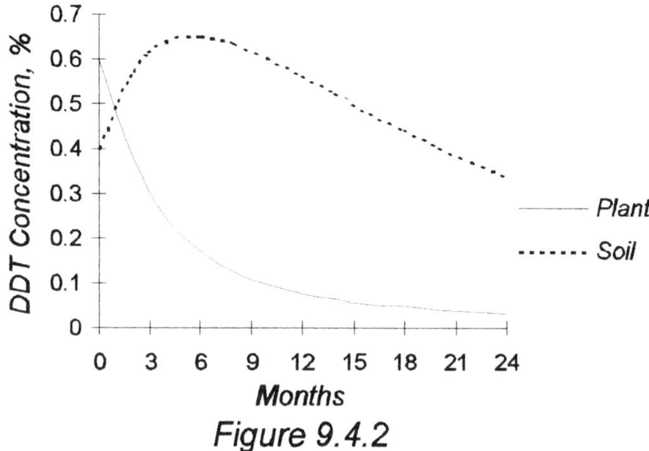

Figure 9.4.2

After rearranging the above terms, the final expression for the state equations becomes

$$Y = \begin{bmatrix} 0.0990 & 0.5010 \\ 1.0122 & -0.6122 \end{bmatrix} \begin{bmatrix} e^{-0.0455t} \\ e^{-0.2745t} \end{bmatrix}$$

Example 9.4.2 A patient with an immunodeficiency problem was dosed with 9.9 grams of gamma globulin intravenously. The following equation describes the blood concentration of the patient Ig globulin[8]:

$$y = 218 + 245e^{-0.0386t}$$

where y is the IgG concentration in mg/dl and t is time in days. Determine the IgG response functions of the system.

Solution: The following is the differential equation of the system:

---

[8]Vohnout, K. Unpublished

$$\frac{dy}{dt} + 0.0386y = 8.415$$

The corresponding Laplace transform is given by the expression

$$G(s) = \frac{g(0)}{s+0.0386} + \frac{8.415}{s(s+0.386)}$$

where the first fraction is related to the free response, the second fraction to the forced response and $g(0) = 463$ is the initial value. Then the state response functions are as follows:

$$y_A = 463e^{-0.0386t}$$
$$y_B = 218\left(1 - e^{0.0386t}\right)$$

The above functions are shown in Fig. 9.4.3.

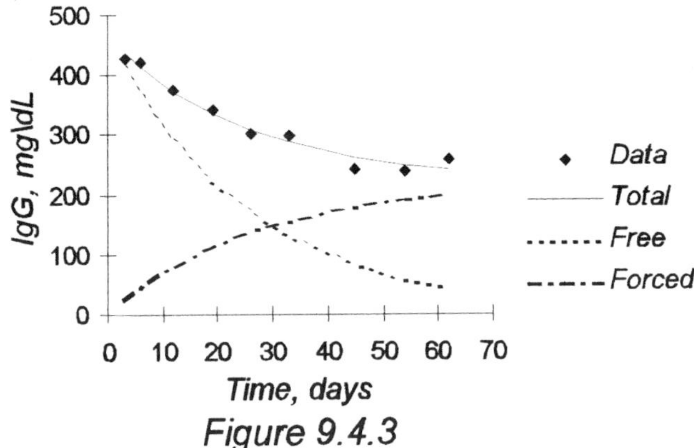

Figure 9.4.3

Compartmental analysis makes it possible determining not just exchange rates, but also distribution volumes and mass of the system compartments. In the above example, the distribution volume is given by the relationship $V = D/g(0)$, where V is the distribution volume in deciliters, D is the gamma globulin dose in milligrams and $g(0)$ is the blood IgG at time zero. Then

$$V = \frac{9900}{463} = 21.4 \text{ dl}$$

By knowing the distribution volume of the marker, it is possible to convert the state equation from concentration of the marker to amount of the marker, such that $y_w = yV$. Then

$$y_w = 4661 + 5239 e^{-0.0386t}$$

where y_w is now milligrams of IgG. The corresponding differential equation is here

$$\frac{dy_w}{dt} = 180 - 0.0386 y_w$$

where 180 is an input and $-0.0386 y_w$ is the output. The system is represented in Fig. 9.4.4

Figure 9.4.4

Compartmental modeling and analysis is used mainly in tracer kinetic studies and a vast literature is available on the subject. The following definitions apply:

- y_i is specific activity or the tracer per unit of volume in compartment i
- v_i is distribution volume of the tracer in compartment i
- r_i is total amount of the tracer in compartment i, $r_i = v_i y_i$
- K_{ij} is exchange rate per unit of volume between compartments i and j
- C_{ij} is exchange rate between compartments i and j, $C_{ij} = v_i K_{ij}$
- λ_i are roots of the ith degree characteristic equation of the system

A model of the I^{131}-thyroxine kinetics is presented in the next example.

Example 9.4.3 As shown in the following diagram, the kinetics of I^{131}-thyroxine is modeled as a three-compartment system. Solve the system for the distribution volumes of the trace in each compartment and for tracer exchange rates between compartments.

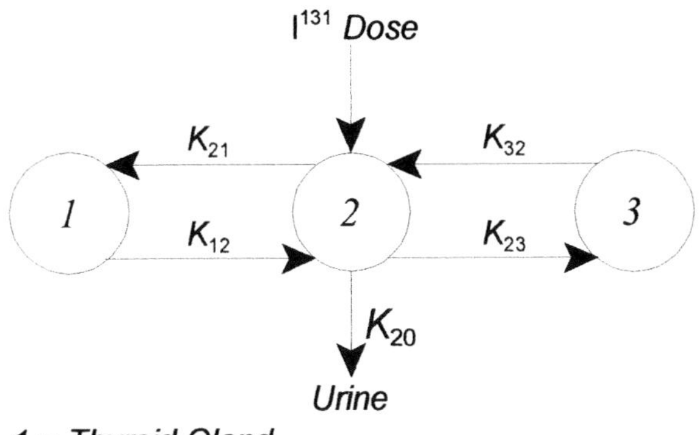

Figure 9.4.5

Solution: The following is the set of differential equations of the system:

$$\frac{dR}{dt} = \begin{bmatrix} -K_{12} & K_{21} & 0 \\ K_{12} & -(K_{21}+K_{23}+K_{20}) & K_{32} \\ 0 & K_{23} & -K_{32} \end{bmatrix} R$$

for $R = (r_1, r_2, r_3)$, where r_1, r_2 and r_3 are the amount of the tracer in compartment 1, 2 and 3 and are the rate constants related to the amount of the tracer. Note that the urine output K_{20} is included in matrix A. The above equation should be converted to specific activity, that is $Y = R/V$, because specific activity is the variable measured by sampling the plasma compartment 2. Then

$$V\frac{dY}{dt} = \begin{bmatrix} -K_{12} & K_{21} & 0 \\ K_{12} & -(K_{21}+K_{23}+K_{20}) & K_{32} \\ 0 & K_{23} & -K_{32} \end{bmatrix} VY = \begin{bmatrix} -C_{12} & C_{21} & 0 \\ C_{12} & -(C_{21}+C_{23}+C_{20}) & C_{32} \\ 0 & C_{23} & -C_{32} \end{bmatrix} Y$$

where $C_{ij} = v_i K_{ij}$ are now the new rate constants related to the specific activity of the tracer. By rearranging terms, the differential system becomes

$$\frac{dY}{dt} = \begin{bmatrix} -C_{12}/v_1 & C_{21}/v_1 & 0 \\ C_{12}/v_2 & -(C_{21}+C_{23}+C_{20})/v_2 & C_{32}/v_2 \\ 0 & C_{23}/v_3 & -C_{32}/v_3 \end{bmatrix} Y$$

The following is the corresponding Laplace transform of the above system of equations:

$$\begin{bmatrix} s+C_{12}/v_1 & -C_{21}/v_1 & 0 \\ -C_{12}/v_2 & s+(C_{21}+C_{23}+C_{20})/v_2 & -C_{32}/v_2 \\ 0 & -C_{23}/v_3 & s+C_{32}/v_3 \end{bmatrix} G(s) = G(0)$$

where $G(0)$ is the set of the initial specific activities in the three compartments of the system. The characteristic equation of the system is here

$$|sI - A)| = \begin{vmatrix} s+C_{12}/v_1 & -C_{21}/v_1 & 0 \\ -C_{12}/v_2 & s+(C_{21}+C_{23}+C_{20})/v_2 & -C_{32}/v_2 \\ 0 & -C_{23}/v_3 & s+C_{32}/v_3 \end{vmatrix}$$

The expansion of this determinant results in the following expression:

$$|sI - A| = s^3 + \left[\frac{C_{32}}{v_3} + \frac{C_{21}+C_{23}+C_{20}}{v_2} + \frac{C_{12}}{v_1}\right]s^2$$
$$+ \left[\frac{C_{32}(C_{21}+C_{20})}{v_2 v_3} + \frac{C_{12}(C_{23}+C_{20})}{v_1 v_2} + \frac{C_{12}C_{32}}{v_1 v_3}\right]s + \frac{C_{12}C_{20}C_{32}}{v_1 v_2 v_3} = s^3 + As^2 + Bs + C$$

This equation is equivalent to

$$(s+\lambda_1)(s+\lambda_2)(s+\lambda_3) = s^3 + (\lambda_1+\lambda_2+\lambda_3)s^2 + (\lambda_1\lambda_2+\lambda_1\lambda_3+\lambda_2\lambda_3)s + \lambda_1\lambda_2\lambda_2$$

where

$$A = \frac{C_{12}}{v_1} + \frac{C_{21}+C_{23}+C_{20}}{v_2} + \frac{C_{32}}{v_3} = \lambda_1+\lambda_2+\lambda_3$$

$$B = \frac{C_{12}(C_{23}+C_{20})}{v_1 v_2} + \frac{C_{12}C_{32}}{v_1 v_3} + \frac{C_{32}(C_{21}+C_{20})}{v_2 v_3} = \lambda_1\lambda_2+\lambda_1\lambda_3+\lambda_2\lambda_3$$

$$C = \frac{C_{12}C_{20}C_{32}}{v_1 v_2 v_3} = \lambda_1\lambda_2\lambda_3$$

Note that the I^{131} dosing and blood sampling takes place at the plasma compartment 2. The Laplace expression for compartment 2 is

$$G_2(s) = \frac{1}{(s+\lambda_2)(s+\lambda_2)(s+\lambda_3)}\begin{vmatrix} s+C_{12}/v_1 & 0 & 0 \\ -C_{12}/v_2 & g_2(0) & -C_{32}/v_2 \\ 0 & 0 & s+C_{32}/v_3 \end{vmatrix} = \frac{g_2(0)(s+C_{12}/v_1)(s+C_{32}/v_3)}{(s+\lambda_2)(s+\lambda_2)(s+\lambda_3)}$$

Then

$$\frac{G_2(s)}{g_2(0)} = \frac{s^2 + (C_{12}/v_1+C_{23}/v_3)/s + C_{12}C_{23}/v_1 v_3}{(s+\lambda_1)(s+\lambda_2)(s+\lambda_3)}$$

$$= \frac{D}{s+\lambda_1} + \frac{E}{s+\lambda_2} + \frac{F}{s+\lambda_3}$$

The state equation for the sampling compartment 2 is the inverse of the above transform:

$$y_2 = g_2(0)\left[De^{-\lambda_1 t} + Ee^{-\lambda_2 t} + Fe^{-\lambda_3 t}\right]$$

where

$$D = \frac{\lambda_1^2 - (C_{12}/v_1 + C_{32}/v_3)\lambda_1 + C_{12}C_{23}/v_1 v_3}{(\lambda_2 - \lambda_1)(\lambda_3 - \lambda_1)}$$

$$E = \frac{\lambda_2^2 - (C_{12}/v_1 + C_{32}/v_3)\lambda_2 + C_{12}C_{32}/v_1 v_3}{(\lambda_1 - \lambda_2)(\lambda_3 - \lambda_2)}$$

$$F = \frac{\lambda_3^2 - (C_{12}/v_1 + C_{32}/v_3)\lambda_3 + C_{12}C_{32}/v_1 v_3}{(\lambda_1 - \lambda_3)(\lambda_2 - \lambda_3)}$$

Note that there are eight unknowns in the system, C_{12}, C_{21}, C_{23}, C_{32}, C_{20} and v_1, v_2, v_3. The C_{20} rate is determined by collecting urine samples and the v_2 distribution volume is determined from the relationship $v_2 = I^{131}\text{Dose}/g_2(0)$. The D, E, F and the λ values are determined from the nonlinear curve fitting process of the state equation corresponding to the sampling compartment. The remaining six unknowns are determined from the six equations, namely A, B, C, D, E, and F. If the unknowns are solved, the Laplace transforms and the state equations for the remaining compartments are easily defined. The following are the corresponding Laplace expressions for compartment 1 and compartment 3:

$$G_1(s) = \frac{1}{(s+\lambda_2)(s+\lambda_2)(s+\lambda_3)} \begin{vmatrix} 0 & -C_{21}/v_1 & 0 \\ g_2(0) & s+(c_{21}+C_{23}+C_{20})/v_2 & -C_{32}/v_2 \\ 0 & -C_{23}/v_3 & s+C_{32}/v_3 \end{vmatrix} = \frac{g_2(0)\,C_{21}/v_1(s+C_{32}/v_3)}{(s+\lambda_2)(s+\lambda_2)(s+\lambda_3)}$$

and

$$G_3(s) = \frac{1}{(s+\lambda_2)(s+\lambda_2)(s+\lambda_3)} \begin{vmatrix} s+C_{12}/v_1 & -C_{21}/v_1 & 0 \\ -C_{12}/v_2 & s+(c_{21}+C_{23}+C_{20})/v_2 & g_2(0) \\ 0 & -C_{23}/v_3 & 0 \end{vmatrix} = \frac{g_2(0)\,C_{23}/v_3(s+C_{12}/v_1)}{(s+\lambda_2)(s+\lambda_2)(s+\lambda_3)}$$

Summary

Compartmental first order linear models are represented by differential equations reducible to the form $dY/dt = (A+B)Y+X$, where Y is the set of state variables, A is a matrix of constant coefficients determining exchange rates between compartments, B is a matrix defining the system outputs to the outside environment and X is the set of input functions of the system. The sum of the coefficients of each column of matrix A should always add up to zero. The model represents a closed system if B is a null matrix, otherwise the system is open.

9.5 FITTING MODELS TO DATA OF CONTINUOUS SYSTEMS

Data of continuous systems are seldom recorded continuously. Most frequently, the data is recorded at regular or at irregular intervals. As disclosed in Chapter 3, always $dt = \Delta t$. Conversely, except for the particular case of the straight line, $dy \neq \Delta y$. However, if Δt is small enough, dy could be an acceptable approximation to the increment Δy of the function. Therefore, the same procedure for fitting models to discrete data is feasible for continuous systems, provided the data is collected and organized in a discrete arrangement. The procedure was presented in the previous chapter and is now illustrated with continuous systems in the examples that follow. The procedure implies the discretization of the continuous system that was already discretized by the data collection method.

Example 9.5.1 The following data corresponds to Example 9.4.2 and is related to a patient with an immunodeficiency problem, dosed with 9.9 grams of gamma globulin intravenously. The following is the blood concentration of the patient Ig globulin:

t	3	6	12	19	26	33	45	54	62
y	427	420	374	340	301	298	241	238	259

where t is days and y is the IgG concentration in mg/dl. Define an appropriate linear model for the data.

Solution: The following is an adjusted difference table for the above data:

Table 9.5.1

t	y	Δy	$\Delta y/\Delta t$	$\Delta^2 y$	$\Delta^2 y/\Delta t^2$
3	427	-7	-2.333	-39	-4.333
6	420	-46	-7.667	12	0.333
12	374	-34	-4.857	-5	-0.102
19	340	-39	-5.571	36	0.745
26	301	-3	-0.429	-54	-1.102
33	298	-57	-4.750	54	0.375
45	241	-3	-0.333	24	0.296
54	238	21	2.625		
62	259				

The following is the second order equation, obtained by linear regression from this table:

$$\frac{\Delta^2 y}{\Delta t^2} + 0.6764\frac{\Delta y}{\Delta t} + 0.0522y = 17.1574 - 0.1113t$$

The statistical evaluation of this equation is shown in the next table. The coefficient of determination and the standard error are here $R^2 = 0.916$ and $s = 0.725$. Note that the coefficient related to the time variable is not significant.

Table 9.5.2

Variable	Coefficient	Standard Error	"t"
Δy	0.6764	0.1364	4.960
y	0.0522	0.0275	1.898
t	-0.1113	0.1200	-0.927
Constant	17.1575	11.7113	1.465

As shown below, a new expression, with the non significant coefficient deleted, greatly improves the stability of the other coefficients:

$$\frac{\Delta^2 y}{\Delta t^2} + 0.6431\frac{\Delta y}{\Delta t} + 0.0272y = 6.4009$$

The statistics of this new equation, with $R^2 = 0.892$ and $s = 0.713$, is shown in the next table:

Table 9.5.3

Variable	Coefficient	Standard Error	"t"
Δy	0.6431	0.1292	4.976
y	0.0272	0.0052	5.227
Constant	6.4010	1.5808	4.049

The following is the Laplace expression of the new equations:

$$G(s) = \frac{g(0)(s+0.6431)+g'(0)}{(s+0.0456)(s+0.5976)} + \frac{6.4010}{s(s+0.0456)(s+0.5976)}$$

where the $g(0)$ and $g'(0)$ are initial conditions. Note that the zero time values are not available from the data. These initial values were guessed from a graph of the data as $g(0)=460$ and $g'(0)=13$. After solving the above transform, rearranging terms and fine tunning the total response by non linear regression, the following are the resulting system responses:

$$y_A = 523.0e^{-0.0456t} - 62.0e^{-0.5976t}$$
$$y_B = 235.4 - 254.8e^{-0.0456t} + 19.4e^{-0.5976t}$$
$$y = 235.4 + 268.2e^{-0.0456t} - 42.6e^{-0.5976t}$$

where y_A, is the free response, y_B is the forced response and yt is the total response of the system. The "goodness" of fit is shown in Fig. 9.5.1. Because the initial values and the discretization procedure affect the state equations, some fine tunning by non linear curve fitting methods is frequently needed. Non linear curve fitting methods were discussed in Chapter 5.

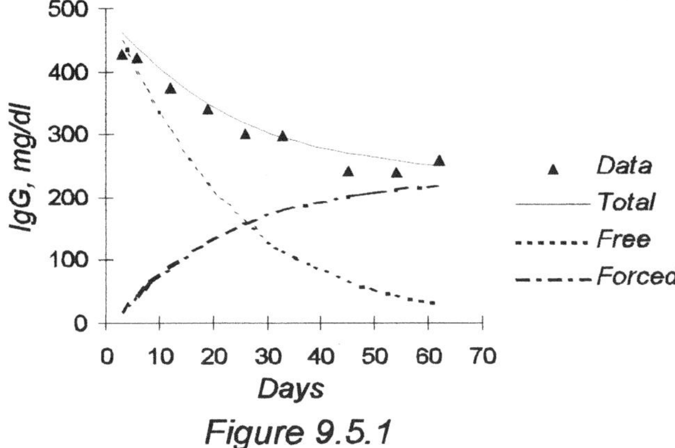

Figure 9.5.1

Example 9.5.2 The following is the data corresponding to Example 9.3.2, related to the nitrogen fractions in the rumen of steers fed a soy-meal diet:

Table 9.5.4

t	y_1	y_2	t	y_1	y_2
0	15.0	127.0	7	19.1	140.5
1	24.5	154.0	8	18.2	142.0
2	26.0	152.0	9	17.1	141.5
3	26.5	142.0	10	16.0	141.0
4	24.3	144.0	11	15.5	140.5
5	21.5	140.0	12	15.0	140.0
6	20.0	139.5			

where y_1 is protein nitrogen, y_2 is ammonia nitrogen in Mg/100 Ml of ruminal fluid and t is days. Determine the differential equations of the system.

Solution: The following is the adjusted difference table corresponding to the above data:

Table 9.5.5

t	y_1	Δy_1	y_2	Δy_2
0	15.0	9.5	127.0	27.0
1	24.5	1.5	154.0	-2.0
2	26.0	0.5	152.0	-10.0
3	26.5	-2.2	142.0	2.0
4	24.3	-2.8	144.0	-4.0
5	21.5	-1.5	140.0	-1.0
6	20.0	-0.9	139.0	1.5
7	19.1	-0.9	140.5	1.5
8	18.2	-1.1	142.0	-0.5
9	17.1	-1.1	141.5	-0.5
10	16.0	-0.5	141.0	-0.5
11	15.5	-0.5	140.5	-0.5
12	15.0		140.0	

The set of equations obtained by linear regression from this difference table is as follows:

$$\frac{\Delta Y}{\Delta t} = \begin{bmatrix} -0.9197 & 0.0964 \\ -1.1692 & -0.6931 \end{bmatrix} Y + \begin{bmatrix} 10.7430 \\ 132.4006 \end{bmatrix} - \begin{bmatrix} -1.0445 \\ 1.6694 \end{bmatrix} t$$

The statistical evaluation of the regression coefficients is shown in Table 9.5.6.

Table 9.5.6

Variable	Coefficient	Standard Error	"t"
y_1	-0.9197	0.1116	-8.238
	-1.1692	0.3341	-3.499
y_2	0.0964	0.0571	1.687
	0.6931	0.1710	-4.054
t	-1.0445	0.0955	-10.938
	-1.6694	0.2858	-5.841
Constant	10.7430	6.3745	1.685
	132.4006	19.0786	6.940

The above set of equations is quite acceptable. The coefficients of determination are $R^2 = 0.953$ for protein nitrogen and $R^2 = 0.944$ for ammonia nitrogen. Standard errors are $s = 0.813$ and $s = 2.432$ for each of the two variables. However, as will be shown, the stability of the coefficients is improved if the non significant coefficient 10.7430 is deleted. The new set of equations is the one defined in Example 9.3.2:

$$\frac{\Delta Y}{\Delta t} = \begin{bmatrix} -1.0249 & 0.1888 \\ -1.1692 & -0.6931 \end{bmatrix} Y + \begin{bmatrix} 0 \\ 132.4006 \end{bmatrix} - \begin{bmatrix} 1.0911 \\ 1.6694 \end{bmatrix} t$$

The set of state equations corresponding to the above difference equations is shown below and the corresponding graph is shown in Fig. 9.5.2:

$$Y = \begin{bmatrix} 27.8 \\ 144.8 \end{bmatrix} - \begin{bmatrix} 1.151 \\ 0.468 \end{bmatrix} t - e^{-0.859t} \begin{bmatrix} 12.80 & 2.82 \\ 17.81 & -27.32 \end{bmatrix} \begin{bmatrix} \cos(0.440t) \\ \sin(0.440t) \end{bmatrix}$$

The statistics for these equations is shown in Table 9.5.7:

Table 9.5.7

Variable	Coefficient	Standard Error	"t"
y_1	-1.0249	0.1016	-10.089
	-1.1692	0.3341	-3.499
y_2	0.1888	0.0175	10.766
	0.6931	0.1710	-4.054
t	-1.0911	0.1032	-10.876
	-1.6694	0.2858	-5.841
Constant	10.7430	6.3745	1.685

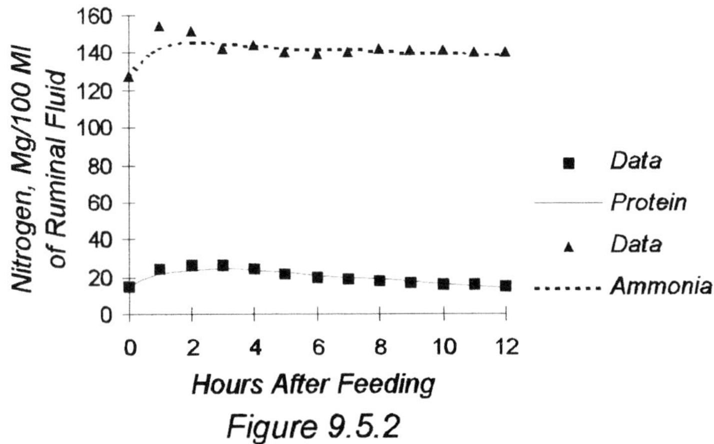

Figure 9.5.2

Note that the response curves do not match the data very accurately at some critical points. Therefore, a fine tuning of the state equations by a non linear regression procedure is needed. The following equation represents the system when only those coefficients related to the initial conditions were allowed to change:

$$Y = \begin{bmatrix} 27.80 \\ 144.81 \end{bmatrix} - \begin{bmatrix} 1.151 \\ 0.468 \end{bmatrix} t - e^{-0.859t} \begin{bmatrix} 12.90 & -16.77 \\ 16.70 & -72.60 \end{bmatrix} \begin{bmatrix} \cos 0.440t \\ \sin 0.440t \end{bmatrix}$$

As shown in Fig. 9.5.3, the new set of state equations greatly reduced the mismatch:

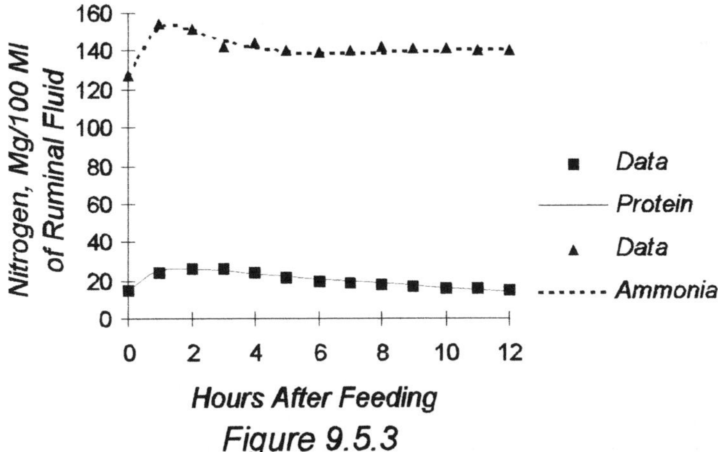

Figure 9.5.3

Summary

The following procedure is recommended for fitting linear models to the data of continuous systems:

- Express the data as a difference table
- Use a linear regression procedure to determine the most appropriate model
- Define the set of differential equation of the system
- Determine the state equations
- Use a non linear curve fitting procedure for fine tuning the state equations

EXPERIMENTAL TESTS FOR A SYSTEM ANALYSIS PROBLEM

The purpose of experimental tests in system analysis is to generate an abstract model of the system. Before the experimental tests take place, the model of the system exists only as a hypothesis. Depending on the statistical outcomes of the experiments, the hypothesis may then be accepted or rejected.

This chapter is related to procedures for modeling and selecting the working hypotheses, in a manner consistent with the concept of a system, as defined in Chapter 1. It is also related with procedures for matching experimental treatments to the mathematical model of the hypothesis.

10.1 THE EXPERIMENTAL HYPOTHESIS

A *hypothesis* is a speculation or conjecture about something that is not proven. Therefore, an *experimental hypothesis* is, in agricultural research, a speculation about a particular population related to agriculture. The researcher has often in mind a definite notion about the population. Then, the purpose of experimentation is to get evidence concerning such belief. Specifically, the following definition for an experimental hypothesis applies in system analysis:

Definition 10.1.1 An experimental hypothesis is a pre experimental proposal of mathematical models for the response functions of the system.

Statement of the Research Problem

A research problem may be defined as a set of questions on the cause-effect relationships among variables. When no acceptable answers to the questions are known to exist, such questions determine the existence of unknown quantities. Finding values of the unknowns may require experimental tests.

A question may be defined as a function assigning a set of unknowns to a set of cause-effect relationships. The domain of the question is the set of all factors affecting the system. It includes factors related and factors not related to the research problem. The codomain of the question is the set of all possible acceptable solutions to the unknowns. It includes only solutions related to the research problem. The main constraints of the solutions are the mathematical model of the hypothesis, the experimental design and the quality of experimental data.

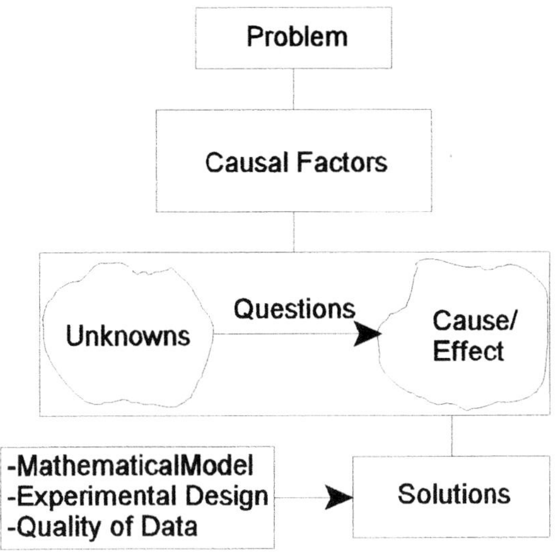

Figure 10.1.1

Example 10.1.1 An experiment is required to evaluate the effects of nitrogen fertilization on the forage yield, as dry matter and as crude protein, of African Star grass and guinea grass pastures. Define the research problem.

The question is here: "How is nitrogen fertilization affecting the dry matter and crude protein yields of Star grass and Guinea grass?" Pasture yield is here a function of the grass species and of nitrogen fertilization. Note that grass species and nitrogen fertilization are two different types of variables. The grass species, s
Star grass and guinea grass, are component variables with no interface relationships. The reader is reminded that component variables were defined in Chapter 6 as components of a conjunctive coupled system. The different levels of nitrogen fertilization are input variables. Thus, the following are the problem related factors that may affect the system:

>	Component variables - Grass species
>	Input variables	- Nitrogen fertilization

The above factors may affect the following variables:

>	State variables	- Available pasture and crude protein content
>	Output variables - Dry matter yield and crude protein yield

<u>Unknowns of the research problem</u>:

- Available pasture as a function of the grass species
- Crude protein content as a function of the grass species
- Available pasture as a function of nitrogen fertilization
- Crude protein content as a function of nitrogen fertilization
- Dry matter yield as a function of available pasture
- Dry matter yield as a function of the crude protein content
- Crude protein yield as a function of available pasture
- Crude protein yield as a function of crude protein content

As shown above, the component variables generate some unknowns and the input variables within components generate others.

<u>Domain</u> - {{Grass species, Nitrogen fertilization},{All other factors affecting the system}}

<u>Codomain</u> - {Solutions}

Note that, before the experiment takes place, the set of solutions is only a set of hypotheses. Depending on the outcomes of the experiment, a hypothesis may then be accepted to become an actual solution or may be rejected.

The graphic representation of the research problem, defined as a conjunctive coupled system, is shown in Fig. 10.1.2.

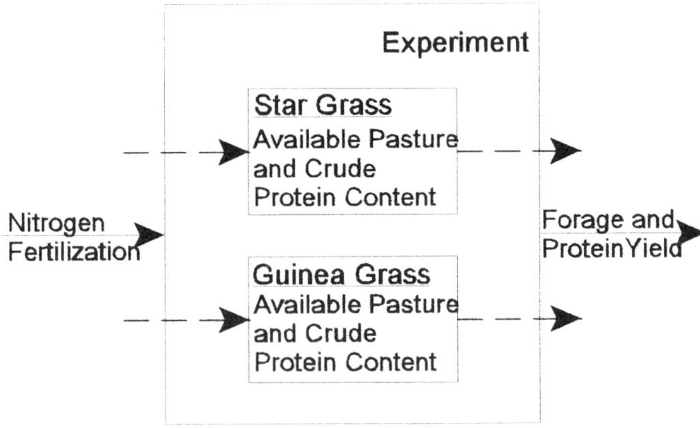

Figure 10.1.2

The statement of the research problem should be a simplified image of the system. In the above example, the system has two component systems, called Star grass and Guinea grass. Each component accepts nitrogen fertilization as an input. Nitrogen fertilization affects the states of the system, namely available pasture and crude protein content. The system states affect the output, namely dry matter and protein yields. The complete picture of the system should include the proposed mathematical models of the response functions.

The problem related factors, namely grass species and nitrogen fertilization in the example, determine the experimental treatments. All other possible factors affecting the system would determine the type of experimental design needed for managing the experimental error.

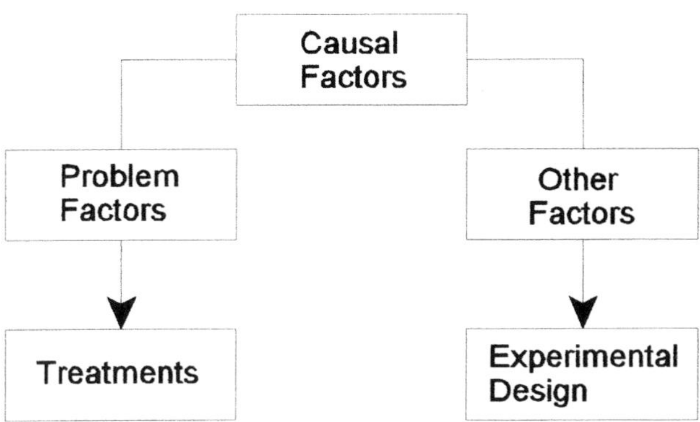

Figure 10.1.3

The statement of the research problem is the foundation for the formal definition of the experiment as a system, as was explained in Chapter 6. It is also the foundation for the mathematical models of experimental hypotheses.

The Null Hypothesis

As indicated before, component variables generate some unknowns in the statement of the problem and input variables generate others. Thus, the notion of experimental hypotheses should include both criteria, the hypothesis on the effects of component variables and the hypothesis related to the state transition function or the output function.

The fundamental hypothesis in experimental statistics is the statement that there are no differences between hypothetical parameters or figures in the experimental sample and the corresponding parameters or figures in the population. This statement is called the *null hypothesis*. Thus, the null hypothesis may be defined as follows:

Definition 10.1.2 A null hypothesis is a statement that there are no differences between hypothetical parameters in a sample and the corresponding parameters in the population

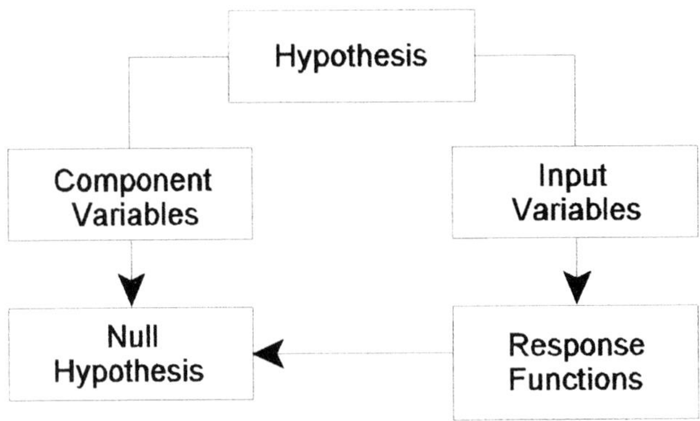

Figure 10.1.4

Depending on the experimental outcomes, the null hypothesis may be accepted or rejected. The most common parameters tested by a null hypothesis are sample means or a particular ratio. Testing differences among component systems only by their sample means can be misleading. In system analysis, averages or ratios are not sufficient criteria for testing a hypothesis. System analysis requires testing the coefficients of the mathematical models of the response functions of the system. Each coefficient of the mathematical model should be evaluated by a "t" test. If k represents a real coefficient of the state transition function and k_0 represents the corresponding hypothetical value, then

$$t = \frac{k - k_0}{S_k}$$

The null hypothesis is here $k - k_0 = 0$, where S_k is the standard error associated with parameter k.

If the system has more than one component, k_i represents a coefficient of the state transition function of the i component and k_j represents the corresponding coefficient of the state transition function of the j component, for $i = 1,2,...,n$ and $j = 1,2,...,n$, then

$$t = \frac{k_i - k_j}{\sqrt{S_{k_i}^2 + S_{k_j}^2}}$$

The null hypothesis is now $k_i - k_j = 0$, where S_{k_i} and S_{k_j} are the standard errors associated with parameters k_i and k_j.

Example 10.1.2 An experiment was designed to study how starch in the diet of steers affects the digestibility of roughage. The experimental roughage were stems of the banana plant, sugarcane leafs and African Star grass hay. Different amounts of green bananas provided the starch. *In vivo* digestibility procedures were carried out with fistulated steers. The following differential equation was proposed as experimental hypothesis:

$$\frac{dy}{dt} = uy - au \quad ; \quad u = be^{cx}$$

where y is percent digestibility of crude protein, t is hours and x is percent of dried bananas in the diet. Define the null hypothesis for the constant coefficients and for differences between roughages.

Solution: Each roughage is a component of the experiment as a system and must have its own differential equation and the corresponding constant coefficients should be compared by a "t" test between roughages. Then, the following is the set of null hypotheses:

$$a_{banana} = a_{sugarcane} = a_{Stargrass} = 0$$
$$b_{banana} = b_{sugarcane} = b_{Stargrass} = 0$$
$$c_{banana} = c_{sugarcane} = c_{Stargrass} = 0$$

Table 10.1.1

Coefficient	Banana Stems	Sugarcane Leafs	Star grass Hay
a	64.25	54.35	64.33
S_a	8.92	5.74	12.41
b	-0.06196	-0.03979	-0.04175
S_b	0.00634	0.00992	0.00872
c	-0.01312	-0.01010	-0.01009
S_c	0.00298	0.00666	0.00574

The a, b and c coefficients and the corresponding errors are displayed in Table10.1.1[1]. The reader may wish to check if the differences are significant, with 20

---

[1] Computed from R.I. Medina-Certad

degrees of freedom for within roughages and 40 for between roughages.

Summary

A research problem may be defined as a set of questions assigning unknowns to the cause-effect relationships among variables. The domain of the question is the set of all factors affecting the system. The codomain is the set of acceptable solutions to the unknowns, as defined in the problem. In system analysis, solutions are expressed as mathematical models of the response functions of the system. Before the experiment takes place, the set of solutions is only a set of hypotheses, defined as proposals of mathematical models of the response functions. Testing these models requires testing the coefficients of the mathematical models using the null hypothesis criteria.

10.2 MATHEMATICAL MODELS OF THE RESPONSE FUNCTIONS

The agricultural scientist has often some notion or image about the relationships between variables in the population that he is dealing with. Mathematical models of the experimental hypothesis must reflect this image. Several choices of mathematical models are often available.

Selecting the Model

The existence of some patterns of the expected response functions are useful indicators in determining an appropriate mathematical model:

- Maximum and minimum values
- Asymptotic values
- Inflection points
- Initial values

However, several empirical models may represent a particular response curve, sometimes sharing all of the above indicators. Then, inspecting additional properties of the mathematical model may be necessary. The following example illustrates this statement.

Example 10.2.1 Determine the most appropriate mathematical model for an experimental hypothesis of the growth curve of steers.

Solution: A growth model is represented by an S shaped curve, meaning that the curve has an inflection point and the rate equation has a maximum. In this example, the curve is expected to have also an asymptotic value. Many models would satisfy these requirements and some are listed in Table 10.2.1. Except for the polynomial, all these models conform to curves having an inflection point and an asymptotic value representing the mature weight of the steers. The task is now selecting the most appropriate experimental hypothesis.

Model 1 in the list is a third degree polynomial. As such, this polynomial has a maximum and a minimum, a feature that is inconsistent with growth curves. Note that using polynomial models is always an option and a temptation. Polynomials are also called "the poor man's equations," because they do not require too much pre experimental thinking and, depending on the degree of the polynomial, they may fit all kind of data. Sometimes, these convenient features may just be what the agricultural scientist needs, especially if the available data corresponds to only a segment of the response curve. The major inconvenience is that finding statistical significance and a geometrical meaning to the constant coefficients of polynomials is not always feasible. In addition, minor extrapolations of conclusions may be very risky.

Table 10.2.1

State Equation	Differential Equation	Inflection Point
1) $y = a + bt + ct^2 + dt^3$	$\dfrac{dy}{dt} = b + 2ct + 3dt^2$	$\left(-\dfrac{c}{3d},\ a - \dfrac{bc}{3d} + \dfrac{2c^3}{27d^2} \right)$
2) $y = \dfrac{1}{a - be^{-ct}}$	$\dfrac{dy}{dt} + acy^2 - cy = 0$	$\left(-\dfrac{1}{c}\ln(-a/b),\ \dfrac{1}{2a} \right)$
3) $y = ae^{b(1-e^{-ct})/c}$	$\dfrac{dy}{dt} - be^{-ct} = 0$	$\left(-\dfrac{1}{c}\ln\dfrac{c}{b},\ ae^{(b-c)/c} \right)$
4) $y = a - e^{-bt}(c + dt)$	$\dfrac{dy}{dt} + by = ab - de^{-bt}$	$\left(\dfrac{d-bc}{bd},\ a - e^{(d-bc)/d}(d/b) \right)$

Finding a geometrical meaning for the constant coefficients of the mathematical model is important for determining the appropriate hypothesis and for determining the appropriate experimental treatments. This statement will be discussed later.

Model 2 is the well known and widely used logistic growth equation. The ordinate of the inflection point of this equation is exactly half the asymptotic value $1/a$. Assuming that the inflection point is determined only by the mature weight of the steers is a severe constraint for the data. Physiological facts do not support such assumptions. In addition, model 2 is represented by a non linear differential equation, which is an inconvenience.

Model 3, known as the Gompertz equation, and model 4 are better choices. Thus, the selection process is confined now to only model 3 and model 4. Note that model 3 is represented by a homogeneous differential equation. This means that the system has only the free response due to initial conditions. This feature of model 3 gives to model 4 some advantage. Data support this advantage. The following are the state equations for both models, using actual data from birth to seven years of age[2]. The third degree polynomial is also included for comparison.

$$y = 18.67 + 293.48t - 38.32t^2 + 1.58t^3$$
$$y = 45.4e^{2.55(1-e^{-0.913t})/0.913}$$
$$y = 749 - e^{-0.948t}(722 + 560t)$$

where y is weight in kilos and t is age in years.

The coefficient of determination for the polynomial is a shining $R^2 = 0.998$ with a standard deviation of $S_{y.t} = 17.02$. However, the intercept 18.67 is not statistically significant. In addition, note that the maximum and the minimum have as coordinates (6.23, 742) and (9.94, 701). Thus, the inflection point is outside the segment corresponding to the seven years of data. A second degree polynomial would be a better choice. Clearly, a polynomial does not represent a growth curve.

The statistics $R^2 = 0.998$ and $S_{y.t} = 13.50$ confirm the accuracy of model 3. Additional statistics are shown in the following table:

Table 10.2.2

Coefficient	Error	"t"
45.37	9.06	5.01
2.548	0.307	8.30
0.9125	0.0511	17.86

A smaller standard deviation of $S_{y.t} = 7.54$ and a better stability of the constant coefficients prove the advantage of model 4, as shown in Table 10.2.3. Note that model 3 overestimates the birth weight of the steers by roughly 50%. The estimate of birth weight by model 3 is 45.4 Kg, as compared with 27 Kg in model 4, a value within expectations. Both models remain good choices. As shown in Fig. 10.2.1, both curves virtually overlap.

---

[2]Vohnout, K. Unpublished

Table 10.2.3

Coefficient	Error	"t"
749.33	6.03	124.27
0.9481	0.0449	21.12
721.70	9.83	73.42
559.64	64.73	8.65

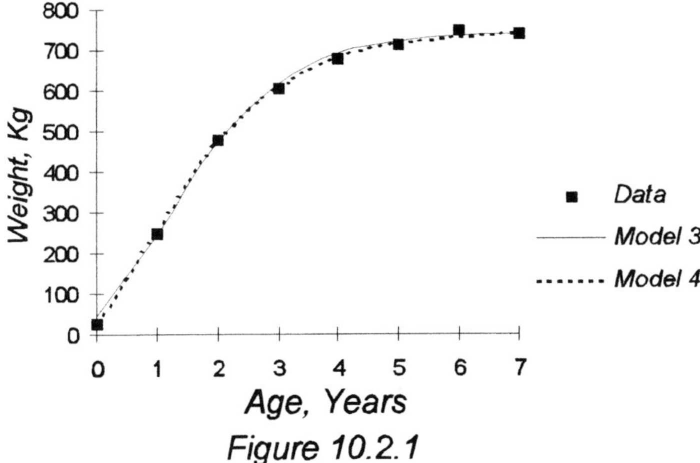

Figure 10.2.1

Geometric Interpretation of the Expected Response

Some common response curves found in agricultural systems include diminishing returns, positive and negative growth, rate equations, periodic functions, among others. However, as new variables are added to the problem, mathematical modeling may be more complex than these basic functions. Some geometrical analysis of the expected response curves of the system may help with the task of modeling and may also help avoiding the dependence on polynomials. The following examples illustrate this statement.

Example 10.2.2 Determine an appropriate mathematical model for the response curve of pasture production, as affected by nitrogen fertilization.

Solution: Pasture production displays a cyclical response due to climatic conditions. Thus, an appropriate option is a second order periodic linear model, with a state equation of the form

$$y = a + be^{\alpha t}\cos[\beta(t - c)]$$

where y is pasture production and t is time.

If the above model is selected, the geometrical meaning of the five constant coefficients should be first determined. The following definitions apply to this model:

- Coefficient a is the distance between the abscissa and the axes of the response curve
- Expression $be^{\alpha t}$ modulates the amplitude response
- Coefficient β modulates the frequency response, such that a cycle is equal to $2\pi/\beta$
- Coefficient c is an out-of-phase parameter

The above geometrical definitions are shown in Fig. 10.2.2:

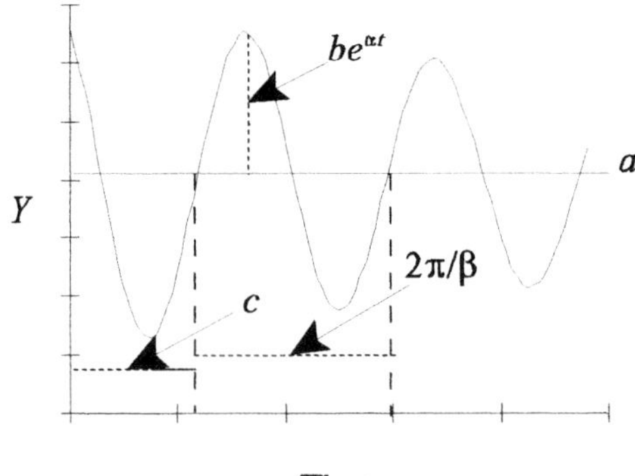

Time

Figure 10.2.2

Note that when $\alpha < 0$ the amplitude decreases over time, when $\alpha > 0$ the amplitude increases and when $\alpha = 0$ the amplitude is only determined by coefficient b.

For determining which coefficients are most likely affected by nitrogen fertilization, is helpful to look at the input term $\lambda_1\lambda_2 a$ in the system differential equation

$$\frac{d^2y}{dt^2} + (\lambda_1 + \lambda_2)\frac{dy}{dt} + \lambda_1\lambda_2 y = \lambda_1\lambda_2 a$$

where $\lambda = \alpha \pm i\beta$. Then, the input term becomes $(\alpha^2 + \beta^2)a$. As shown, the coefficients related to the input term in the differential equation are a, α and β.

Nitrogen fertilization would most likely affect coefficient *a*, the distance between the time abscissa and the axes of the response curve, because *a* represents the average pasture production. Nitrogen may also affect coefficient α, because α changes the amplitude of the response curve over the time variable. It is unlikely that nitrogen fertilization would affect the frequency coefficient β. The proposed model is shown in Fig. 10.2.2 for an $\alpha < 0$ value.

The task is now to determine appropriate equations for the presumable nitrogen dependent coefficients *a* and α. If these coefficients are no longer considered constants, they should be renamed as $a = u \in U$ and $\alpha = v \in V$. It is safe to assume that nitrogen would increase pasture production. Then, as indicated in Fig. 10.2.3, the response of variable *u* to fertilization would probably be a curve of diminishing returns with an initial value of $k_1 - k_2$ and an asymptotic value of k_1.

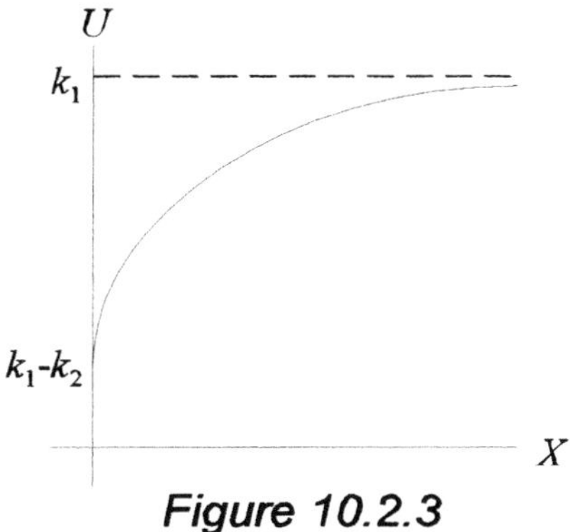

Figure 10.2.3

Equation $u = k_1 - k_2 e^{-k_3 x}$ represents the above relationship, where *u* is average pasture production and *x* is nitrogen fertilization.

It is also save to assume that pasture production over time would decrease due to the nutrient depletion of the soil. Then, variable *v* would have negative values.

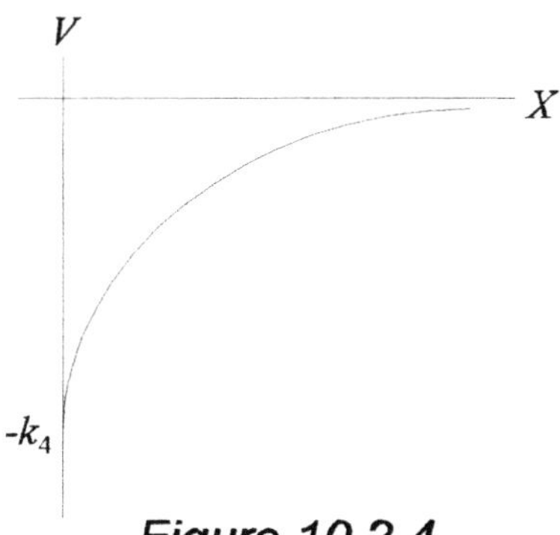

Figure 10.2.4

If the goal is maintaining a constant pasture production with fertilization, v would have to grow to an asymptotic value of zero, such that $v = -k_4 e^{-k_5 x}$, where v determines the changes of amplitude of the response function over time. The picture of this expression is shown in Fig. 10.2.4.

After replacing the above expressions in the original state equation, the following new equation represents now the system experimental hypothesis:

$$y = k_1 - k_2 e^{-k_3 x} + be^{k_4 t e^{-k_5 x}} \cos[\beta(t - c)]$$

Example 10.2.3 Determine a mathematical model for the response curve of pasture production, as affected by nitrogen fertilization and by stocking rate.

Solution: It was assumed in the previous example that only coefficients a and α, in the state equation for pasture production, where $a = u$ and $\alpha = v$, are affected by nitrogen fertilization, such that

$$y = u + be^{vt}\cos[\beta(t-c)]$$
$$u = k_1 - k_2 e^{-k_3 x_1}$$
$$v = -k_4 e^{-k_5 x_1}$$

where u is the average pasture production, v determines the changes in amplitude of the response curve over time, x_1 is nitrogen fertilization and t is time.

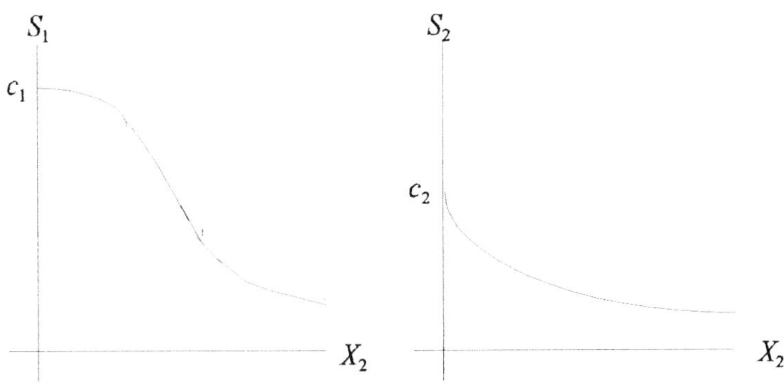

Figure 10.2.5

It is now valid to assume that the same coefficients affected by fertilization are also affected by stocking rate. Then, as stocking rate grows very large, coefficients k_1 and k_2 in variable u should approach zero. Thus, these coefficients may be renamed as $k_1 = s_1 \in S_1$ and $k_2 = s_2 \in S_2$, such that $s_1 = c_1(1 - e^{-k6/x_2})$ and $s_2 = c_2 e^{k_7 x_2}$, where x_2 is stocking rate. The graph of these expressions is shown in Fig. 10.2.5.

After replacing the above expressions in u, the new equation for average pasture production is now

$$u = c_1(1 - e^{-k_6/x_2}) - c_2 e^{-(k_3 x_1 + k_7 x_2)}$$

showing that nitrogen fertilization would increase pasture production by making u larger, from an initial value of $c_1 - c_2$ to an asymptotic value of c_1. Conversely, increasing stocking rate would decrease pasture production by diminishing the value of u. These relations are shown in Fig. 10.2.6.

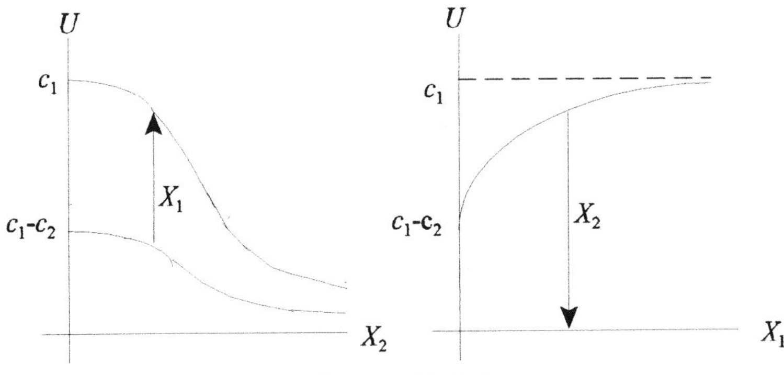

Figure 10.2.6

Increasing stocking rate would probably deplete the soil progressively faster. Therefore, assuming that stocking rate affects coefficient k_4 in the equation for variable v, this coefficient would grow gradually larger as stocking rate increases. Then, coefficient k_4 should be renamed as variable $k_4 = s_3 \in S_3$, such that $s_3 = c_3 x_2^{k_8}$. After replacing the above expression in v, the new equation is now $v = -c_3 x_2^{k_8} e^{-k_5 x_1}$, showing that nitrogen fertilization would make the negative values of v progressively smaller, from the initial value of $c_3 x_2^{k_8}$ to an asymptotic value of zero. Conversely, from an initial value of zero, stocking rate would make the negative values of v progressively larger. The above relationships are shown in Fig. 10.2.7:

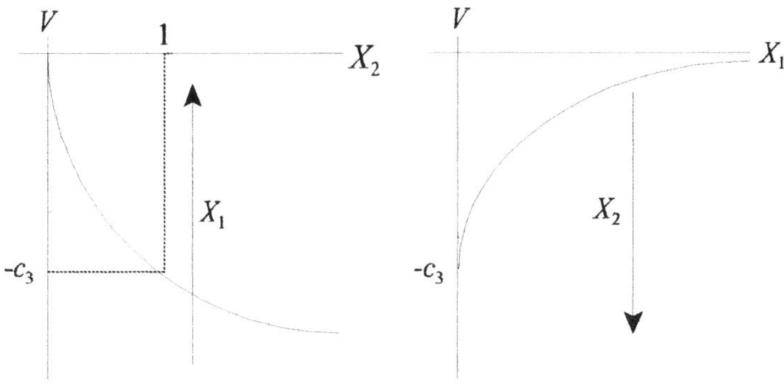

Figure 10.2.7

After replacing the *u* and *v* equations in the initial state equation, the experimental hypothesis becomes

$$y = c_1(1 - e^{-k_6/x_2}) - c_2 e^{-(k_3 x_1 + k_7 x_2)} + b e^{-c_3 x_2^{k_8} e^{-k_5 x_1}} \cos[\beta(t-c)]$$

In general terms, mathematical modeling may take the following steps:

First step: Select a mathematical model for the experimental hypothesis using critical points indicators

Second step: Determine the geometrical meaning of the constant coefficients of the selected state equation

Third step: Determine which constant coefficients may be affected by the input variables of the system

Fourth step: Define mathematical expressions for the possible relationships between input variables and coefficients

Experimental design and treatments must be consistent with the experimental hypothesis. After the data are collected, all the constant coefficients of the model must be tested by null hypothesis criteria.

Summary

Critical points of the response curve, such as maximum and minimum values, inflection points and asymptotic or initial values, are useful indicators for defining a mathematical model of the experimental hypothesis. Understanding the geometrical meaning of the constant coefficients of the state equation, determining which coefficients are most likely affected by input variables and determining expressions for the presumable relationships between the input variables and the selected coefficients, are also useful modeling procedures.

10.3 GENERATION OF EQUATIONS BY GEOMETRIC ANALYSIS

The models of the expected response curves must reflect what is assumed that the response curves might be and treatments must provide the data points for the mathematical expression. Therefore, the pre-experimental selection of the mathematical model is essential for determining the proper experimental treatments.

If the researcher can figure them out, several choices of mathematical models are often available for a given system. As indicated before, some geometric analysis is required for determining those choices.

As a complement of the previous section and to help the reader defining mathematical models, this section gives some guidelines in analytic geometry, as it applies to assembling the most common response curves found in agricultural research. Several examples are provided to show how theoretical considerations, related to the expected response of the system, match the observed response.

Equations Related to the Straight Line

The simplest equation to assemble is the straight line. Without understanding how a straight line is born, building more complicated mathematical models is hardly possible. Consider the points $P_1(t_1, y_1)$ and $P_2(t_2, y_2)$ in the TY plane. The two points determine a line segment with the following slope:

$$\tan \theta = \frac{y_2 - y_1}{t_2 - t_1} = b$$

By rearranging terms, the above expression may be written as the straight line equation $y_2 = y_1 + b(t_2 - t_1)$. If $t_1 = 0$ and $y_1 = a$, this equation becomes

$$y = a + bt$$

This is the most common expression for the straight line. The above relation is shown in Fig. 10.3.1.

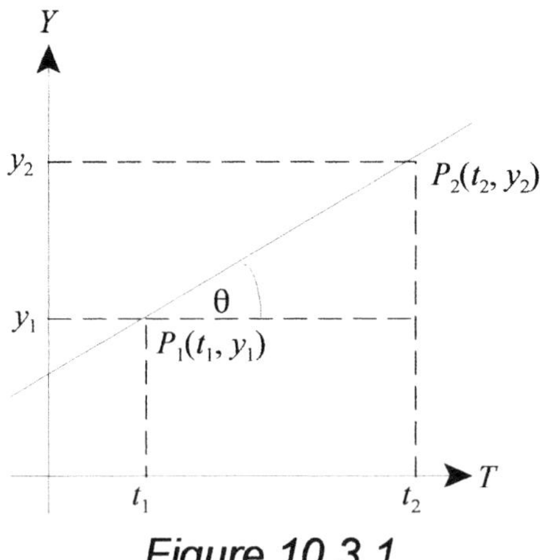

Figure 10.3.1

Consider now equation $y = at^{\,b}$. By a logarithmic transformation, this expression is the straight line $\ln y = b \ln t + k$, where $k = \ln a$. This mathematical model has been applied mainly to problems of animal metabolism and growth of body parts relative to the whole body.

Example 10.3.1 The following equation represents the relationship between body weight and metabolic rate in 26 animal species, from mouse to cow [3]:

$$y_1 = 67.4 y_2^{0.756}$$
$$\ln y_1 = 4.21 + 0.756 \ln y_2$$

where y_1 is metabolic rate in kcal/day and y_2 is body weight in kilograms. The corresponding graphs are shown in Fig. 10.3.2. Note that this is not a state equation but a relationship between two state variables.

A straight line is also obtained from a logarithmic transformation of the state equation $y = ae^{\,bt}$. Then $\ln y = k + bt$, where a is the initial value and $k = \ln a$. If $b>0$, the equation represents exponential growth. If $b<0$, this expression represents exponential decay. The following is the corresponding differential equation:

$$\frac{dy}{dt} - by = 0$$

---

[3]Kleiber, M.

This is a first order homogeneous equation, representing only the free response due to initial conditions.

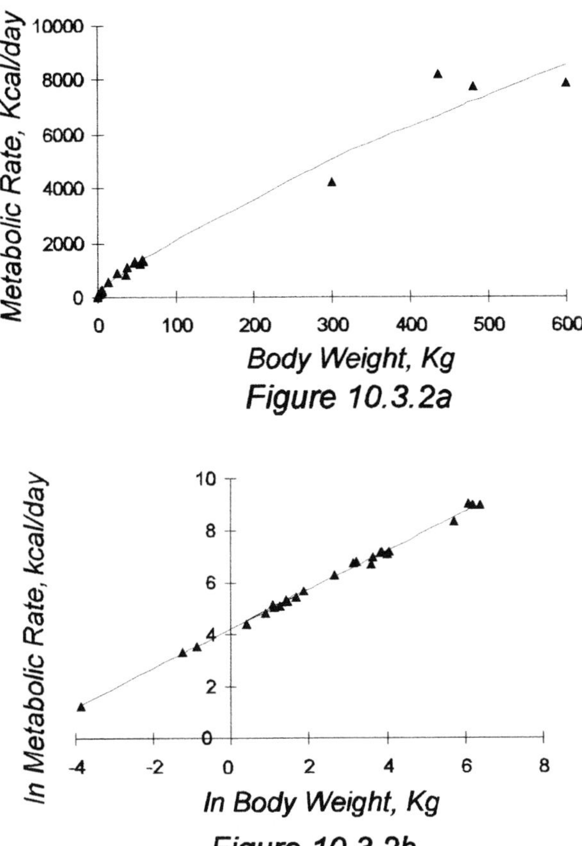

Figure 10.3.2a

Figure 10.3.2b

Example 10.3.2 The following is the fitted equation for the residual of *in vivo* digestibility of cell walls of sugarcane leaves [4]:

$$y = 102e^{-0.00751t}$$
$$\ln y = 4.628 - 0.00751t$$

where y is percent of undigested residual and t is days. The graphs of this function are

---

[4]Computed from San Martin, F.A.

shown in Fig. 10.3.3:

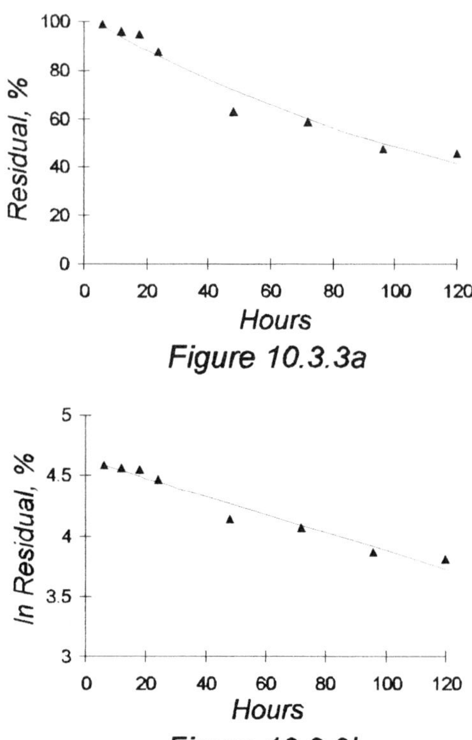

Figure 10.3.3a

Figure 10.3.3b

If a constant c is added to the exponential decay equation, expression $y = ae^{-bt} + c$ is obtained, where c is an asymptotic value and $a+c$ is the initial value. Note that when $c<a$, the asymptotic value is negative. Conversely, when $c>a$ the asymptote is a positive value. The following is the corresponding differential equation:

$$\frac{dy}{dt} + by = bc$$

This is a non homogeneous equation and represents the total response of the system. Thus, by adding the constant c to the exponential decay expression, a first order model for the total response was obtained.

Example 10.3.3 The following equation was fitted to the energy content of milk from a group of cows[5]:

$$y = 2.821 + 0.965e^{-0.423t}$$

where y is the energy content of milk in Mjoules/Kg and t is days after calving. The graph of the response functions of this system is shown in Fig. 10.3.4:

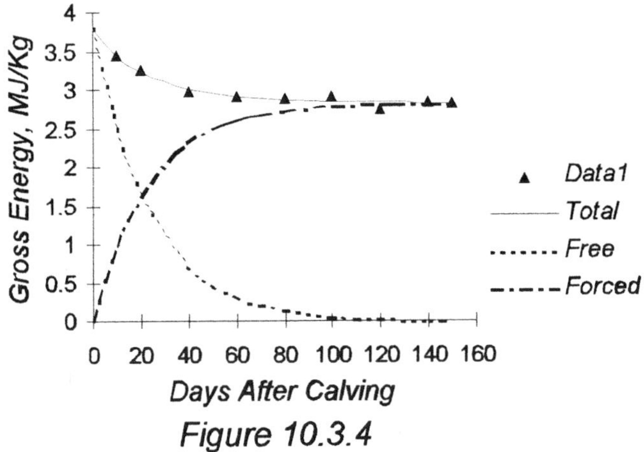

Figure 10.3.4

    If the expression of exponential decay is subtracted from a constant c, equation $y = c - ae^{-bt}$ of diminishing returns is obtained, where c is the asymptotic value and $c - a$ is the initial value. Note that when $c=a$, the above equation becomes $y = a(1 - e^{-bt})$ and the initial value is zero. Then, there is no free response, because the initial value is zero. Note also that when $c<a$, the initial value of the function is negative. Conversely, when $c>a$ the initial value is positive.

Example 10.3.4 The following is the fitted equation for the bacteria count of the rumen of a calve[6]:

$$y = 4.59 - 3.74e^{-0.151t}$$

where y is the bacteria count in millions/gram x 10^4 and t is weeks. Fig. 10.3.5 shows the graph of the response curves of the system.

---

[5]Computed from Lowman, B.G. et.al.

[6]Computed from Lengemann, F.W. and N.N. Allen

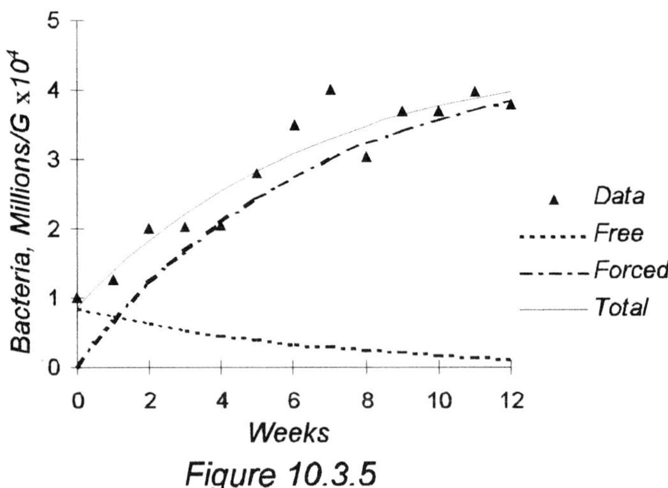

Figure 10.3.5

None of the models assembled so far provide maximum, minimum or inflection points for the response curves. For such, combining two or more different terms in the time variable is needed. Two terms determine a second order system, three terms determine a third order system, and so on.

Polynomials are always an option but, as was pointed out in the previous section, determining a geometrical meaning of the polynomial constant coefficients is often difficult. Since assembling response curves by the geometrical meaning of the constant coefficients is what this section is all about, polynomials are here excluded from consideration.

Example 10.3.5 An insect control program was tested in a pasture field and the following is the corresponding fitted equation:

$$y_n = 2193(0.6686)^n - 1943(0.5359)^n$$

where y is the number of insects per square meter and n is months. This equation has two terms in the time variable. Therefore, it represents a second order system. The following is the second order difference equation of the system:

$$y_{n+2} - 1.2045y_{n+1} + 0.3583y_n = 0$$

This is a homogeneous time invariant equation. As discussed in a previous chapter, a second order time invariant homogeneous equation is equivalent to a first order time variant non homogeneous equation. The following is the first order difference equation:

$$y_{n+1} - 0.5359y_n = 290.77(0.6686)^n$$

The response curves of the system are shown in Fig. 10.3.6.

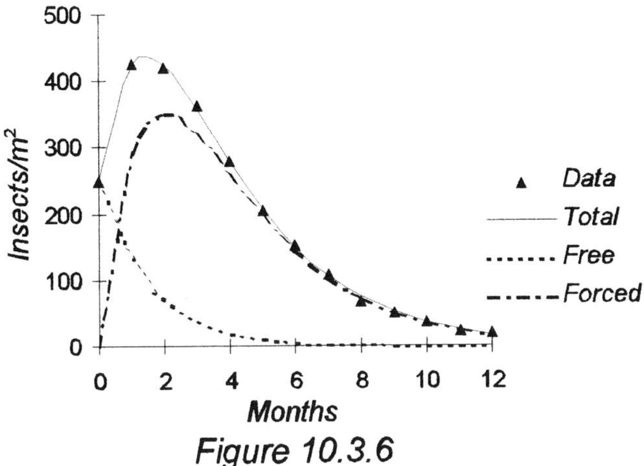

Figure 10.3.6

Time delay is another feature worth considering. The following example illustrates the manipulation for including the time delay in the model.

Example 10.3.6 The following is the fitted equation to the *in vivo* digestion of the cell walls of corn plant stubs[7]:

$$y = 0.783\left[e^{-0.00883(t-17.98)} - e^{-0.115(t-17.98)}\right]$$

where y is digestion rate, as percent per hour and t is hours. Note that there is a time lag of 17.8 hours in the digestion process. The graph of the total response of the system is shown in Fig. 10.3.7.

The following equation was also fitted to the data:

$$y = 0.0453(t-16.7)e^{-0.0280(t-16.7)}$$

Note that the time delay coefficient always goes with the time variable. This equation is

---

[7]Computed from San Martin, F.A

less accurate than the first one.

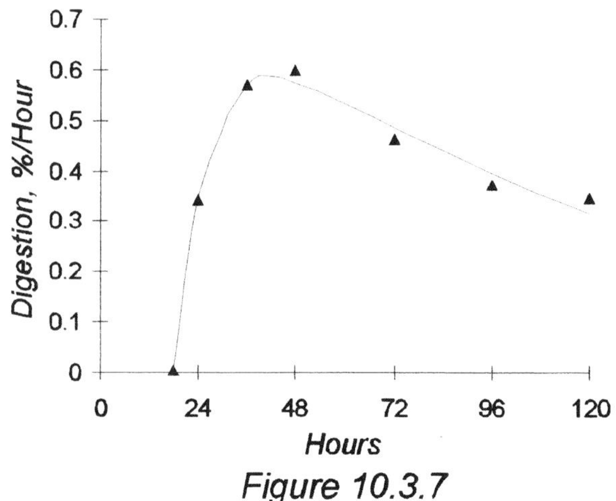

Figure 10.3.7

Periodic Functions

Periodic functions are mathematical expressions defined in polar coordinates. As such, they are functions of the radius r of the circle. Then, a point P is defined over parameters r and θ, such that $P(r,\theta)$, where θ is the angle between two vectors whose length is r.

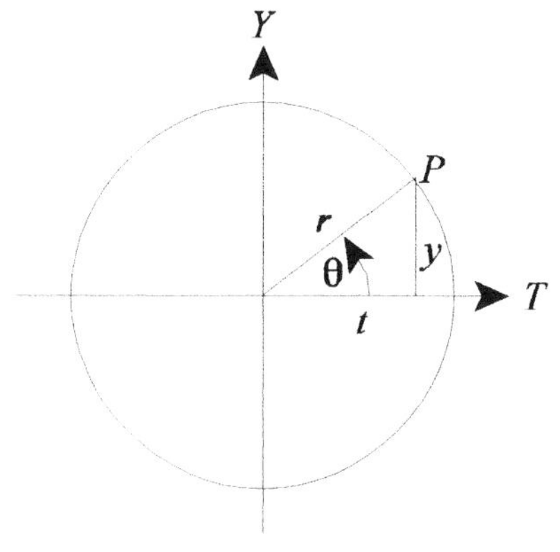

Figure 10.3.8

As shown in Fig 10.3.8, $\sin\theta = y/r$ and $\cos\theta = t/r$. Then

$$y = r\sin\theta$$
$$t = r\cos\theta$$

The above expressions are called *parametric equations of the curve*.

If the angle θ is expressed in radians, such that $\theta = at$, then the parametric equations may be written as

$$y = r\sin at$$
$$t = r\cos at$$

where the radius r modulates the amplitude of the function, such that $-r \leq y \leq r$ and parameter a modulates the frequency. Note that these functions are defined in periods or cycles of magnitude 2π. A cycle is completed when $at = 2\pi$. Then

$$\text{Cycle} = \frac{2\pi}{a}$$

The graph of these functions is shown in Fig. 10.3.9.

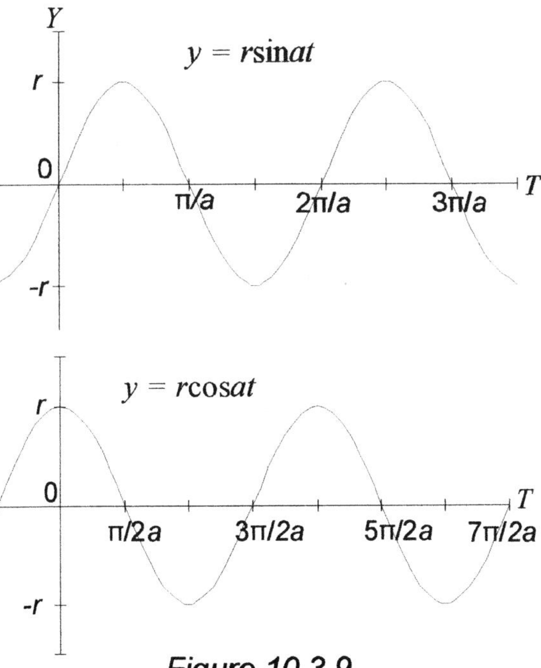

Figure 10.3.9

An important equation representing response curves that occur frequently in biological problems is $y = r_1\cos at + r_2\sin at$. As discussed in the previous chapter, an equivalent expression has the form $y = r\cos[a(t - b)]$, where a modulates the frequency, b is a time lag or out of phase coefficient and r modulates the amplitude. Coupling the above equation to an exponential term could change the amplitude of the response curve over the time variable. If the exponent α is negative, the amplitude decreases over time. Conversely, if the exponent is positive, the amplitude increases. The distance between the abscissa and the axes of the response curve could also be changed by adding a constant c to the equation. The following is the resulting expression:

$$y = re^{\alpha t}\cos[a(t - b)] + c$$

Example 10.3.7 The population of an animal species decreases each generation roughly by half the number of animals of the previous one. To prevent extinction new animals are introduced with each generation. The following is the equation representing the total response function of the system:

$$y_n = 309 + 957(0.5)^n\cos\left[\frac{2\pi}{3}(n - 0.365)\right]$$

where y is number of animals and n is generations. Note that the axes of the response curve has a value of 309. The radius is 957 and the cycle is $2\pi/3$. The curve is 0.365 out of phase. The graph of the above function is shown in Fig. 10.3.10.

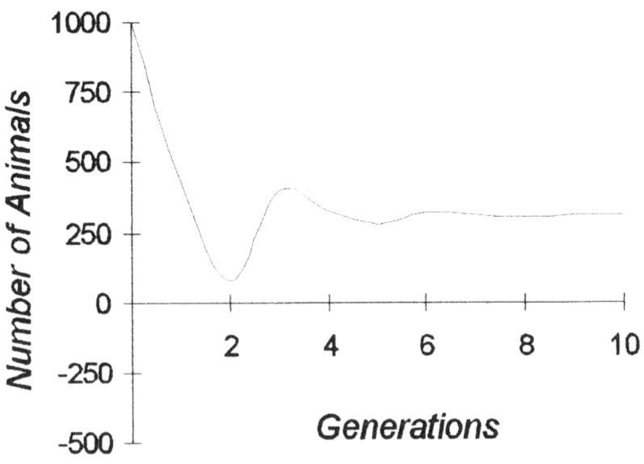

Figure 10.3.10

As disclosed in the previous section, input variables may affect the constant coefficients of a state equation. The task for the researcher is proposing also appropriate expressions for such relationships.

Summary

Many equations in Cartesian coordinates are related to the straight line by a logarithmic transformation and assembling complicated equations is possible by simple manipulations of these transformations. Periodic functions are expressions defined in polar coordinates and assembling many useful equations is possible by manipulation of the sinus, cosines and exponential functions. A pre experimental selection of mathematical models of the expected response curves of the system is essential for determining the proper experimental treatments.

10.4 ASSIGNMENT AND ARRANGEMENT OF TREATMENTS

The assignment and arrangement of *experimental treatments* must be consistent with the statement of the research problem and with the hypothesis representing the state transition function of the system. Note that an input trajectory f and the initial state y_0 determine the state y of the system at any time t, such that

$$y = u(f, y_0, t)$$

This is the fundamental expression determining the experimental treatments. Note also that input variables are related to the mode of operation of the system and that component variables are related to the structure. Thus, the following variables determine the categories of treatments that must be considered:

- Input Variables
- Initial state of the system
- Component variables

The reader should not confuse the notion of assignment and arrangement of experimental treatments with the concept of *experimental design*. Experimental design is the procedure for increasing the accuracy of experiments by grouping the sources of variation and determining the variability that is not due to treatments. Such variability is then subtracted from the experimental error. Therefore, experimental design is the management of the variability that is not related to the experimental problem and is not within the scope of this book.

Treatments Related to Input Variables

For continuous systems, an input trajectory is a set of ordered pairs of the form (t, x), where $x \in X$ is an input value and $t \in T$ is a value defined over the time scale T of the system. In agricultural experiments, an input trajectory is usually a constant value for the duration of the experiment. Thus, input related treatments may generate a factorial arrangement of the form

$$X = X_1 \times ... \times X_i \times ... \times X_m$$

where $X_i = (x_{i1},...,x_{in})$ is an input variable. Then, a treatment is an element of the factorial, such that

$$X = \left\{ x = (x_{1j},...,x_{mj}) : x_{ij} \in X_i \right\}$$

where the m-tuple x is a treatment, X_i for $i = 1,2,...,m$ is an input variable and X_{ji} for $j = 1,2,...,n$ is a value within a treatment. This factorial may be nested in any experimental design.

The following example illustrates the concept of a factorial arrangement of treatments as defined above.

Example 10.4.1 An experiment is designed with three levels of nitrogen fertilization and two levels of phosphoric acid. Define the factorial arrangement of treatments.

Solution: If nitrogen is denoted by X_1 and phosphoric acid by X_2, then

$$X = X_1 \times X_2 = \left\{ (x_{11},x_{21}),(x_{11},x_{22}),(x_{12},x_{21}),(x_{12},x_{22}),(x_{13},x_{21}),(x_{13},x_{22}) \right\}$$

where each ordered pair is a treatment that includes a nitrogen value and a phosphoric acid value. Therefore, there are six treatments in the experiment.

Given the input variables in the research problem, the following factors determine the experimental treatments:

- A capacity factor related to the number of treatments
- A potential factor related to the values assigned to each treatment

The number of treatments affects the significance test for the constant coefficients of the response functions, because significance tests are related to the degrees of freedom. The values assigned to each treatment modulate the response functions, affecting the coefficient of determination.

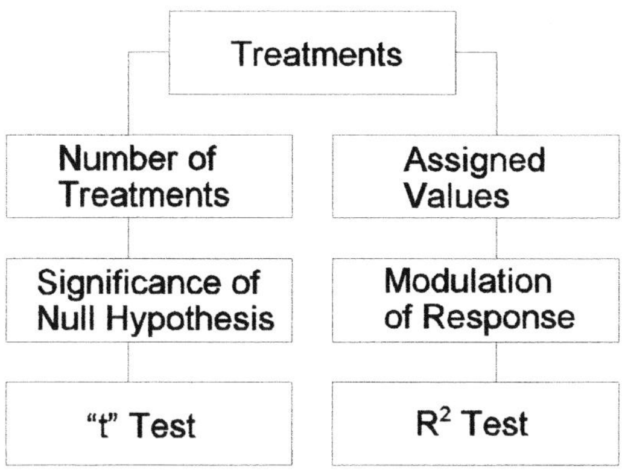

Figure 10.4.1

The relationship between degrees of freedom and the probability of a larger value in the "t" test is shown in Fig. 10.4.2.

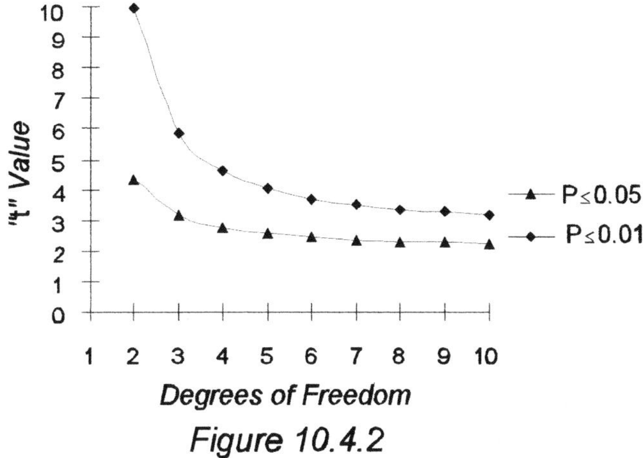

Figure 10.4.2

Note that, as the number of degrees of freedom increases, the "t" value approaches 2.576 for $P \leq 0.01$ and 1.960 for $P \leq 0.05$. The researcher would have to compromise between the accuracy of the "t" test and the affordable number of treatments in the experiment. Note also that more than four or five degrees of freedom no longer provide dramatic reductions of the "t" values. Therefore, a minimum of four or five degrees of freedom for the error term in the "t" test seems an acceptable settlement. Without previous knowledge of the variability expected for a research problem, the chart of Fig. 10.4.2 or a corresponding table, may prove useful in choosing the number of treatments for an experiment.

Example 10.4.2 The following is the proposed mathematical model for the expected milk production response of dairy cows to a low protein supplement:

$$y = e^{-ax}(b + cx) + d$$

where y is milk production and x is the percentage of a low protein supplement in the diet. Determine the number of treatments if five degrees of freedom are chosen.

Solution: This equation has four constant coefficients. If five degrees of freedom are chosen, then nine treatments are required because $n=4+5$, where n is the number of treatments. Nine treatments would fulfill the chosen degrees of freedom and also the modulation of the response equation.

Note that this function increases from an initial value of $b+d$, to a maximum and then decreases to an asymptotic value of d. Thus, the expected shape of the milk production response to supplementation is as shown in Fig. 10.4.3. Note also that the treatments were not spaced equally, but placed more closely where the expected maximum would most likely happen. Such distribution of treatments was decided for modulation of the response curve.

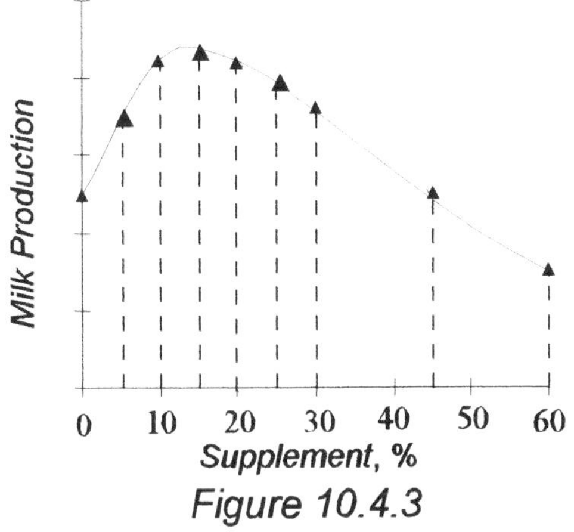

Figure 10.4.3

Usually, for experiments with a single independent variable, providing degrees of freedom for the error term is often more critical that providing treatments for modulation of the response curve. Conversely, when the research problem has more than one variable, providing treatments for modulation of the response curve may be more critical than providing degrees of freedom for the error term.

Example 10.4.3 Determine an appropriate number and distribution of treatments for the pasture production problem in Example 10.2.3.

Solution: The following was the mathematical model proposed for the experimental hypothesis of Example 10.2.3:

$$y = u + be^{vt}\cos\left[\beta(t - c)\right]$$
$$u = c_1(1 - e^{-k_6/x_2}) - c_2 e^{-(k_3 x_1 + k_7 x_2)}$$
$$v = -c_3 x_2^{k_8} e^{-k_5 x_1}$$

where u is average pasture production, v determines the amplitude of the response function over time, x_1 is nitrogen fertilization, x_2 is stocking rate and t is time. There are three independent variables and 11 coefficients in the above equation. With 11 coefficients, if five degrees of freedom are chosen for the error term, a minimum of 16 data points would be required. A 3X3 factorial would provide nine data points from nine treatments. If data is collected monthly during one year, then a total of 108 data points would be available, giving a generous 97 degrees of freedom to the error term. Clearly, obtaining degrees of freedom for the error term is not critical here. The graph of the average pasture production u, for a 3X3 factorial arrangement, is shown in Fig. 10.4.4:

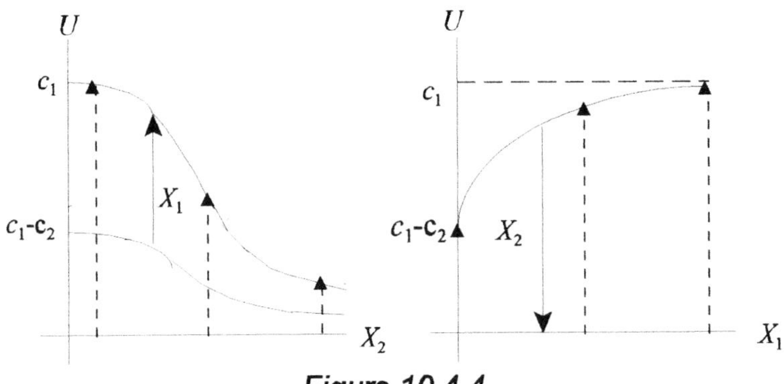

Figure 10.4.4

The question is whether a 3X3 factorial would provide sufficient data points for the modulation of the expected response curve. By observing Fig.10.4.4, it seems very unlikely that three data points would modulate properly the expected curves. As shown in Fig.10.4.5, a 4X4 factorial, representing 16 treatments, seems more suitable here:

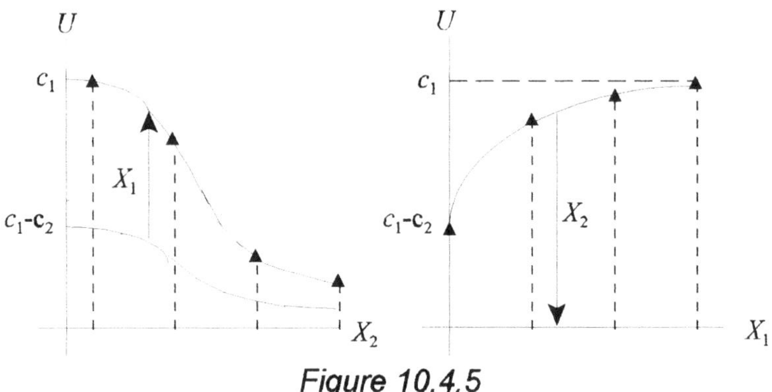

Figure 10.4.5

Clearly, the number of treatments, determined to provide degrees of freedom for the error term, is not a sufficient criterion for defining the experimental treatments. Treatment distribution should be determined only after an evaluation of where the system response is expected to have extreme values and the largest changes, as affected by the independent variables.

Modulation of the response curve should be determined also for the system response to the time variable. Therefore, data collected over time should meet the same criteria and requirements as data collected over the input variables.

Example 10.4.4 The following is the graph of the expected state equation of an insect control program:

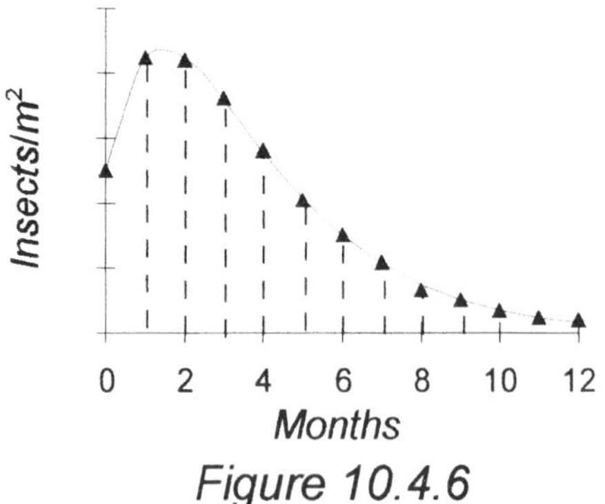

Figure 10.4.6

The number of data points in the above response curve may look as too many. However, as often happens, it is assumed here that data are collected monthly and that the time scale of this system is discrete.

Example 10.4.5 The following is the graph of the expected response of the pasture production problem in Example 10.2.2.

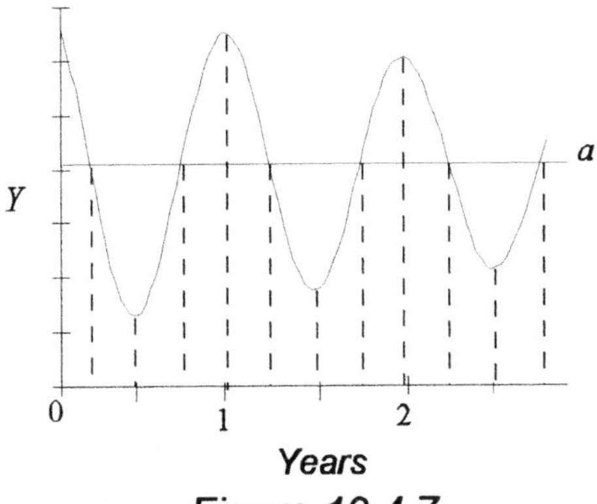

Figure 10.4.7

As shown in the graph, four data points per year may yield an adequate modulation of the response curve. However, a bimonthly collection of data would probably be more appropriate.

Researchers that are more concerned with parameters such as the mean, rather than with the dynamic condition of the system, may question the importance placed here on modulation of the system response. Systems change over time and the name of the game is forecasting the system response. Much effort and resources are wasted in small experiments, designed with mass production mentality, that are irrelevant because conclusions are based on the traditional A versus B comparison of means. The A versus B comparison of means has often very little predictive value.

The coefficient of determination estimates the predictive value of the mathematical model of the system response. As mentioned earlier, modulation of the system response affects the coefficient of determination. The coefficient of determination is the proportion of the total variability that is attributable to regression, that is

$$R^2 = \frac{\sum \hat{y}^2}{\sum \hat{y}^2 + \sum d^2}$$

where $\sum \hat{y}_2$ is the variability due to the mathematical model that is, the variability due to regression and $\sum d^2$ is the variability of the deviations from regression. Thus, the following factors affect the coefficient of determination:

- The mathematical model of the system response
- The modulation of the system response
- Variability not due to regression

The variability that is not due to regression is estimated by the corresponding standard deviation. The standard deviation from regression represents the size of the deviations of the system response in relation to the mathematical model of the system response, that is

$$S_{Y.12\ldots} = \sqrt{\frac{\sum d^2}{n-m}}$$

where n is the total number of data points and m is the total number of constant coefficients. Thus, the following factors affect the standard deviation from regression:

- The mathematical model of the system response
- Degrees of freedom
- Variability of the sample

The reader is reminded that, as the number of constant coefficients increases, R^2 gets larger, approaching 1.0 as the number of coefficients approaches the number of data points. Thus, how much the coefficient of determination can be trusted depends on the standard deviation of the regression coefficients. Non significant coefficients must be deleted from the regression equation or an alternative mathematical model should be determined. Multiple options of mathematical models for the expected response of a system are often available. The subject was presented and discussed several times before. The reader is also reminded that the experimental design affects the variability of the sample.

Treatments Related to Initial States of the System

As previously indicated, the dynamic condition of a system is represented by the expression $y = u(f, y_0, t)$, where y is a state trajectory, u is the state transition function, f is an input trajectory, y_0 is the initial state and t is time. Note that the initial value in the above expression is an independent variable but is not an input. Therefore, initial values are not a part of the factorial arrangement of treatments.

The set of initial states is defined as

$$Y_0 = \{y_{0,1}, y_{0,2}, \ldots y_{0,n}\} \quad ; \quad Y_0 \subset Y$$

Each initial value $y_{0,j}$ may generate a different state trajectory. Thus, an initial state related treatment is a subclass within the experiment. As such, initial states determine a set of component systems coupled conjunctively in the experiment as a system. This coupling is shown in Fig. 10.4.8:

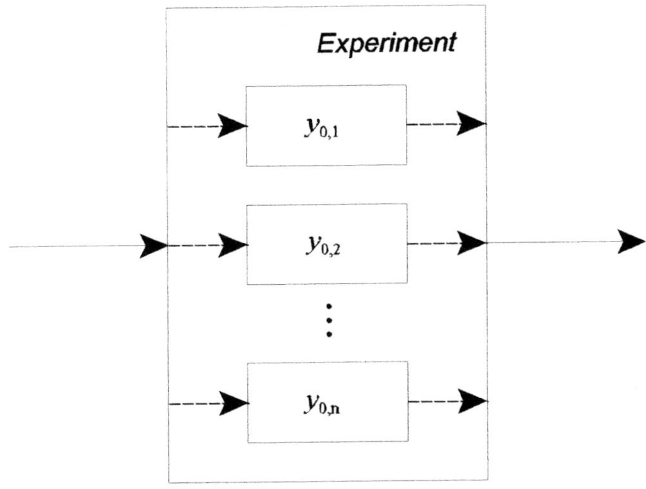

Figure 10.4.8

The factorial arrangement of treatments should be nested in each of the above subclasses and treatments should be randomized over each subclass. This conception is illustrated in Fig. 10.4.9:

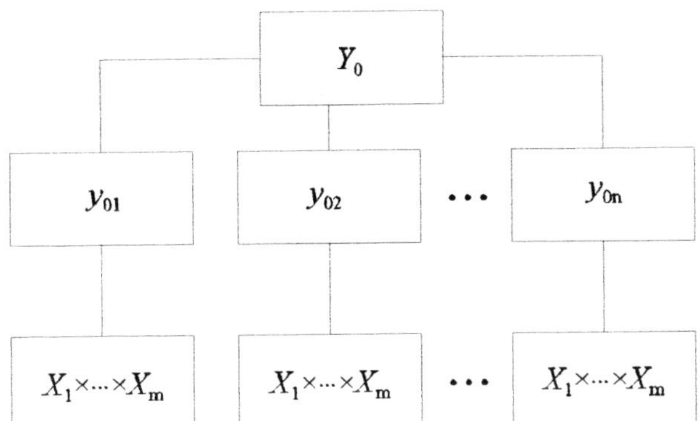

Figure 10.4.9

Example 10.4.6 The following is the mathematical model proposed for the experimental hypothesis of Example 10.2.3:

$$y = u + be^{vt}\cos\left[\beta(t - c)\right]$$
$$u = c_1(1-e^{-k_6/x_2}) - c_2 e^{-(k_3 x_1 + k_7 x_2)}$$
$$v = -c_3 x_2^{k_8} e^{-k_5 x_1}$$
$$y_{0,j} = c_{1,j} - c_{2,j} - b_j\cos(\beta_j c_j)$$

where u is average pasture production, v determines the amplitude of the response function over time, x_1 is nitrogen fertilization, x_2 is stocking rate, $y_{0,j}$ is an initial state and t is time. It is assumed here that the maximum production is obtained during the rainy season of a tropical environment and the minimum production during the dry season. The experiment is conceived as to start the system at three different stages of the production cycle: three months before the peak production of the rainy season, at the peak production and three months after the peak production. Determine the subclasses and the nesting of the factorial.

Solution: The following is the graph of the expected state trajectories for the three initial states defined for the system. Note that each of the three state trajectories corresponds to a different component system, determined by a different initial state.

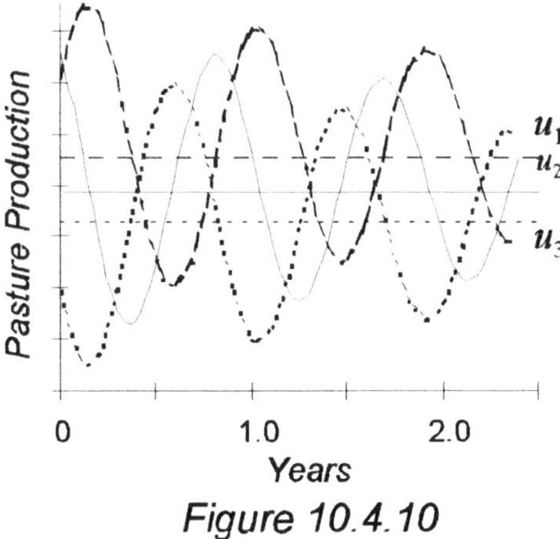

Figure 10.4.10

As shown in Fig. 10.4.11, the $X_1 \times X_2$ factorial should be nested and treatments randomized over each of the component systems.

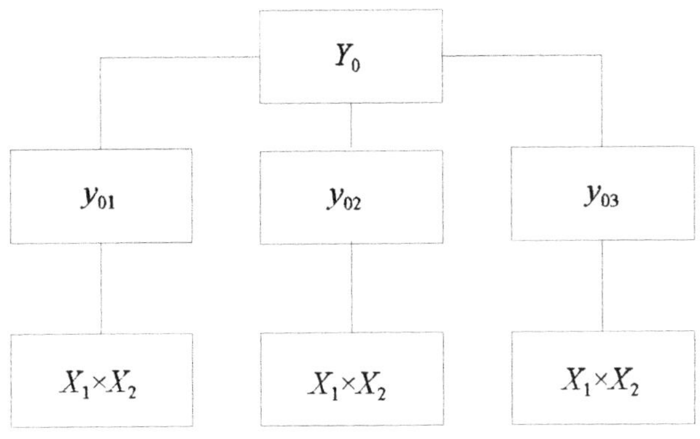

Figure 10.4.11

Treatments Related to Structural Components

The causal factors related to an experimental problem may include input variables and variables that cannot be defined as functions of time. Variables that cannot be defined as functions of time are usually qualitative traits and were called components in Chapter 6.

As defined previously, treatments related to component variables are not inputs of the system, but may determine component systems coupled conjunctively within the experiment as a system. It was also defined previously that treatments related to initial states may also generate component systems coupled conjunctively. However, systems determined by component variables and systems determined by initial states are at different structural levels within the experiment. Component variables are at a higher level of the pyramid. Then, if treatments related to initial states are subclasses, treatments related to component variables should be classes. This stratification of treatments is known as a *split-plot* arrangement. A class is a *main plot* that is being split into smaller *subplots* or subclasses. Subplots should be nested randomly over a main plot.

The set C of component variables is denoted here as $C = \{c_1, c_2, ..., c_m\}$, where each component $c_i \in C$ should nest the set of treatments Y_0 related to initial states. The total number of split-plot treatments SP is the product $SP = C \times Y_0$. Each component $y_{0,j} \in Y_0$ should nest the factorial arrangement of input related treatments. Thus, the total number of experimental treatments E is $E = C \times Y_0 \times X$, where $X = X_1 \times ... \times X_m$. This structural arrangement of treatments is shown in Fig. 10.4.12.

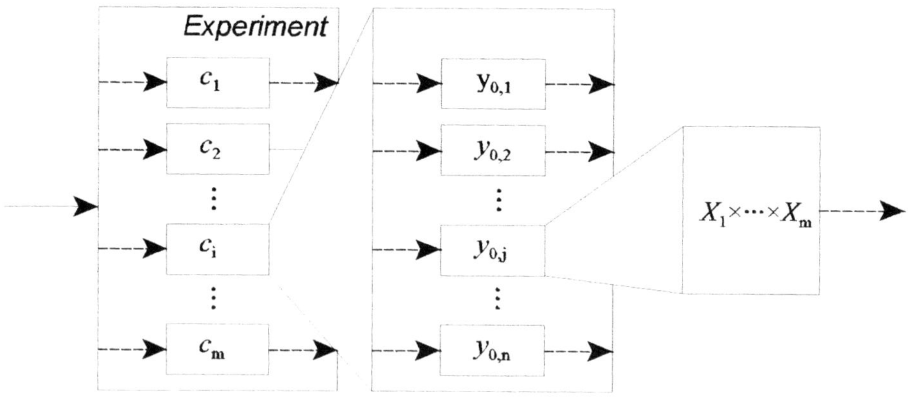

Figure 10.4.12

Note that the sequence shown in Fig. 10.4.12, component variables - initial states - input variables, should never be changed. This sequence may be nested in any experimental design. Note also that the component variables may also determine a factorial arrangement of treatments at the upper level of the pyramid.

Example 10.4.7 Determine the arrangement of treatments in a pasture production experiment with three blocks, three pasture species, three initial states, four levels of nitrogen fertilization and four levels of stocking rate.

Solution: This experiment has four variables: pastures, the initial state, nitrogen fertilization and stocking rate. It has four structural levels: blocks, pastures, initial states and the factorial. The split-plot arrangement has nine treatments nested in each block for a total of $9 \times 3 = 27$ plots. The 16 factorial treatments are nested in each initial state for a total of $16 \times 9 \times 3 = 432$ plots. An experiment with 420 plots is very large, mainly for the necessity of modulating the system response and for the inclusion of the initial state variable. Thus, the researcher may be facing the challenge of reducing the size of the experiment by cutting down the number of plots with the minimum loss of information. Deleting the initial state variable is one choice. Manipulation of the factorial arrangement of treatments is another available choice.

The following summarized plan for the analysis of variance reflects the picture of the arrangement of treatments:

Table 10.4.1

Sources of Variation		Degrees of Freedom	
Blocks (B)			2
Split-plot			24
Pastures (P)		6	
P	2		
P×B	4		
Initial States (I)		18	
I	2		
I×P			4
I×B			4
I×P×B	8		
Factorial			405
Nitrogen	3		
Stocking Rate	3		
Interactions	399		
Total			431

Manipulation of the Factorial Arrangement of Treatments

As indicated before, if the problem has more than one independent variable, providing treatments for modulation of the response functions of the system is often critical. For such, four or five and sometimes more data points are needed for each input variable. For a 5×5 factorial, that means 25 treatments. Three variables would result in 125 treatments. Central composite rotatable designs have been among the most widely proposed schemes for reducing the number of treatments. The concept of *rotatability* is related to the distribution of the standard error of the regression estimate. In a rotatable scheme, the standard error is the same for all points that are at the same distance from the center of the response curve[8]. A central composite rotatable arrangement combines a 2^k factorial, a $2(k)$ star part and central points, where k is the number of input variables.

Example 10.4.8 Determine a central composite rotatable arrangement of treatments for two input variables.

Solution: With two input variables, a central composite rotatable design is obtained by placing four treatments equally spaced around a circumference of a circle in the X_1, X_2 plane with center (0,0), plus one or more points at the center. The number of treatment replications at the center of the circle is chosen so that the standard error of the regression estimate is approximately the same at the center as it is at all points on the circle with

---

[8]Box, G.E.P. and J.S. Hunter

radius 1. In addition, four more treatments are placed at the vertices of a square inscribed in the circle. This arrangement is shown in Fig. 10.4.13.

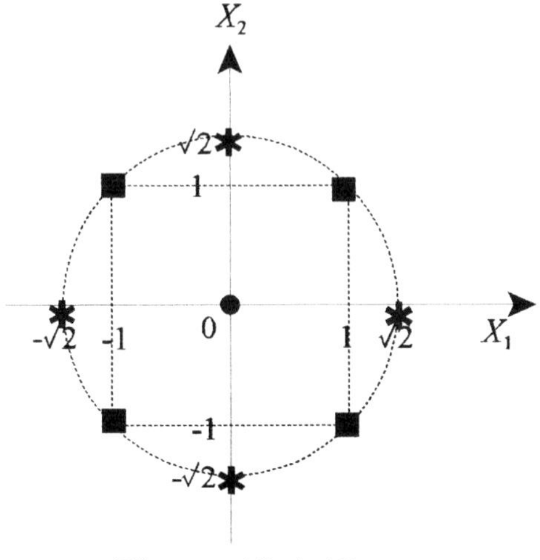

Figure 10.4.13

As shown above, the total number of treatments is nine and the total number of data points is 13, including the five central points, as compared with 25 in a regular 5×5 factorial. As with the regular factorial, there are also five data points for the modulation of the response of each of the two input variables.

Note in Fig.10.4.13 that the coordinate values are coded. The rotatable designs were developed at a time where the most sophisticated tools were the IBM 1620 computer and the Monroe electric desk calculator. Thus, coding of the coordinate values was very helpful for the computation of regression polynomials. After computations were completed, regression coefficients had to be decoded. Coding is unnecessary with the computer facilities available today.

If blocking is used in central composite rotatable schemes, the factorial and some central points form one block or sometimes two blocks in larger designs. The star part plus the remaining central points form an additional block. These are called incomplete blocks.

An inconvenience of central composite rotatable designs is their rigidity, because the distance between data points is fixed. Such distribution of treatments makes modulation of the response function of the system more difficult. Another inconvenience is the exclusion of the corner data points of the 5×5 grid. Corner points represent extreme values, which are often required for an efficient modulation of the response curve. An optional compromise would be inserting those corner treatments in the layout. Thus, this modified arrangement includes the 2^k factorial, the $2(k)$ star points, one central point and the additional 2^k corner points.

Example 10.4.9 Determine a modified central composite rotatable arrangement with two input variables.

Solution: A modified central composite design with two factors should have four factorial points, four star points, one central point and four corner points, totaling 13 treatments. The modified design is shown in Fig. 10.4.14.

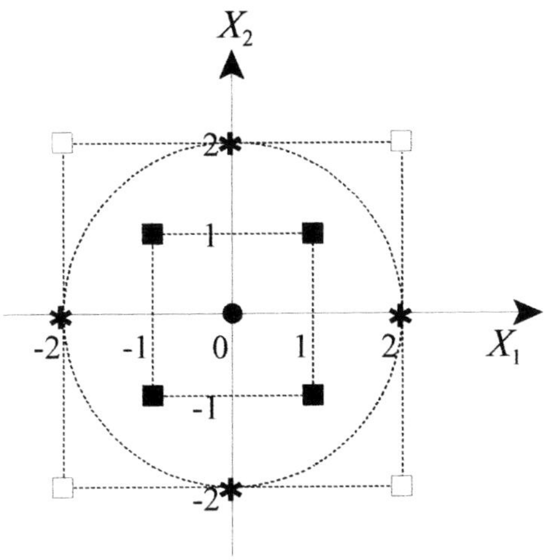

Figure 10.4.14

Example 10.4.10 Define a modified central composite design with three factors.

Solution: The modified central composite design with three variables should include eight factorial points, six star points, one central point and eight corner points, totaling 23 treatments. This arrangement is shown in Fig. 10.4.15.

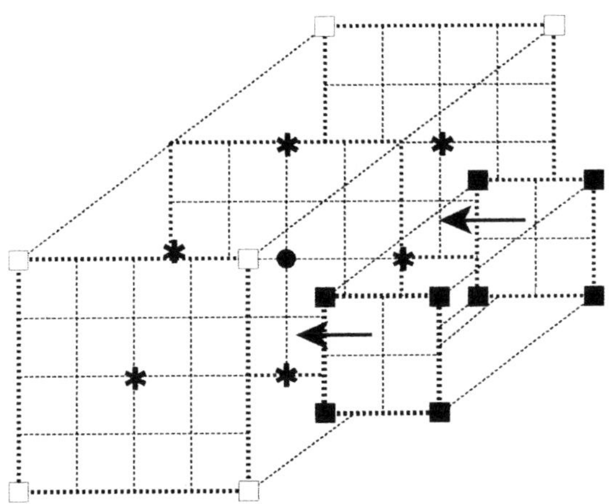

Figure 10.4.15

Note that the central composite rotatable scheme with three input variables has only 15 treatments, making a total of 20 data points with the replications of the central points, as compared with 23 for the modified design. Thus, the trade-off for the convenience of having the corner treatments is three more data points.

A less restrictive arrangement of treatments is a combination of factorials, as shown in the next example.

Example 10.8.11 Combine a 3^3 and a 2^3 factorial.

Solution: Combining a 3^3 and a 2^3 factorial, as in Fig. 10.4.16, requires a total of 35 treatments.

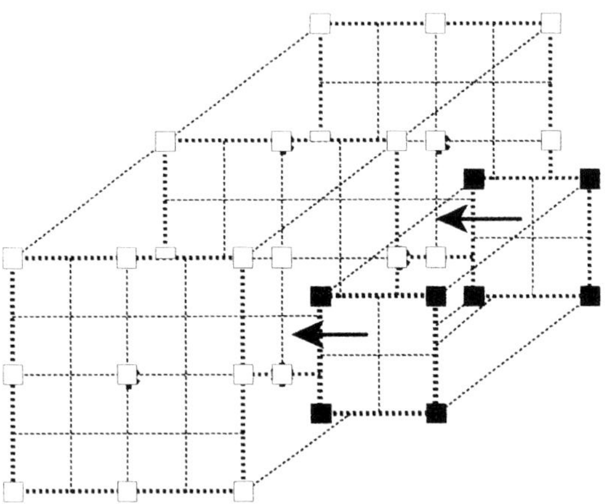

Figure 10.4.16

Options for the arrangement of treatments depend on and the ability of the research team to figure them out, on the research needs or on the availability of resources. Thus, in general terms, determining a design for the distribution of treatments may take the following steps:

First step: Define the degrees of freedom required for the error term and the data points for modulation of the response functions of the system

Second step: Determine if the size of the experiment is compatible with available resources

Third step: Adjust the number and distribution of treatments with a design compatible with available resources

Summary

Assignment of treatments should be consistent with the research problem and with the model of the expected state transition function of the system. As such, treatments should be related to input variables, to initial states and to component variables. Treatments related to input variables may generate a factorial arrangement. The number and distribution of treatments in the factorial must be chosen to attain the best modulation of the system response functions. When appropriate, the factorial should be nested in the initial states related treatments, which in turn are nested in the treatments related to component variables. Component variables are here the main plots and initial states are the subplots. An experiment with this structure may become extremely large. Central composite rotatable designs and modifications of these schemes are available for negotiating between the accuracy and the feasibility of experiments.

A

MISCELLANEOUS MATRIX
CONCEPTS AND PROCEDURES

Definition of a Matrix. Matrix is a rectangular array of numbers arranged in rows and columns. In general terms, a matrix A is represented as follows:

$$
A = \begin{bmatrix}
a_{11} & a_{12} & \cdots & a_{1j} & \cdots & a_{1n} \\
a_{21} & a_{22} & \cdots & a_{2j} & \cdots & a_{2n} \\
\vdots & & & & & \vdots \\
a_{i1} & a_{i2} & \cdots & a_{ij} & \cdots & a_{in} \\
\vdots & & & & & \vdots \\
a_{m1} & a_{m2} & \cdots & a_{mj} & \cdots & a_{mn}
\end{bmatrix}
$$

where a_{ij} denotes the element in the i th row and the j th column for a matrix of r rows and c columns.

An abbreviated form is here

$$A = \{a_{ij}\}, \quad \text{for} \quad i = 1, 2, \ldots, m, \quad \text{and} \quad j = 1, 2, \ldots, n$$

Matrix addition. If $A = \{a_{ij}\}$ and $B = \{b_{ij}\}$, then

$$A + B = \{ a_{ij} + b_{ij} \}, \quad \text{for} \quad i = 1, 2, \ldots, m \quad \text{and} \quad j = 1, 2, \ldots, n$$

The above operation can take place only if the matrices involved have the same number of rows and the same number of columns, in other words, if they are of the same order. The following definition is here set forth:

• When matrices are of the same order, it is said that they are *conformable for addition.*
• Matrices that are conformable for addition are also *conformable for subtraction*

Matrix addition and subtraction are *commutative*, that is

$$A + B = B + A$$

Matrix addition is *associative*, that is

$$(A + B) + C = A + (B + C)$$

Matrix Multiplication. Addition of two equal matrices can be written as follows:

$$A + A = \{a_{ij}\} + \{a_{ij}\} = \{2a_{ij}\} = 2A \quad \text{for} \quad i = 1,2,...,m \text{ and } j = 1,2,...,n$$

This result can be extended to the addition of k equal matrices:

$$A + A + ... + A = kA$$

Thus, a matrix A multiplied by a constant k is the matrix A with each of its elements multiplied by k.

The product of two vectors is the sum of the product of each element a_i of the row vector a' multiplied by the corresponding element x_i of the column vector x. Then, if

$$a' = (a_1 \; a_2 \; ... \; a_n) \quad \text{and} \quad x = \begin{bmatrix} x_1 \\ x_2 \\ \vdots \\ x_n \end{bmatrix}$$

the product $a'x$ must be

$$a'x = a_1x_1 + a_2x_2 + ... + a_nx_n = \sum_{i=1}^{n} a_ix_i$$
$$i = 1,2,...,n$$

If each row in matrix A is multiplied by vector x. Then the product Ax is as follows:

$$
Ax = \begin{bmatrix} a_{11} & a_{12} & \cdots & a_{1n} \\ a_{21} & a_{22} & \cdots & a_{2n} \\ \vdots & \vdots & & \vdots \\ a_{m1} & a_{m2} & \cdots & a_{mn} \end{bmatrix} \begin{bmatrix} x_1 \\ x_2 \\ \vdots \\ x_n \end{bmatrix} = \begin{bmatrix} \sum_{j=1}^{n} a_{1j} x_j \\ \sum_{j=1}^{n} a_{2j} x_j \\ \vdots \\ \sum_{j=1}^{n} a_{mj} x_j \end{bmatrix}
$$

Thus,   if $A = \{ a_{ij} \}$ and $x = \{ x_j \}$ then

$$
Ax = \left\{ \sum_{j=1}^{n} a_{ij} x_j \right\}, \qquad \text{for} \qquad i = 1,2,...,m \; ; \; j = 1,2,...,n
$$

In multiplying two matrices, each row vector of one matrix is multiplied by a column vector of the other matrix. This is a repetitive operation of multiplying one matrix by each column vector of the other matrix.

For a product AB to exist, it is required that the number of columns in matrix A to be the same as the number of rows in matrix B, such that

$$
A_{m \times n} B_{n \times s} = AB_{m \times s}
$$

The following definition applies here:

• When the number of columns in matrix A is the same as the number of rows in matrix B, the matrices are said to be *conformable for multiplication.*

Thus, if $A = \{ a_{ij} \}$ for $i = 1,2,...,m$ and $j = 1,2,...,n$ and $B = \{ b_{ij} \}$ for $i = 1,2,...,n$ and $j = 1,2,...,s$, such that $A_{m \times n} B_{n \times s} = AB_{m \times s}$ then:

The ij th element of AB is $\sum_{k=1}^{n} a_{ik} b_{kj}$ and $AB = \left\{ \sum_{k=1}^{n} a_{ik} b_{kj} \right\}$

If the matrices are conformable for multiplication, the following properties are here set forth:

- Matrix multiplication is *associative*, that is $A(BC) = (AB)C$

- Matrix multiplication is *distributive,* that is

$$A(B + C) = (AB + AC)$$
$$(A + B)\,C = AC + BC$$
$$A(kB) = k(AB) = (kA)B$$

- Matrix multiplication is not *commutative*, unless all the matrices are of the same order, that is $AB \neq BA$

Transpose of a Matrix. By interchanging rows and columns, matrix A' becomes the transpose of matrix A. Thus

$$A = \{\, a_{ij} \,\} \neq A' = \{\, a_{ji} \,\} \quad ; \quad i = 1,2,...,m \quad ; \quad j = 1,2,...,n$$

The following properties for the transpose operation are here set forth without proof:

- The transpose operation is *reflexive,* that is $(A')' = A$

- The transpose of a product matrix is the product of the transposed matrices taken in reverse order, that is $(AB)' = B'A'$

- The transpose of the addition of matrices is the addition of the transposed matrices, that is $(A + B)' = A'B'$

Elementary Operations. The following are called elementary operations:

- Exchanging two rows or two columns
- Adding or subtracting a multiple of a row or column to another row or column
- Multiplying a row or column by a constant $k \neq 0$

Matrices related by elementary operations are called *equivalent matrices*. Thus, matrix A and matrix B are said to be equivalent, if matrix B can be obtained from matrix A by elementary operations. Then $A \approx B$

Expansion of Determinants. A determinant $|A|$ is a polynomial of the elements of a square matrix A. It is a scalar value. Thus, if A is a second order matrix, then $|A|$ is defined as follows:

$$|A| = \begin{vmatrix} a_{11} & a_{12} \\ a_{21} & a_{22} \end{vmatrix} = a_{11}(a_{22}) + (-1)\, a_{21}(a_{12})$$

The process of obtaining the value of a determinant $|A|$ is known as *evaluation* or *expansion* of the determinant.

Elementary expansion. The expansion of a second order determinant is the product of the diagonal terms minus the product of the off-diagonal terms. A third order determinant can be expanded to three second order determinants:

$$A = \begin{vmatrix} a_{11} & a_{12} & a_{13} \\ a_{21} & a_{22} & a_{23} \\ a_{31} & a_{32} & a_{33} \end{vmatrix}$$

$$= a_{11} \begin{vmatrix} a_{22} & a_{23} \\ a_{32} & a_{33} \end{vmatrix} + (-1)\, a_{12} \begin{vmatrix} a_{21} & a_{23} \\ a_{31} & a_{33} \end{vmatrix} + a_{13} \begin{vmatrix} a_{21} & a_{22} \\ a_{31} & a_{32} \end{vmatrix}$$

$$= a_{11}a_{22}a_{33} - a_{11}a_{32}a_{23} - a_{12}a_{21}a_{33} + a_{12}a_{31}a_{23} + a_{13}a_{21}a_{32} - a_{13}a_{31}a_{22}$$

Note that the three second order determinants are multiplied by three coefficients that are elements either of a row or from a column of the third order determinant. Each second order determinant is multiplied, in an alternate way, by $(+1)$ and by (-1). Thus, the third order determinant is a linear function of three second order determinants, whose coefficients are either elements of a row or elements of a column. The second order determinants are called *minor determinants* or simply *minors*.

By making this process general, the expansion of a determinant of a square matrix of order n is as follows:

$$|A| = \begin{vmatrix} a_{11} & a_{12} & \cdots & a_{1n} \\ a_{21} & a_{22} & \cdots & a_{2n} \\ \vdots & \vdots & \vdots & \\ a_{n1} & a_{n2} & \cdots & a_{nn} \end{vmatrix} = \sum_{j=1}^{n} a_{ij}(-1)^{i+j}|A_{ij}| = \sum_{i=1}^{n} a_{ij}(-1)^{i+j}|A_{ij}| \quad ; \quad i, j = 1,2,\ldots,n$$

Note that, when j (or i) is even, $(-1)^{i+j} = -1$ and when j (or i) is odd, $(-1)^{i+j} = 1$. Note also that a_{ij} is an element of matrix A and that $|A_{ij}|$ is its *minor*.

The product $(-1)^{i+j}|A_{ij}|$ is known as the *cofactor* of coefficient a_{ij} and is written u_{ij}. The minor $|A_{ij}|$ is written as $|M_{ij}|$. Thus

$$|A| = \sum_{j=1}^{n} a_{ij}(-1)^{i+j}|M_{ij}| = \sum_{j=1}^{n} a_{ij}u_{ij} = \sum_{i=1}^{n} a_{ij}u_{ij} \quad ; \quad i, j = 1,2,...,n$$

Diagonal Expansion. A matrix can be expressed as the sum of two matrices, one of which is a *diagonal* matrix. A square matrix is called diagonal when all the non diagonal elements are zero. Such a matrix is shown below:

$$D = \begin{bmatrix} d_{11} & 0 & ... & 0 \\ 0 & d_{22} & ... & 0 \\ \vdots & \vdots & \vdots & \vdots \\ 0 & 0 & ... & d_{nn} \end{bmatrix}$$

Then, given $A = \{a_{ij}\}$, for $i,j = 1,2,...,n$, a matrix $(A+D)$ can be defined. The determinant of such a matrix can be obtained as a polynomial of the elements of D.

The diagonal expansion is useful, because the determinant form $|A+D|$ occurs often. It is also useful when some minors of $|A|$ are zeros or can be made zeros by adopting a $|A+D|$ form or when all the elements of the diagonal matrix D are the same.

For a second order matrix, the determinant $|A+D|$ can be expanded as follows:

$$|A+D| = \begin{bmatrix} a_{11} + d_1 & a_{12} \\ a_{21} & a_{22} + d2 \end{bmatrix} = (a_{11} + d_1)(a_{22} + d_2) - a_{12}a_{21}$$

This expression can be written as a function of d_1 and d_2:

$$|A+D| = d_1 d_2 + d_1 a_{22} + d_2 a_{11} + \begin{vmatrix} a_{11} & a_{12} \\ a_{21} & a_{22} \end{vmatrix}$$

For a third order determinant the expansion is as follows:

$$|A+D| = \begin{vmatrix} a_{11} + d_1 & a_{12} & a_{13} \\ a_{21} & a_{22} + d_2 & a_{23} \\ a_{31} & a_{32} & a_{33} + d_3 \end{vmatrix}$$

$$= d_1 d_2 d_3 + d_1 d_2 a_{33} + d_1 d_3 a_{22} + d_2 d_3 a_{11} +$$

$$d_1 \begin{vmatrix} a_{22} & a_{23} \\ a_{32} & a_{33} \end{vmatrix} + d_2 \begin{vmatrix} a_{11} & a_{13} \\ a_{31} & a_{33} \end{vmatrix} + d_3 \begin{vmatrix} a_{11} & a_{12} \\ a_{21} & a_{22} \end{vmatrix} + \begin{vmatrix} a_{11} & a_{12} & a_{13} \\ a_{21} & a_{22} & a_{23} \\ a_{31} & a_{32} & a_{33} \end{vmatrix}$$

The following are some additional important properties of determinants:

• The determinant of the transpose of a matrix is the same as the determinant of the matrix, that is $|A'| = |A|$

• If λ is a scalar and a factor of a row, it is also a factor of the determinant, such that $\lambda|A|$

• If λ is a scalar and a factor of an $n \times n$ matrix, then $|\lambda A| = \lambda^n |A|$

• If one row of a determinant is a multiple of another row, the value of the determinant is zero. The same rule applies for columns.

• If a determinant has a row or column of zeros, the value of the determinant is zero.

• The determinant of the product of two squared matrices of the same order is the product of the determinants of the individual matrices, that is $|AB| = |A||B|$

Matrix Inversion. Given a *square* matrix A, defining a matrix A^{-1} is possible, such that $AA^{-1}=I$, where I is the *identity* matrix. Matrix A^{-1} is called the *inverse* matrix of A. The process of finding matrix A^{-1} is called *matrix inversion*. If A has an inverse then, A is called an *invertible* or *non singular* matrix.

$$
I = \begin{bmatrix} 1 & 0 & ... & 0 \\ 0 & 1 & ... & 0 \\ \vdots & \vdots & & \vdots \\ 0 & 0 & ... & 1 \end{bmatrix}
$$

Matrix inversion is related to the process of division of a matrix by another matrix. In a strict sense, however, division does not exist in matrix algebra. The process of dividing a matrix B by a matrix A is, actually, a multiplication of B by the inverse matrix A^{-1}.

The following properties of the inverse are here set:

• An inverse matrix A^{-1} is *commutative* with A, that is $A^{-1}A = AA^{-1}$

• The inverse of A is unique

• The determinant of the inverse of A is the reciprocal of the determinant of A, that is

$$
|A^{-1}| = \frac{1}{|A|}
$$

• The inverse of A^{-1} is A, that is $(A^{-1})^{-1} = A$

• The inverse of a transpose is the transpose of the inverse, that is $(A')^{-1} = (A^{-1})'$

• The inverse of a product is the product of the inverses taken in reverse order, that is $(AB)^{-1} = B^{-1}A^{-1}$

• If a matrix A is such that its inverse equals its transpose, A is said to be *orthogonal*. Then, the product of the two matrices is the identity matrix I, that is $AA' = I$

• A squared matrix is invertible when determinant $|A| \neq 0$

B

BASIC CONCEPTS AND
PROCEDURES IN CALCULUS

Definition of a Derivative. A derivative is defined as follows:

- The derivative of a function at a point is the limit of the ratio of the increment Δy of the dependent variable, to the increment Δx of the independent variable, when the latter increment approaches zero as a limit.

When such a limit exists, the function is said to be differentiable at that point. The process of obtaining the derivative of a function is called differentiation.

The above definition can be expressed mathematically as follows:

$$
\begin{aligned}
y &= f(x) \\
y + \Delta y &= f(x + \Delta x) \\
\Delta y &= f(x + \Delta x) - f(x) \\
\frac{\Delta y}{\Delta x} &= \frac{f(x + \Delta x) - f(x)}{\Delta x} \\
\lim_{\Delta x \to 0} \frac{\Delta y}{\Delta x} &= \lim_{\Delta x \to 0} \frac{f(x + \Delta x) - f(x)}{\Delta x} = \frac{dy}{dx}
\end{aligned}
$$

A derivative of $f(x)$ is denoted by the $\dfrac{dy}{dx}$ symbol. It is also a function of the variable x. Therefore, if $y = f(x)$ then

$$
\frac{dy}{dx} = \frac{d}{dx} f(x)
$$

The symbol $\dfrac{d}{dx}$ is called the *derivative operator*. Other commonly used symbols are D, y' and $f'(x)$.

All derivatives can be found by applying the above definition, which is called *the general rule of differentiation*. The above process, for a function $y = f(x)$, is shown in Fig. B1:

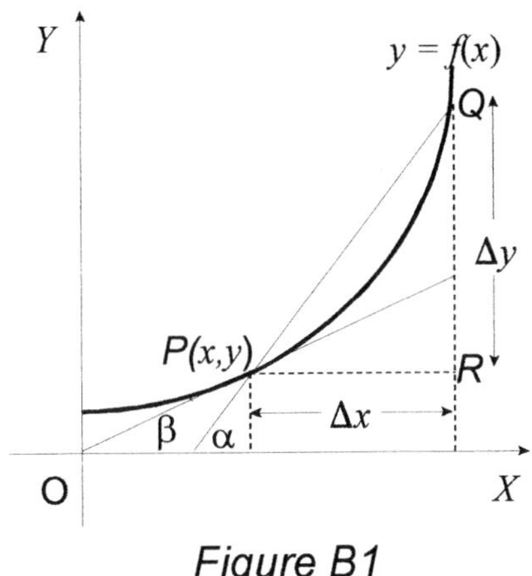

Figure B1

According to Fig. B1 and by the third step of the general rule of differentiation, the following relationship is obtained:

$$\frac{\Delta y}{\Delta x} = \frac{f(x + \Delta x) - f(x)}{\Delta x}$$

$$= \frac{RQ}{PR} = \frac{\sin\alpha}{\cos\alpha}$$

Note that, as Δx approaches zero as a limit, the secant line PQ approaches the tangent line at point $P(x, y)$ and angle α approaches angle β. Then

$$\lim_{\Delta x \to 0} \alpha = \beta$$

By the fourth step of the differentiation process, the following is obtained:

$$\lim_{\Delta x \to 0} \frac{\Delta y}{\Delta x} = \lim_{\Delta x \to 0} \frac{\sin\alpha}{\cos\alpha} = \lim_{\Delta x \to 0} \tan\alpha = \tan\beta$$

Then, the value of the derivative becomes

$$\frac{dy}{dx} = \tan\beta$$

The above definition can be stated as follows:

• The value of the derivative at any point of a curve is equal to the slope of the tangent line to the curve at that point.

Thus, the direction of the curve at any point can be determined by the value of the tangent line at that point. *Maximum* and *minimum* values of a function are obtained by making zero the value of the first derivative. *Inflection points* are determined by making zero the value of the second derivative.

Partial derivatives. Many biological problems are determined by more than one independent variable. Consider the function $z=f(x, y)$. If x is held constant, such that $x=a$, then

$$z = f(a, y) \text{ is a function of } y, \text{ then, } \frac{\partial z}{\partial y} = \frac{d}{dy}f(a, y)$$

where $\partial z/\partial y$ is called the *partial derivative* of z with respect to y. If y is now hold constant, such that $y=b$, then

$$z = f(x, b) \text{ is a function of } x \text{ then, } \frac{\partial z}{\partial x} = \frac{d}{dx}f(x, b)$$

where $\partial z/\partial x$ is called the *partial derivative* of z with respect to x.

As shown in Fig. B2, function $z = f(x, y)$ determines a response surface defined over the *XY* plain with values in coordinate Z. If a point $P(x, y, z)$ is defined on this response surface, such a point, projected on the *XY* plain, determines a point $P'(a, b)$ on the *XY* plain. Note that, by cutting the response surface through point P with a plane TPS, parallel to the plane *YZ*, a function $z=f(a, y)$ is defined. Note also that by cutting the response surface through point P with a plane QPR, parallel to the plane *XZ*, a function $z=f(x, b)$ is also determined.

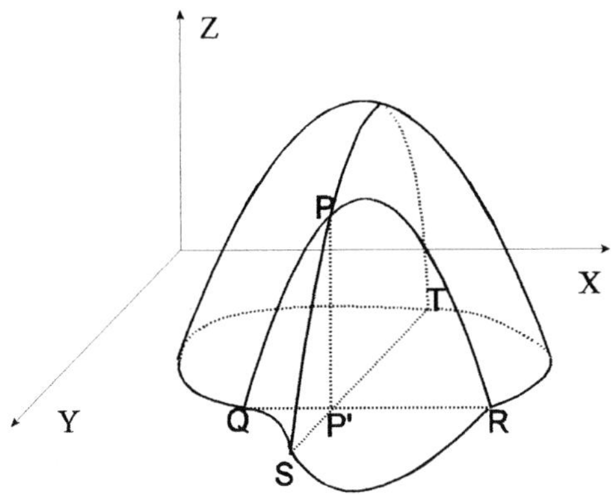

Figure B2

By holding one independent variable as a constant, all the rules developed for function $y = f(x)$ apply for each of the partial derivatives of function $z = f(x, y)$. It should be pointed out here that

$$\frac{d}{dy} f(a, y) \text{ and } \frac{d}{dx} f(x, b)$$

are the tangent lines at P in function $z = f(a, y)$ and in function $z = f(a, y)$, respectively. These relationships are illustrated in Fig. B3.

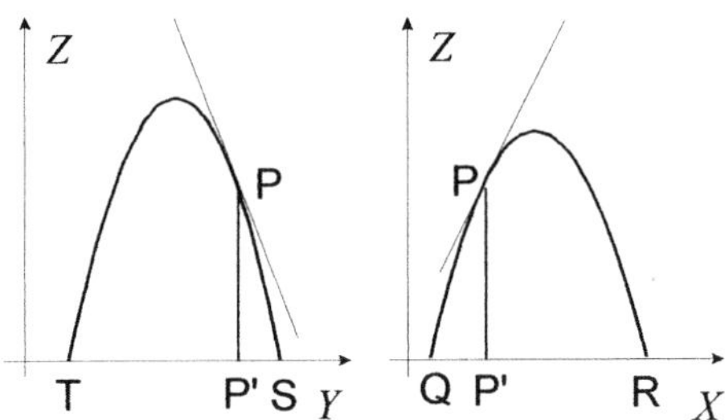

Figure B3

Definition of an Integral. An *integral* is the *antiderivative* of a function, that is

$$\frac{d}{dx} f(x) = f'(x) \text{ is the derivative of function } f(x)$$

$$\int f'(x)\ dx\ =\ f(x)\ +\ C \text{ is the integral or anti derivative of function } f'(x)$$

The symbol $\int (...)\ dx$ is called the *integral operator* and C is called the *constant of integration*. The process of finding the integral of $f'(x)$ is called integration. Note that, upon integration, C is unknown and indefinite. Thus

$$f(x)\ +\ C \quad \text{is the } \textit{indefinite integral} \text{ of } f'(x)dx$$

The indefinite integral of $f'(x)\ dx$ determines a family of curves whose difference is only the value of the constant C.

Definite Integrals. Consider the function $y = g(x)$ in Fig B4. Let u be the area CMPD. Note that when x takes an increment Δx, the area y takes an increment Δu, where Δu is the area MNQP.

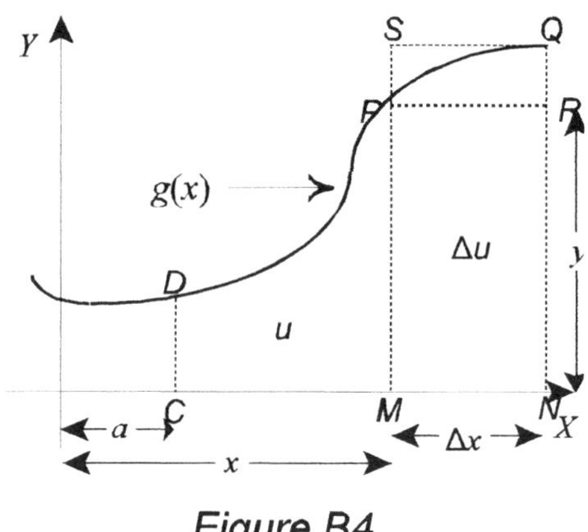

Figure B4

Note also the following relationships among areas:

$$\text{Area MNRP} < \text{area MNQP} < \text{area MNQS}$$

Thus

$$\text{MP}(\Delta x) < \Delta u < \text{NQ}(\Delta x) \text{ or, MP} < \frac{\Delta u}{\Delta x} < \text{NQ}$$

Now let Δx approach now zero as a limit:

$$\lim_{\Delta x \to 0} \frac{\Delta u}{\Delta x} = \frac{du}{dx} = MP = y$$

Then, using differentials, the following relation is determined:

$$du = y\,dx$$

The above is translated into the following definition:

• The differential of an area bounded by a curve, the x-axis, a fixed ordinate value and a variable ordinate value, is equal to the product of the variable ordinate and the differential of the corresponding abscissa.

It follows that, if $y = g(x)$ then

$$du = g(x)\,dx$$
$$u = \int g(x)\,dx$$
$$= f(x) + C$$

Note in Fig. B4 that, when $x = a$, then $u = 0$. Thus

$$f(a) + C = 0$$
$$C = -f(a)$$
$$u = f(x) - f(a)$$

Now, if $x = b$, the area $CMPD$ becomes

$$\text{Area } CMPD = f(b) - f(a)$$

The following new definition is now possible:

• The area bounded by any curve whose ordinate is y, by the x-axis and by two fixed values for the ordinate, corresponding to $x = a$ and $x = b$, is equal to the difference of the values of $\int y\,dx$ for $x = a$ and $x = b$.

This area is represented by the symbol

$$\int_a^b y \; dx \quad \text{or} \quad \int_a^b g(x) \; dx$$

If between a and b the curve does not rise or fall to infinity or cross the x-axis, an integration between the finite limits a and b has always a definite value. That is why integral $\int_a^b y \; dx$ is called a *definite integral*. The constant C has been canceled out here.

From the above, the definite integral clearly represents, in geometrical terms, the area under the curve and between two limits in the independent variable. The process of determining the definite integral is summarized as follows:

First Step: Integrate the differential expression.

Second Step: Substitute in the indefinite integral the value of the independent variable, first by the value of the upper limit and then by value of the lower limit.

Third Step: Subtract the last result from the first.

Improper Integrals. A special case of the definite integral is the *improper integral*. An improper integral is determined when one or both of the limits are infinity. Three cases are here possible provided that the limits exist: the upper limit is infinite, the lower limit is infinite or both limits are infinite, that is

$$\int_a^\infty g(x) \; dx \; = \; \lim_{b \to \infty} \int_a^b g(x) \; dx$$

$$\int_{-\infty}^b g(x) \; dx \; = \; \lim_{a \to -\infty} \int_a^b g(x) \; dx$$

$$\int_{-\infty}^\infty g(x) \; dx \; = \; \lim_{\substack{a \to -\infty \\ b \to \infty}} \int_a^b g(x) \; dx$$

C

PROBABILITY DEFINITIONS AND FORMULAS

Probability of Events in a Finite Sample Space. A *sample space* is defined as follows:

• A sample space S is the set of all possible outcomes of an experiment.

The outcomes of the sample space are called *sample points* or *elementary events*. Then

$$S = \{e_1, e_2, ..., e_n\}$$

where e is an outcome.

When the number of outcomes is finite, the sample space is also finite. Conversely, when the number of outcomes is infinite, then the sample space is infinite.

Any union of outcomes is a subset of the sample space and is called an *event*. Then, an event may be defined as follows:

• An event E is any subset of the sample space S.

Then

$$E = \{e; e \in S\}$$

Since events are sets, using set operations is possible.

Events that are complementary are defined as follows:

• Given an event E in the sample space S, the complementary event $\overline{E}$ is the set of outcomes in S but not in E.

Then

$$\overline{E} = \{e; e \in S; e \notin E\}$$

In some situations the probability of an event can be obtained simply by common sense. Common sense suggests that, if a coin is perfectly balanced, then the probability of heads would be ½. Frequently, however, obtaining probability values may require gathering experimental data. If an experiment is performed to test the coin, in the early stages of tossing a coin the proportion of heads varies considerable. As the experiment continues, the proportion of heads approaches the expected value of 0.5. Then, the following generalization is possible. The probability p for an event E to occur m times, is the limit of the *relative frequency* m/n of the event, when the number of repetitions n of the experiment approaches infinity. Thus

$$\lim_{n \to \infty} \frac{m}{n} = p(E)$$

where p is the *probability function* and $p(E)$ is the *probability of the event E*. If n is large enough, the probability of an event may be defined as

$$p(E) = \frac{m}{n} = \frac{\text{number of elementary events of } E}{\text{number of elementary events of } S}$$

An event E was defined previously as a subset of the sample space S and as a set of elementary events. Then, the probability $p(E)$ of the event is the sum of the probabilities of the elementary events. Thus, if

$$E = \{e_1, e_2, ..., e_m\}$$

then

$$p(E) = p(e_1) + p(e_2) + ... + p(e_m)$$

Note that given $p(E) = m/n$, when $m = $ n then $E = S$. Thus

$$p(S) = 1$$

where S is the sample space. Expression $p(S)$ is called the *probability space*. Conversely, when $m = 0$, then

$$p(\varnothing) = 0$$

where $\varnothing$ is called the *impossible event*. The above implies that the probability of event E is some value between 0 and 1. Thus

$$0 \le p(E) \le 1 \; ; E \subset S$$

It also implies that the probability of the complementary event $\overline{E}$ must be $1 - p(E)$:

$$p(\overline{E}) = p(S) - p(E) = 1 - p(E)$$

Mutually Exclusive Events. The following definition applies for *mutually exclusive events*:

• Mutually exclusive events are events that cannot occur simultaneously because they have no sample points in common.

If two events E_1 and E_2 are mutually exclusive, then $E_1 \cap E_2 = \varnothing$

If E_1 and E_2 are two mutually exclusive events, the probability that either E_1 or E_2 occurs is the sum of the individual probabilities of the two events, that is

$$p(E_1 \cup E_2) = p(E_1) + p(E_2) \text{ if } E_1 \cap E_2 = \varnothing$$

This result can be extended to n mutually exclusive events, that is

$$p(E_1 \cup E_2 \cup ... \cup E_n) = p(E_1) + p(E_2) + ... + p(E_n)$$

The above relationship does not apply when events are not mutually exclusive. For any two events not mutually exclusive, the following relationship applies:

$$p(E_1 \cup E_2) = p(E_1) + p(E_2) - p(E_1 \cap E_2)$$

Note that, for mutually exclusive events, $p(E_1 \cap E_2) = \varnothing$. Therefore, this formula applies equally for not mutually exclusive events and for mutually exclusive events.

The following probability applies for three not mutually exclusive events:

$$p(E_1 \cup E_2 \cup E_3) = p(E_1) + p(E_2) + p(E_3)$$
$$- p(E_1 \cap E_2) - p(E_1 \cap E_3) - p(E_2 \cap E_3)$$
$$+ p(E_1 \cap E_2 \cap E_3)$$

Conditional Probability. Between no information and complete information on the outcomes of an event, there may be many levels of partial information. If known, this partial information may be a condition that can affect the probabilities of occurrence of events and must be taken into account. Then, *conditional probability* may be defined as follows:

• For a sample space S and events E_1 and E_2 in S, conditional probability is the probability that event E_2 occurs given the condition that event E_1 also happened.

Conditional probability has the following notation:

$$p(E_1 | E_2)$$

The probability $p(E_1 | E_2)$ is called the conditional probability of E_1 given E_2.

To calculate $p(E_1 | E_2)$ it is necessary to know the probability that E_2 occurs and the probability that E_1 and E_2 occur together on a trial. If an event E_2 contains l sample points and an event $E_1 \cap E_2$ contains m sample points, then

$$p(E_2) = \frac{l}{n} \text{ and } p(E_1 \cap E_2) = \frac{m}{n}$$

where n is the total number of sample points in the space S. Since the condition is that the event E_2 occurs, the outcome of the experiment must be one of the l sample points in E_2. Among these l points, there are sample points for which E_1 also occurs. These are the m sample points in $E_1 \cap E_2$. Thus

$$p(E_1 | E_2) = \frac{m}{l} = \frac{\text{number of sample points of } E_1 \cap E_2}{\text{number of sample points of } E_2}$$

Note that $E_1 \cap E_2$ is considered as an event in the sample space $E_2 \subset S$. This concept is illustrated in Fig. C1.

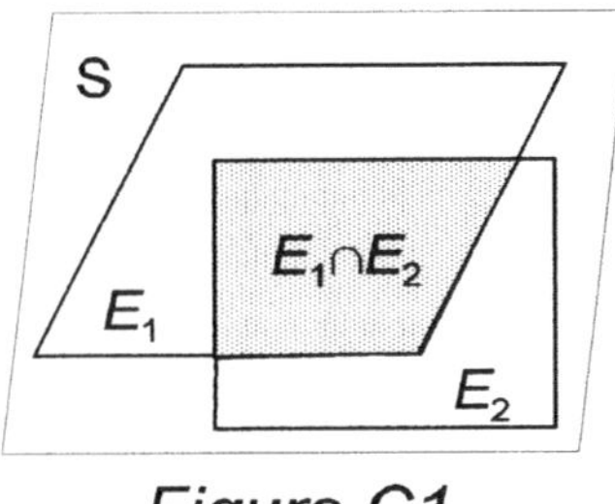

Figure C1

If numerator and denominator are divided by n, the following is obtained:

$$p(E_1|E_2) = \frac{m/n}{l/n} = \frac{p(E_1 \cap E_2)}{p(E_2)}$$

From the above, if E_1 and E_2 are events in the sample space S and $p(E_2) \neq 0$, the conditional probability of E_1 given E_2 is

$$p(E_1|E_2) = \frac{p(E_1 \cap E_2)}{p(E_2)}$$

Frequently, the problem is to compute the probability of events that occur together in a trial. The following is obtained by rearranging the conditional probability expression for two events:

$$p(E_1 \cap E_2) = p(E_1)p(E_2|E_1) = p(E_2)p(E_1|E_2)$$

This result is known as the *multiplication theorem of conditional probability*. The multiplication theorem of conditional probability can be generalized for n events:

- Given $E_1, E_2, ..., E_n$ events in the sample space S, and $p(E_1 \cap E_2 \cap ... \cap E_{n-1}) \neq 0$, then

$$p(E_1 \cap E_2 \cap ... \cap E_n) = p(E_1)\,p(E_2|E_1)\,p(E_3|E_1 \cap E_2)...\,p(E_n|E_1 \cap E_2 \cap ... \cap E_{n-1})$$

In many problems, partitioning the sample space S into subsets may be more meaningful. This would be the case, for example, if a herd of cattle is partitioned into subgroups according to breed, sex, age or any other criteria. As indicated in Fig. C2, an event E can be expressed as the following disjoint union:

$$E = (E \cap S_1) \cup (E \cap S_2) \cup \ldots \cup (E \cap S_n)$$

with probability

$$p(E) = p(E \cap S_1) + p(E \cap S_2) + \ldots + p(E \cap S_n)$$

If $p(E \cap S_i) = p(E|S_i)p(S_i)$, then

$$p(E) = p(E|S_1)p(S_1) + p(E|S_2)p(S_2) + \ldots + p(E|S_n)p(S_n)$$

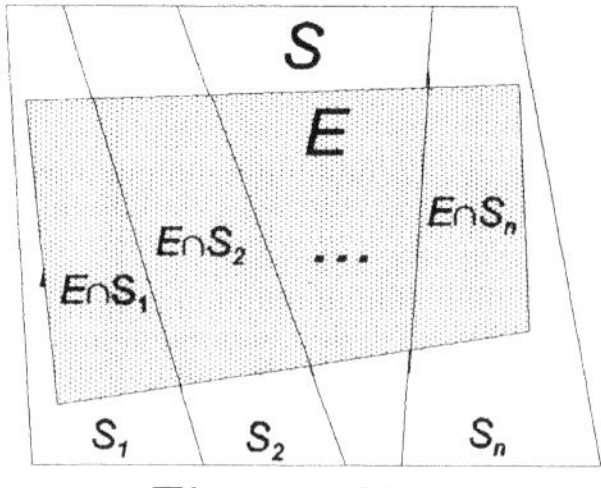

Figure C2

Independent Events. *Independent events* are defined as follows:

• Two events are said to be *independent* when the probability of an event E_1 is not affected by the knowledge that an event E_2 occurs.

Then

$$p(E_1|E_2) = p(E_1)$$

It was shown, however, that

$$p(E_1|E_2) = \frac{p(E_1 \cap E_2)}{p(E_2)}$$

By equating these two expressions, the following is the mathematical definition for independent events:

$$p(E_1 \cap E_2) = p(E_1)p(E_2)$$

The definition of independent events can be generalized to n events. If events $A_1, A_2, \ldots, A_n$ are independent, then

$$p(E_1 \cap E_2 \cap \ldots \cap E_n) = p(E_1)p(E_2)\ldots p(E_n)$$

The idea of independent events should not be confused with the concept of mutually exclusive events. For two events to be mutually exclusive, it is required that $E_1 \cap E_2 = \varnothing$. Then $p(E_1 \cap E_2) = 0$, simply because $p(\varnothing) = 0$. In two independent events, it is required that the probability of one event to be zero when $p(E_1 \cap E_2) = p(E_1)p(E_2) = 0$.

D

RULES OF COUNTING

Multiplication Principle. The *multiplication principle* is stated as follows:

• If a set A_1 contains n_1 objects, a set A_2 contains n_2 objects,..., and a set A_k contains n_k objects, the number of ways to choose one object from each of the k sets is $n_1 \times n_2 \times ... \times n_k$

Permutation. A permutation is the act or process of changing the order of a set of objects and to arrange these objects in all possible ways, that is:

• A permutation of n objects is an arrangement of these objects in a definite order and without repetition

Thus, the first rule for permutations can be written as follows:

• The number of permutations of a set with n distinct objects is
$$1(2)...(n-1)(n) = n!$$

Often the interest is in the number of ways of choosing n objects taken k at a time, rather than all possible ways in choosing n objects. Then, the second rule of permutations is as follows:

• The number of permutations in a set of n distinct objects taken k at a time is
$$P_{n,k} = \frac{n!}{(n-k)!}$$

Often, not all the objects that are being permuted need to be distinguished. The following is the third rule of permutations:

• For a set of $n = n_1 + n_2 + ... + n_k$ objects, where the objects within $n_1, n_2, ..., n_k$ are indistinguishable, the number of permutations is the multinomial coefficient
$$\binom{n}{n_1, n_2, ..., n_k} = \frac{n!}{n_1! n_2!, ..., n_k!}$$

Combinations. There are many cases in which the ordering of objects is not important. For example, with three objects that can be selected from the set {1, 2, 3, 4}, there are 24 permutations of objects taken three at a time. There are more permutations than subsets with objects taken three at a time, because permutations distinguish the ordering of objects, while subsets do not. Therefore

$$1\ 2\ 3 \text{ and } 3\ 2\ 1$$

are different permutations of $\{1, 2, 3, 4\}$ with objects taken three at a time, while the sets

$$\{1, 2, 3\} \text{ and } \{3, 2, 1\}$$

are not different because they have the same objects. Thus, the following definition applies here:

• A subset of k objects chosen from a set of n objects is called a *combination* of the n objects taken k at a time.

The rule of combination is the number of permutations divided by $k!$ and is stated as follows:

• The number of combinations of n distinct objects taken k at a time is
$$C_{n,k} = \frac{n!}{k!(n-k)!}$$

Often the notation is used in place of $C_{n,k}$ is $\binom{n}{k}$. This notation is known as the *binomial symbol* and usually is used here in relation to the expansion of the binomial expression $(a + b)^n$.

E

PROBABILITY DISTRIBUTIONS

The Binomial Distribution

In many experiments, the primary interest is whether a certain outcome does or does not occur. These are experiments for which there are only two possible outcomes, *success* and *failure*. On a one trial experiment, the probability of success may be called p and the probability of failure q. Then

$$\text{Probability of success is } p$$
$$\text{Probability of failure is } q = 1 - p$$

Thus, the sample space would have only two elementary events. If the above experiment is repeated twice, the sample space will have four elementary events, that is 2^2. If the outcome of the second trial is not affected by the outcome of the first trial, independence allows the multiplication of probabilities. Then, the probabilities of the four elementary events would be

$$\text{success, success} = p^2$$
$$\text{success, failure} = pq$$
$$\text{failure, success} = qp$$
$$\text{failure, failure} = q^2$$

By adding up the above probabilities, the following expression is obtained:

$$p^2 + pq + qp + q^2 = (p+q)^2 = 1$$

If the experiment is repeated three times, the sample space would have eight outcomes, that is 2^3. Then

$$p^3 + 3p^2q + 3pq^2 + q^3 = (p+q)^3 = 1$$

If the experiment with two outcomes is repeated n times, then it will be 2^n possible outcomes in the sample space. Thus

$$(p+q)^n = p^n + np^{n-1}q + \frac{n(n-1)}{2!}p^{n-2}q^2 + \dots + q^n = \sum_0^n \binom{n}{x}p^x q^{n-x} = 1$$

where $\binom{n}{x} = \dfrac{n!}{x!(n-x)!}$ for $x = 1,2,\dots,n$.

Each element of the above binomial is a *binomial probability* $P(X=x)$. Experiments for which there are only two possible outcomes are called *binomial experiments* or *Bernoulli experiments* and the variable X is called a *binomial random variable*. Thus, the following definition applies here:

- The binomial probability $P(X=x)$ of having x successes in n independent trials of a binomial experiment is given by expression

$$P(X=x) = f(x) = \binom{n}{x}p^x q^{n-x}$$

Given the number of trials n, the number of successes x and the probability of success of an outcome p, the binomial probability can also be written as

$$f(n,x,p) = \binom{n}{x}p^x q^{n-x}$$

Individual terms of the binomial distribution can be found in standard mathematical tables.

The expected value of the mean $E(X)$ and the variance σ^2 of the binomial random variable X are given by the following expressions:

$$E(X) = np$$
$$\sigma^2 = npq = \mu(1-p)$$

where n is the number of trials, p is the probability of success and q is the probability of failure.

The Multinomial Distribution

The binomial distribution was shown to represent repeated trials of an experiment with two random variables, X_1 and X_2 with probabilities $p_1 = P(X_1 = x_1)$ and $p_2 = P(X_2 = x_2)$. Then, expression

$$(p_1 + p_2)^n = \sum_0^n \binom{x}{n} p_1^{x_1} p_2^{x_2}$$

represents the set of all the outcomes in n trials. If x_1 is a success, then x_2 must be a failure. For n trials, it will be n_1 outcomes of type X_1 and n_2 outcomes of type X_2. Therefore, $x_1 + x_2 = n$.

Each element of the above sum is a binomial probability $P(X_1 = x_1, X_2 = x_2)$, such that

$$P(X_1 = x_1, X_2 = x_2) = f(x_1, x_2) = \frac{n!}{x_1! x_2!} p_1^{x_1} p_2^{x_2}$$

This approach can be generalized to experiments with m possible types of outcomes. Let $X_1, X_2, ..., X_m$ be the different types of outcomes of an experimental trial and $p_1 = P(X_1 = x_1)$, $p_2 = P(X_2 = x_2)$, ..., $p_m = P(X_m = x_m)$ be the probabilities of these outcomes where $p_1 + p_2 + ... + p_m = 1$. If the experiment with m random variables is repeated n times, then the total possible number of outcomes is represented by the sum

$$(p_1 + p_2 + ... + p_m)^n = \sum_0^n \frac{n!}{n_1! n_2! ... n_m!} p_1^{x_1} p_2^{x_2} ... p_m^{x_m}$$

where $x_1 + x_2 + ... + x_m = n$ is the total number of outcomes. Each element of the above sum is a *multinomial probability* $P(X_1 = x_1, X_2 = x_2, ..., X_m = x_m)$ and $X_1, X_2, ..., X_m$ are the *multinomial random variables*. The joint distribution of these variables is called the *multinomial distribution*. Thus, the following definition applies here:

• The multinomial probability $P(X_1 = x_1, X_2 = x_2, ..., X_m = x_m)$ is given by the expression

$$f(x_1, x_2, ..., x_m) = \binom{n}{n_1, n_2, ..., n_m} p_1^{x_1} p_2^{x_2} ... p_m^{x_m}$$

where $\displaystyle \binom{n}{n_1, n_2, \ldots, n_m} = \frac{n!}{n_1! \, n_2! \ldots n_m!}$.

As expected, the mean μ and the variance σ^2 of the multinomial distribution are the same as in the binomial distribution, but defined over each of the X_i random variables, that is

$$\mu_i = np_i$$
$$\sigma_i^2 = np_i(1 - p_i) = \mu_i(1 - p_i)$$

where n is the number of trials and p_i is the probability of the outcome X_i.

The Geometric Distribution

The binomial probability $P(X = x)$ was shown to be

$$P(X = x) = \binom{n}{x} p^x q^{n-x}$$

which is the probability of x successes in n trials of the experiment, where p is the probability of a success and q the probability of a failure. If the binomial experiment is performed n times, until the first success occurs, then the above formula becomes

$$P(X = 1) = \binom{n}{1} pq^{n-1} = pq^{n-1}$$

Note that the random variable X becomes now the number of trials required for the first success to occur and not the number of successes. Then, the random variable X is said to have a *geometric distribution* and can be any positive integer value. The *geometric probability function* takes the following form and definition:

• The geometric probability $P(X=x)$ of having a success in x trials is given by the expression $f(x) = pq^{x-1}$

Since the random variable X is any positive integer, the sample space is infinite, with range $S_r = \{x : x \in I^+\}$ where I^+ is the set of positive integers.

The *cumulative geometric probability* $P(X \geq x)$ is the sum

$$F(x) = \sum_0^n pq^{x-1} = p\sum_0^n q^{x-1}$$
$$= p(1 + q + q^2 + ... + q^n)$$
$$= \frac{p(1 - q^n)}{1 - q} = 1 - q^n$$

Note that this sum is a geometric series, which gives the name to the geometric distribution. Note also that

$$\lim_{n \to \infty} (1 - q^n) = 1$$

The expected number of trials for the first success and the variance for the geometric distribution are the following expressions:

$$\mu = \frac{1}{p}$$
$$\sigma^2 = \frac{q}{p^2}$$

where p and q are the probabilities of success and of failure.

The Poisson Distribution

For experiments with probabilities p for success and $q = 1 - p$ for failure, the sample size is often too large or undefined and the probability of success is very small. In such situations, applying the binomial distribution may not be convenient or even feasible. To avoid these problems, an approximation to the binomial distribution, called the *Poisson* distribution, has been developed. The Poisson distribution can be used whenever n is too large and p is too small and is defined as follows:

• The Poisson probability $P(X=x)$ of having a success x is given by the expression

$$f(x) = \frac{\mu^x}{x!} e^{-u}$$

where $\mu = np > 0$ is the mean.

The expected value of the mean for the Poisson distribution was already defined as $\mu = np$ where n is large and p is small. Since p is small, then $q = 1 - p$ is approximately equal to 1. The variance in the binomial distribution is $\sigma^2 = npq$. Since q is approximately 1, then npq is approximately equal to np and the variance in the Poisson distribution becomes equal to np, which is also the mean μ. Thus

$$\mu = np$$
$$\sigma^2 = np$$

A classical example of a Poisson distribution is the radioactive decay. Given the following definitions

n the number of radioactive atoms
p the small probability of decaying during a 1-second period
$\mu = np$ the expected number of decays per second
t the time in seconds

then the probability distribution is

$$f(x) = \frac{(\mu t)^x}{x!} e^{-\mu t}$$

where μt is the expected number of decays in t seconds.

The Normal Curve

The normal distribution may be defined as an approximation to the binomial distribution. As the number n of trials increases, the area under the histograms approaches the area under a bell shaped curve. Then, if p is success and q is failure, as n increases, the binomial probability function

$$f(n,x,p) = \binom{x}{n} p^x q^{n-x}$$

approaches the density function

$$f(x) \; = \; \frac{1}{\sqrt{npq}} \; \frac{1}{\sqrt{2\pi}} \; e^{\,-(x-np)^2/2npq}$$

called the *normal density* function.

Note that *npq* is the variance σ^2 and *np* is the mean μ. Then, the following definition applies:

- The random variable X is said to be normally distributed on the range $-\infty < x < \infty$,

  if its density function is $f(x) \; = \; \dfrac{1}{\sigma\sqrt{2\pi}} \; e^{\,-(x-\mu)^2/2\sigma^2}$, where μ is the mean and σ^2
  is the standard deviation

The graph of the normal curve is shown in Fig. E1. Note that the mean $x = \mu$ is the center of symmetry of the curve and that the inflection points occur at $x = \mu - \sigma$ and at $\square = \mu + \sigma$.

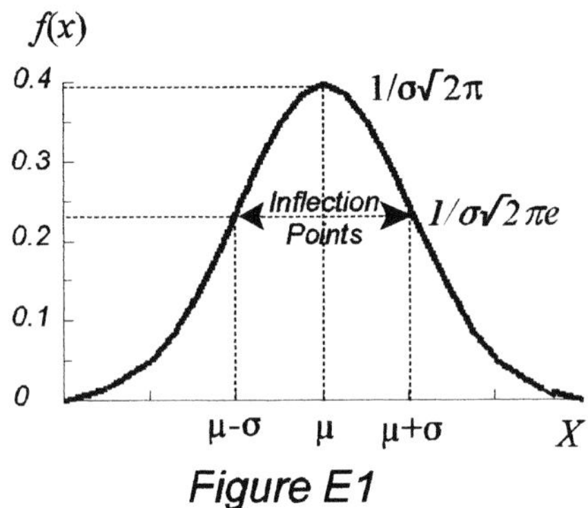

Figure E1

The random variable X has the following distribution function:

$$F(x) \; = \; P(X \le x) \; = \; \frac{1}{\sigma\sqrt{2\pi}} \int_{-\infty}^{x} e^{\,(y-\mu)^2/2\sigma^2} \, dy$$

Of particular importance is the normal curve with $\sigma = 1$ and $\mu = 0$, called *standard normal distribution* or *unit normal distribution* function. By defining a random variable Z, such that $z = (x - \mu)/\sigma$, the normal density function is simplified to the following definition:

• Given the random variable Z, the standard normal density function is determined by

$$\text{the expression } f(z) = \frac{1}{\sqrt{2\pi}} e^{-z^2/2}$$

By avoiding the parameters μ and σ, the mathematical manipulation of the normal distribution function is made more practical. Tables for the standardized random variable Z are available.

The random variable Z has the following standard distribution function:

$$F(z) = P(X \le \mu + \sigma z) = P(Z \le z) = \frac{1}{\sqrt{2\pi}} \int_{-\infty}^{z} e^{-s^2/2} \, ds$$

where $s = (y - \mu)/\sigma$. This means that, if the random variable Z has a standard normal distribution, then $X = \mu + \sigma Z$ is normally distributed with mean μ and variance σ^2. Thus, tables of probability values of the standard density and distribution functions can be used to obtain probabilities for normal random variables with any mean μ and variance σ^2.

Confidence Intervals

In statistical analysis, measuring the deviation of an outcome from the expected mean as standard deviation units, is often convenient. If the variable is the sample mean, it is possible to define a random variable "T", such that

$$t = \frac{\bar{x} - \mu}{s_{\bar{x}}}$$

Note that "t" is the deviation of the sample mean from that of the population mean, measured as standard error units for n observations. The *sampling distribution* of the random variable "T" is known as *Student's* "t" *distribution*. Like the normal, the "t" distribution is symmetrical about the mean. In large samples it is essentially normal, with $\mu = 0$ and $\sigma = 1$. However, for samples of less than 30 the difference becomes evident, the "t" distribution being more peaking at the center than the normal. Note also that μ is seldom known. Without knowing μ, "t" cannot be calculated, but tables of the "t" distribution for the required degrees of freedom are available. Thus, the actual "t" is expected to lie between the negative and the positive tails of the "t" distribution. For example, if the value $t_{0.05}$ is chosen, then

$$-t_{0.05} \leq \frac{\bar{x} - \mu}{s_{\bar{x}}} \leq t_{0.05}$$

Upon rearranging, the above expression becomes

$$\bar{x} - t_{0.05}\, s_{\bar{x}} \leq \mu \leq \bar{x} + t_{0.05}\, s_{\bar{x}}$$

that is, the probability that the above interval will include μ is 0.95 .

     A major application of this analysis is determining the sample size of an experiment. If "t" is redefined, such that

$$t = \frac{\bar{x} - \mu}{s/\sqrt{n}} = \frac{d}{s/\sqrt{n}}$$

where d is the *least significant difference* that the experiment is expected to detect. Then

$$d = \frac{s\,t}{\sqrt{n}}$$

where s is the sample standard deviation, "t" is a tabulated value and n is the sample size. Note that, as a starting value, "t" is not known. Thus, it takes some manipulation to converge to the value n of the sample size.

F

MOST FREQUENTLY USED
STATISTICAL FORMULAS

- Random variable: $X = \{(e, x);\ e \in S\}$

- Discrete probability function: $f = \{(x, P(X=x));\ x \in S_r\}$

- Discrete distribution function: $F = \{(x, P(X \geq x));\ x \in S_r\}$

- Continuous distribution function: $F = \{(x, P(X \leq x));\ -\infty < x < \infty\}$

- Density function: $f(x) = \dfrac{d}{dx} F(x)$

- Expected value: $E(X) = \mu = \sum_{i=1}^{m} x_i f(x_i)$

- Variance: $\sigma^2 = \sum_{i=1}^{m} (x_i - \mu)^2 f(x_i) = \dfrac{1}{n} \sum_{i=1}^{m} (x_i - \mu)^2 r_i = \dfrac{1}{n} \sum_{j=1}^{n} (x_j - \mu)^2$, where n is the number of cases

- Standard deviation: $\sigma = \sqrt{\sigma^2}$

- Standard error: $\sigma_\mu = \sigma/\sqrt{n}$

- Sample mean: $\bar{x} = (x_1 + x_2 + \ldots + x_n)/n = (x_1 r_1 + x_2 r_2 + \ldots + x_m r_m)/n$

- Sample variance: $s^2 = \dfrac{1}{n-1} \sum_{j=1}^{n} (x_j - \bar{x})^2 = \dfrac{1}{n-1} \left[\sum_{j=1}^{n} (x_j^2) - \dfrac{1}{n} \left(\sum_{j=1}^{n} x_j \right)^2 \right]$

- Sample standard deviation: $s = \sqrt{s^2}$

- Sample standard error: $s_{\bar{x}} = s/\sqrt{n}$

- Coefficient of Variation: $C = (s/\bar{x})100$

- The χ^2 distribution test for E expected and O observed values:

$$\chi^2 = \sum \frac{(O - E)^2}{E}$$

- Comparison between two means, paired observations: $t = \dfrac{\bar{d}}{s_{\bar{d}}}$

- Comparison between two means, not paired observations: $t = \dfrac{\bar{x}_1 - \bar{x}_2}{\sqrt{s_1^2 / n_1 + s_2^2 / n_2}}$

- The regression coefficient: $b = \dfrac{\sum xy}{\sum x^2}$

- The multiple regression coefficient: $b_j = \sum c_{jj} \sum x_j y$, where

$$c_{11} \sum x_1^2 + c_{12} \sum x_1 x_2 + \ldots + c_{1m} \sum x_1 x_m = 1,0,\ldots,0$$
$$c_{21} \sum x_2 x_1 + c_{22} \sum x_2^2 + \ldots + c_{2m} \sum x_2 x_m = 0,1,\ldots,0$$
$$\vdots$$
$$c_{m1} \sum x_m x_1 + c_{m2} \sum x_m x_2 + \ldots + c_{mm} \sum x_m^2 = 0,0,\ldots,1$$

- Sum of squares due to regression: $\sum \hat{y} = \sum b_j \sum x_j y$

- Sum of squares not due to regression: $\sum d^2 = \sum y^2 - \sum \hat{y}^2$

- Mean square deviation from regression for p independent variables:

$$s_{Y.1,2,\ldots,p}^2 = \sum \frac{d^2}{n - p}$$

- Sample standard deviation of a regression coefficient: $s_{bj} = s_{Y.12\ldots p} \sqrt{c_{jj}}$

- Comparison between a regression coefficient and hypothetical value b_0 : $t = \dfrac{b - b_0}{s_b}$

• Comparison between a regression coefficient of two different treatments:

$$t = \frac{b_i - b_j}{\sqrt{\sum s_{b_i}^2 + s_{b_j}^2}}$$

• Coefficient of determination: $R^2 = \dfrac{\sum \hat{y}^2}{\sum y^2}$

• Coefficient of determination adjusted for sample size: $R_a^2 = R^2 - \dfrac{p(1 - R^2)}{n - p - 1}$

• Coefficient of determination when a new independent variable is entered in the equation:

$$R_{change} = R^2 - R_j^2$$

where R_j^2 is the coefficient of determination when all independent variables except the jth are in the equation

• Comparison between two coefficients of determination:

$$F_{change} = \frac{R_{change}^2 (n - p - 1)}{q(1 - R^2)}$$

where q is the number of variables entered at this step

TABLE OF LAPLACE TRANSFORMS

Definition of the Laplace Transform

$$L[f(t)] = F(s) = \int_0^\infty e^{-st} f(t)\,dt$$

Selected Properties of the Laplace Transform

$F(s)$	$f(t)$
$a_1 F_1(s) + a_2 F_2(s)$	$a_1 f_1(t) + a_2 f_2(t)$
$F(s/a)$	$af(at)$
$aF(as)$	$f(t/a)$
$F(s+a)$	$e^{-at} f(t)$
$F(s-a)$	$e^{at} f(t)$
$sF(s) - f(0)$	$f'(t)$
$s^2 F(s) - sf(0) - f'(0)$	$f''(t)$
$s^n F(s) - s^{n-1} f(0) - s^{n-2} f'(0) - \cdots - f^{n-1}(0)$	$f^n(t)$
$F'(s)$	$-tf(t)$
$F''(s)$	$t^2 f(t)$
$F^n(s)$	$(-1)^n t^n f(t)$

$F(s)/s$	$\int_0^t f(\tau)d\tau$
$F(s)G(s)$	$\int_0^t f(\tau)g(t-\tau)d\tau = \int_0^t f(t-\tau)g(\tau)d\tau$
$\lim_{s\to\infty} sF(s)$	$\lim_{t\to 0} f(t) = f(0^+)$
$\lim_{s\to 0} sF(s)$	$\lim_{t\to\infty} f(t) = f(\infty)$
$e^{-s\tau}F(s)$	$f(t-\tau)$

Some Useful Laplace Transform Pairs

$F(s)$	$f(t)\quad t>0$
1	$\delta(t)$ unit impulse
$e^{-\tau s}$	$\delta(t-\tau)$ delayed impulse
$\dfrac{1}{s}$	$1(t)$ unit step
$\dfrac{1}{s^2}$	t unit ramp
$\dfrac{2}{s^3}$	t^2
$\dfrac{1}{s^n}\quad n = 1,2,3,\ldots$	$\dfrac{t^{n-1}}{(n-1)!}$

$\dfrac{1}{s}e^{-\tau s}$	$1(t-\tau)$ delayed step
$\dfrac{1}{s+a}$	e^{-at}
$\dfrac{1}{(s-a)^n}\quad n=1,2,3,\ldots$	$\dfrac{1}{(n-1)!}\left(t^{n-1}e^{-at}\right)$
$\dfrac{1}{s(s+a)}$	$\dfrac{1}{a}(1-e^{-at})$
$\dfrac{1}{(s+a)(s+b)}$	$\dfrac{1}{b-a}(e^{-at}-e^{-bt})$
$\dfrac{1}{s(s+a)(s+b)}$	$\dfrac{1}{ab}\left(1-\dfrac{be^{-at}}{b-a}+\dfrac{ae^{-bt}}{b-a}\right)$
$\dfrac{1}{(s+a)(s+b)(s+c)}$	$\dfrac{e^{-at}}{(b-a)(c-a)}+\dfrac{e^{-bt}}{(a-b)c-b)}+\dfrac{e^{-ct}}{(a-c)b-c)}$
$\dfrac{s}{s+a}=1-\dfrac{a}{s+a}$	$\delta(t)-e^{-at}$
$\dfrac{s}{(s+a)(s+b)}$	$\dfrac{1}{a-b}\left(ae^{-at}-be^{-bt}\right)$
$\dfrac{1}{s^2(s+a)}$	$\dfrac{t}{a}-\dfrac{1}{a^2}\left(1-e^{-at}\right)$
$\dfrac{1}{s^3(s+a)}$	$\dfrac{t^2}{2a}-\dfrac{t}{a^2}+\dfrac{1}{a^3}\left(1-e^{-at}\right)$

$\dfrac{1}{s^2 - a^2}$	$\dfrac{1}{2a}\left(e^{at} + e^{-at}\right)$
$\dfrac{s}{s^2 - a^2}$	$\dfrac{1}{2}\left(e^{at} - e^{-at}\right)$
$\dfrac{s}{(s+a)^2}$	$(1 - at)e^{-at}$
$\dfrac{1}{s(s^2 - a^2)}$	$\dfrac{1}{2a^2}\left(e^{at} + e^{-at} - 2\right)$
$\dfrac{a}{s^2 + a^2}$	$\sin at$
$\dfrac{s}{s^2 + a^2}$	$\cos at$
$\dfrac{1}{\left(s^2 + a^2\right)^2}$	$\dfrac{\sin at - at\cos at}{2a^3}$
$\dfrac{s}{\left(s^2 + a^2\right)^2}$	$\dfrac{t\sin at}{2a}$
$\dfrac{s^2}{\left(s^2 + a^2\right)^2}$	$\dfrac{\sin at + at\cos at}{2a}$
$\dfrac{s^2 - a^2}{\left(s^2 + a^2\right)^2}$	$t\cos at$
$\dfrac{a}{(s+b)^2 + a^2}$	$e^{-bt}\sin at$

$\dfrac{a}{\left(s-b\right)^2 + a^2}$	$e^{bt}\sin at$
$\dfrac{s+a}{\left(s+b\right)^2 + a^2}$	$e^{-bt}\cos at$
$\dfrac{s+a}{\left(s-b\right)^2 + a^2}$	$e^{bt}\cos at$
$\dfrac{s^3}{\left(s^2 + a^2\right)^2}$	$\cos at - \dfrac{at\sin at}{2}$

H

TABLE OF Z TRANSFORMS

Definition of The Z Transform

$$Z[f(n)] = F(z) = \sum_0^\infty f(n)z^{-n}$$

Selected Properties of The Z Transform

$F(z)$	$f(n) \quad n=1,2,\ldots$
$a_1 F_1(z) + a_2 F_2(z)$	$a_1 f_1(n) + a_2 f_2(n)$
$zF(z) - zf(0)$	$f(n+1)$
$z^2 F(z) - z^2 f(0)$	$f(n+2)$
$z^n F(z) - z^n f(0) - z^{n-1} f(1) - \cdots - zf(n-1)$	$f(n+k)$
$z^{-k} F(z)$	$f(n-k)$
$\lim_{z \to \infty}(1 - z^{-1})F(z) = F(\infty)$	$f(0)$
$\lim_{z \to 1}(1 - z^{-1})F(z)$	$f(\infty)$
$F_1(z)F_2(z)$	$\sum_{n=0}^\infty \left(\sum_{k=0}^\infty f_1(n-k)f_2(i) \right) z^{-n} = \sum_{n=0}^\infty \left(\sum_{k=0}^\infty f_2(n-k)f_1(i) \right) z^{-n}$

$F(z/a)$	$a^n f(n)$

Some Useful Z Transform Pairs

$F(z)$	$f(n)$ $n=1,2,...$
z^{-n}	1 at n, 0 elsewhere
$\dfrac{z}{z-1}$	1 unit step sequence
$\dfrac{z}{(z-1)^2}$	n unit ramp sequence
$\dfrac{z(z+1)}{(z-1)^3}$	n^2
$\dfrac{z}{z-e^{-a}}$	e^{-an}
$\dfrac{z^n}{(z-1)^n}$	$\dfrac{(n+1)(n+2)\cdots(n+k-1)}{(n-1)!}$
$\dfrac{z}{z-a}$	a^n
$\dfrac{az}{(z-a)^2}$	na^n
$\dfrac{z^n}{(z-a)^n}$	$\dfrac{(n+1)(n+2)\cdots(n+k-1)}{(n-1)!}a^n$

$$\dfrac{1}{z^2 - a^2}$$	$$\dfrac{a^n + (-a)^n}{2a^2}$$
$$\dfrac{z}{(z-a)(z-1)^2}$$	$$\dfrac{a^n}{(a-1)^2} + \dfrac{n}{1-a} - \dfrac{1}{(1-a)^2}$$
$$\dfrac{1}{(z-a)(z-b)}$$	0 for $n = 0$ $\dfrac{1}{a-b}\left(a^{n-1} - b^{n-1}\right)$ for $n > 0$
$$\dfrac{z}{(z-a)(z-b)}$$	$$\dfrac{1}{a-b}\left(a^n - b^n\right)$$
$$\dfrac{z(1-a)}{(z-a)(z-b)}$$	$$1 - a^n$$
$$\dfrac{z\sin a}{z^2 - 2z\cos a + 1}$$	$$\sin an$$
$$\dfrac{z(z-\cos a)}{z^2 - 2z\cos a + 1}$$	$$\cos an$$
$$\dfrac{z}{z+a}$$	$$a^n \cos \pi n$$

THE DELTA FUNCTION

To define and understand the delta function, it is first necessary to define the *unit step function* represented by the expression

$$1(t - t_0) = \begin{cases} 1 \text{ for } t > t_0 \\ 1 \text{ for } t \le t_0 \end{cases}$$

The *unit impulse function* or delta function $\delta(t)$ is defined as follows:

$$\delta(t) = \lim_{\Delta t \to 0^+} \left[\frac{1(t) - 1(t - \Delta t)}{\Delta t} \right]$$

The mathematical interpretation is that of a function representing a spike whose ordinate approaches infinity and the width of the independent variable approaches zero. The area under the curve is equal to one, that is

$$\int_{-\infty}^{\infty} \delta(t)dt = 1$$

meaning that a unit of input is compressed to an infinitesimally small duration of time. Note that the delta function has a value only at $t = 0$. The following is the Laplace transform of the delta function:

$$L[\delta(t)] = \int_0^{\infty} e^{st} \delta(t)dt = 1$$

Then, $L^{-1}[1] = \delta(t)$.

Expression $\delta(t - \tau)$ is called a *delayed impulse*, such that

$$L^{-1}[\delta(t - \tau) = e^{-s\tau}$$

REFERENCES

Beaudouin, J., 1968, Efectos de la melaza sobre el consumo de pasto en bovinos. MS Thesis, Instituto Interamericano de Ciencias Agricolas (IICA).

Box, G.E.P. and J.S.Hunter, 1957, Multi factor Experimental Designs. Ann. Math. Stat. 28.

Davis G.V. and O.T. Stallcup, 1967, Effect of soybean meal, raw soybeans, corn gluten feed and urea on the concentration of rumen fluid components at intervals after feeding. J. Dairy Sci. 50:1638-1644.

Johnson, H.D., 1973, Data presented at "Curso intensivo de fisiología de la producción con énfasis en bovinos". IICA-CTEI, Turrialba, Costa Rica.

Kleiber, M., 1961, The Fire of Life. Wiley & Sons, Inc. New York, London.

Lengeman, F.W. and N.N. Allen, 1959, Development of Rumen Function in the Dairy Calf. II effect of Diet Upon Characteristics of the Rumen Flora and Fauna of Young Calves. J. Dairy Sci. 42:1171-1181.

Lowman, B.G., R.A. Edwards and S.H. Somerville, 1979, The effects of plane of nutrition in early lactation on the performance of beef cows. Anim. Prod. 29:293-303.

Medina-Certad, R.I., 1980, Tasa de Digestion y Digestibilidad Potencial Ruminal de Materiales Fibrosos en Funcion de Niveles de Almidon Suplementario. M.S. Thesis, CATIE-University of Costa Rica.

Miller and N.W. Hooven, Jr., R.H., 1969, Variation in part-lactation and whole-lactation feed efficiency of Holstein cows. J. Dairy Sci. 52:1025-1036.

Murtagh, G.J., A.G. Kaiser, D.O. Huett and R.M. Hughes, 1980, Summer-growing components of a pasture system in a subtropical environment. 1. Pasture growth, carrying capacity and milk production. J. Agric. Sci. Camb. 94:645-663.

Ray D.E. and R.E. Roubicek, 1971, Behavior of feedlot cattle during two seasons. J. Anim. Sci. 33:72-76.

San Martin, F.A., 1980, Digestibilidad, tasas de digestion y consumo de forraje, en funcion de la suplementacion con banano verde. MS Thesis, Universidad de Costa Rica-CATIE.

SPSS Professional Statistics 7.5., 1997, SPSS Inc.

Streeter, C.L., C.O. Little, G.E. Mitchell, Jr. and R.A. Scott, 1973, Influence of rate of ruminal administration of urea on nitrogen utilization in lambs. J. Animal Science 37:796-799.

Van Soest, P.J., 1973, The uniformity and nutritive availability of cellulose. Fed. Proc. 32:1804-1808.

Waymore, A.W. 1976, Systems Engineering Methodology for Interdisciplinary Teams. Wiley & Sons, New York.